Astrophysics of Gaseous Nebulae and Active Galactic Nuclei

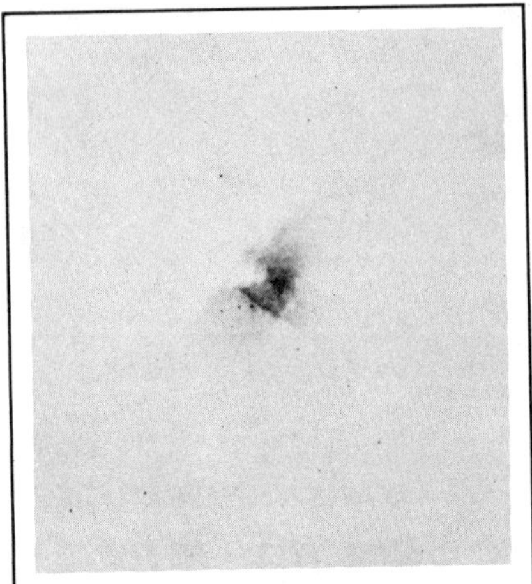

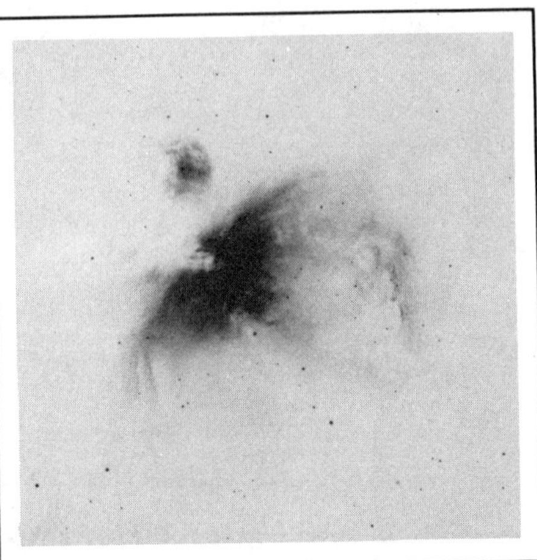

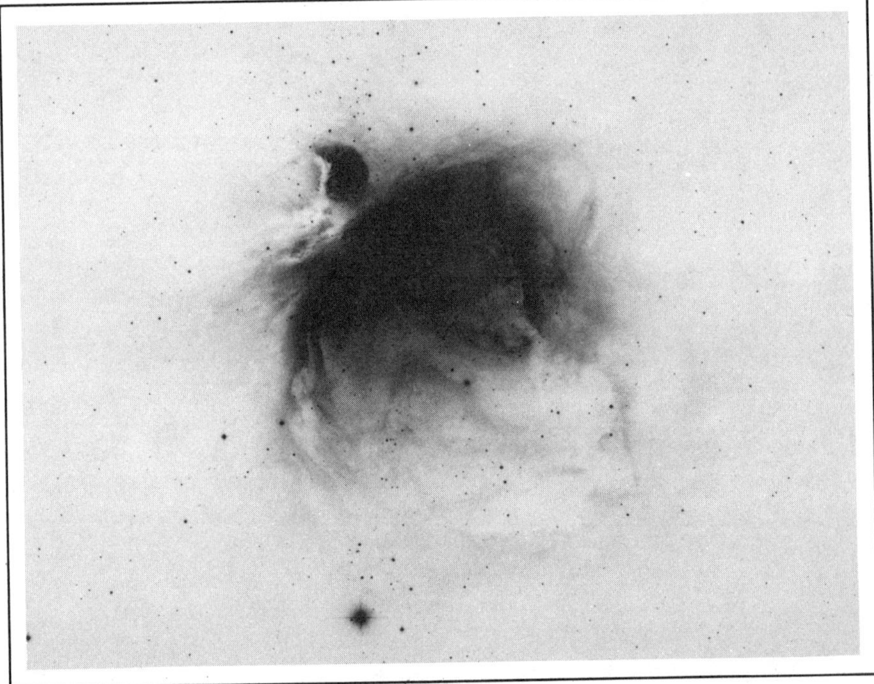

NGC 1976, the Orion Nebula, probably the most-studied H II region in the sky. Three different exposures, taken with the 48-inch Schmidt telescope, red filter and 103a-E plates, emphasizing Hα λ6563, [N II] $\lambda\lambda$6548, 6583. The 5-second exposure (*upper left*) shows only the brightest parts of nebula; θ^1 Ori (the Trapezium), the multiple-exciting star, cannot be seen on even this short an exposure. In the 40-second exposure (*upper right*), the central parts are overexposed. The nebula to the upper left (northeast) is NGC 1982. Though it is partly cut off from NGC 1976 by foreground extinction, radio-frequency measurements show that there is a real density minimum between the two nebulae. The 6-minute exposure (*bottom*) shows outer fainter parts of nebula. Still fainter regions can be recorded on even longer exposures. Compare these photographs with Figure 7.6, which was taken in continuum radiation. (*Hale Observatories photographs.*)

Astrophysics of Gaseous Nebulae and Active Galactic Nuclei

Donald E. Osterbrock
Lick Observatory, University of California, Santa Cruz

UNIVERSITY SCIENCE BOOKS
Mill Valley, California

University Science Books
20 Edgehill Road
Mill Valley, CA 94941

Production managers: Mary Miller and Bernie Scheier
Manuscript editor: Aidan Kelly
Indexer: Irene H. Osterbrock
Text and jacket designer: Robert Ishi
Technical illustrator: John Foster
Compositor: Patricia Shand
Printer and binder: Maple-Vail Book Manufacturing Group

Library of Congress Catalog Number: 87-050026

ISBN 0-935702-22-9

Printed in the United States of America

10 9 8 7 6 5 4 3 2 1

Dedication

To my friends at Lick Observatory, University of California, Santa Cruz – faculty members, postdocs, and graduate students – who have worked with me on problems connected with active galactic nuclei during the past fifteen years. Together we learned so much of what went into this book – and of how much more remains to be learned.

Contents

Preface

Fifteen years ago I sent to the publisher my book on *Astrophysics of Gaseous Nebulae.* It was a graduate-level text and research monograph that evidently filled a need, for it soon became widely used and quoted. Over the years since then the book has found increasing use, not only in nebular research, but also in problems connected with quasars, Seyfert galaxies, quasistellar objects, and all the other fascinating types of active galactic nuclei whose emission-line spectra are similar, in general terms, to those of gaseous nebulae. My own research had turned in those directions since I came to Lick Observatory in 1973 and began obtaining data with its superbly instrumented 3-m Shane reflecting telescope, as it now is named.

Hence as *AGN* (for so my first book is often referred to) gradually became dated, particularly in its tables of observational results and theoretical calculations, it was natural for me to think of revising it, and of extending it to *Astrophysics of Gaseous Nebulae and Active Galactic Nuclei* at the same time. Many of my friends and colleagues urged me to do so. Thus the present *(AGN)*^{2} came about.

Like the earlier *AGN*, it is both a graduate-level text and an introduction to nebular and AGN research. The first nine chapters are based upon the first nine chapters of the earlier book, but have been heavily revised and updated. The last three chapters are completely new, one on nova and supernova remnants, and the final two chapters on active galactic nuclei. The emphasis is very strongly on the ionized gas in AGNs and the emission-line spectra they emit; their X-ray and radio-frequency radiations are only briefly mentioned.

The book is based upon graduate courses that I have given often at the University of California, Santa Cruz. It represents the material I consider necessary to understand research papers that are now being published in its fields. So much is known today, and so many new results are pouring out, that it is probably impossible to go straight from studying any book to doing frontier research oneself. But I believe that this book will enable the reader to get up to speed, so that he or she will be able to read and understand current research, and then begin to add to it.

The reader for whom $(AGN)^2$ was written is assumed to have a reasonably good preparation in physics, and some knowledge of astronomy and astrophysics. The simplest concepts of radiative transfer are used without explanation, since the reader almost invariably has studied stellar atmospheres before gaseous nebulae and active galactic nuclei. Physical parameters, such as collision cross sections, transition probabilities, and energy levels, are taken as known quantities; no attempt is made to derive them. When I teach this material I usually include some of these derivations, linking them to the quantum-mechanics textbooks with which the students are most familiar. Omitting this material from the book left room to include more interpretation and results on gaseous nebulae and active galactic nuclei.

References are given at the end of each chapter, in a separate section. They are not inserted in the text, partly so that they will not break up the continuity of the discussions, and partly because the text is a complicated amalgam of many papers, with no obvious single place at which to refer to many of them. Almost all the references are to the American, English, and European astronomical literature, with which I am most familiar; it is also the literature that will be most accessible to the readers of this book.

I would like to express my deep gratitude to my teachers at the University of Chicago, who introduced me to the study of gaseous nebulae: Thornton L. Page, S. Chandrasekhar, W. W. Morgan, and the late Bengt Strömgren. I am also very grateful to my colleagues and mentors at the Mount Wilson and Palomar Observatories, as it was then named, the late Walter Baade and the late Rudolph Minkowski, who encouraged me to apply what I knew of nebular astrophysics to the study of galaxies. I owe much to all these men, and I am grateful to them all for their continued encouragement, support, and stimulation.

I am extremely grateful to my colleagues and friends who read early drafts of various chapters in this book and sent me their suggestions, comments, and criticisms on them: Donald P. Cox, Gary J. Ferland, William G. Mathews, John S. Mathis, Manuel Peimbert, Richard A. Shaw, Gregory A. Shields, Sidney van den Bergh, Robert E. Williams, and Stanford E. Woosley. In addition, my two current graduate students, Richard W. Pogge and Sylvain Veilleux, carefully read the entire manuscript; their comments and corrections greatly improved it, as did those of Dieter Hartmann and Philip A. Pinto, both of whom carefully read the supernova material. I am most grateful to them all.

Though these readers found many misprints and errors, corrected many misstatements, and clarified many obscurities, the ultimate responsibility for the book is mine. I have tried very hard to find and remove *all* the errors, but some must surely remain, to be discovered only after publication. I can do no better than repeat once again the words of a great physicist, Richard P. Feynman, "Listen to what I mean, not to what I say." If the reader finds an error, I am sorry I did not catch it, but he or she will have proved his or her real understanding of the material, and I shall be very pleased to receive a correction.

I am greatly indebted to Gerri McLellan, who entered on the word processor the first drafts of all the chapters, and all the successive revisions of the manuscript, and to Pat Shand, who made the final editorial revisions and prepared the camera-ready copy for publication. I deeply appreciate the skill, accuracy, and dedication with which they worked on this book. I am also most grateful to my wife, Irene H. Osterbrock, who prepared the index for the book.

My research on gaseous nebulae and active galactic nuclei has been supported over the past fifteen years by the University of California, the John Simon Guggenheim Memorial Foundation, the University of Minnesota, the University of Chicago, the Institute for Advanced Study, the Ohio State University, and especially by the National Science Foundation. I am grateful to all of these organizations for their generous support. Much of my own research, and of the research of the graduate students and post-docs who have worked and are working with me, has gone into this book; I could never have written it without doing that research myself.

I am especially grateful to my friends George H. Herbig, Paul W. Hodge, Guido Münch, and Robert E. Williams, who provided original photographs included in this book. I am grateful to them and also to Palomar Observatory, Lick Observatory, and the National Optical Astronomy Observatories for permission to use the photographs (which are all credited individually) in this book. Publication of the photographs from NOAO does not imply the endorsement by NOAO, or by any NOAO employee, of this book! Many of the other figures are derived from published papers, and I am grateful to their authors for permission to modify and use their figures in this book.

Lastly, I wish to express my sincere thanks to my friends Bruce Armbruster, president of University Science Books, and Joseph S. Miller, my colleague, former student, and astronomy co-editor with me for USB, both of whom encouraged me time after time to go on with revising *AGN* and writing the additional new chapters for $(AGN)^2$. Bruce was the astronomy editor for W. H. Freeman and Company when I wrote the earlier book, and he helped me greatly with it then, as he has helped me with $(AGN)^2$ now. It was a great pleasure for me to work with him on both these books. I am also grateful to W. H. Freeman and Company for releasing me from my obligation to them, and allowing me to publish this book with USB.

Donald E. Osterbrock

Astrophysics of Gaseous Nebulae and Active Galactic Nuclei

1

General Introduction

1.1 Introduction

Many important topics in astrophysics involve the physics of ionized gases and the interpretation of their emission-line spectra. The subject is fascinating in itself. In addition, H II regions allow us to probe the chemical evolution and the star-formation history of the far reaches of our own Galaxy, and of distant galaxies. Planetary nebulae let us see the outer remaining envelopes of dying stars. Supernova remnants allow us to observe material from the burned-out deep interiors of exploded, massive stars. Quasars and QSOs are the most luminous objects in the universe, and hence the most distant that we can observe. All of these are subjects we shall cover in this book. Further applications, such as the properties of intergalactic material, X-ray flows, and primordial galaxies, though not treated here, are straightforward extensions of the physics that forms the spline of this volume.

1.2 Gaseous Nebulae

Gaseous nebulae are observed as bright extended objects in the sky. Those with the highest surface brightness, such as the Orion Nebula (NGC 1976) or the Ring Nebula (NGC 6720), are easily observed on direct photographs, or even at the eyepiece of a telescope. Many other nebulae that are intrinsically less luminous or that are more strongly affected by interstellar extinction are faint on ordinary photographs, but can be photographed on long exposures with filters that isolate a narrow wavelength region around a prominent nebular emission line, so that the background and foreground stellar and sky radiations are suppressed. The largest gaseous nebula in the sky is the Gum Nebula, which has an angular diameter of the order of 30°, while many familiar nebulae have sizes of the order of one degree, ranging down to the smallest objects at the limit of resolution of the largest telescopes. The surface brightness of a nebula is independent of its distance, but more distant nebulae have (on the average) smaller angular size and greater interstellar extinction; so the nearest members of any particular type of nebula tend to be the most-studied objects.

Gaseous nebulae have an emission-line spectrum. This spectrum is dominated by forbidden lines of ions of common elements, such as [O III] $\lambda\lambda$4959, 5007, the famous green nebular lines once thought to indicate the presence of the hypothetical element nebulium; [N II] $\lambda\lambda$6548, 6583 in the red; and [O II] $\lambda\lambda$3726, 3729, the ultraviolet doublet which appears as a blended λ3727 line on low-dispersion spectrograms of almost every nebula. In addition, the permitted lines of hydrogen, Hα λ6563 in the red, Hβ λ4861 in the blue, Hγ λ4340 in the violet, and so on, are characteristic features of every nebular spectrum, as is He I λ5876, which is considerably weaker, while He II λ4686 occurs only in higher-ionization nebulae. Long-exposure spectrograms, or photoelectric spectrophotometric observations extending to faint intensities, show progressively weaker forbidden lines, as well as faint permitted lines of common elements, such as C II, C III, C IV, O II, and so on. The emission-line spectrum, of course, extends into the infrared, where [Ne II] $\lambda 12.8\mu$ and [O III] $\lambda 88.4\mu$ are among the lines measured, and into the ultraviolet, where Mg II $\lambda\lambda$2796, 2803, C III] $\lambda\lambda$1907, 1909 and C IV $\lambda\lambda$1548, 1551 are also observed.

Gaseous nebulae have weak continuous spectra, consisting of atomic and reflection components. The atomic continuum is emitted chiefly by free-bound transitions, mainly in the Paschen continuum of H I at $\lambda > 3646$ Å, and the Balmer continuum at $\lambda < 3646$ Å. In addition, many nebulae have reflection continua consisting of starlight scattered by dust. The amount of dust varies from nebula to nebula, and the strength of this continuum fluctuates correspondingly. In the infrared, the nebular continuum is largely thermal radiation emitted by the dust.

In the radio-frequency region, emission nebulae have a reasonably strong continuous spectrum, mostly due to free-free emission or bremsstrahlung of thermal electrons accelerated on Coulomb collisions with protons. Superimposed on this continuum are weak emission lines of H, such as 109α at $\lambda =$ 6 cm, resulting from bound-bound transitions between very high levels of H. Weaker radio recombination lines of He and still weaker lines of other elements can also be observed in the radio region, slightly shifted from the H lines by the isotope effect.

1.3 Observational Material

Practically every observational tool of astronomy can be and has been applied to the study of gaseous nebulae. Because nebulae are low-surface-brightness, extended objects, the most effective instruments for studying them are fast, wide-field optical systems. For instance, large Schmidt cameras are ideal for direct photography of gaseous nebulae, and many of the most familiar pictures of nebulae, including several of the illustrations in this book, were taken with the 48-inch Schmidt telescope at Palomar Observatory. The finest small-scale detail in bright nebulae is best shown on photographs taken with longer focal-length instruments, as we can see in other illustrations taken with the 5-m F/3.67 Hale telescope, the 4-m F/2.7 Mayall telescope, and the 3-m F/5 Shane telescope.

Though the brighter nebulae are known from the early visual observations, many fainter-emission nebulae have been discovered by systematic programs of direct photography, comparing an exposure taken in a narrow wavelength region around prominent nebular lines (most often Hα $\lambda6563$ + [N II] $\lambda\lambda6548$, 6583) with an exposure taken in another wavelength region that suppresses the nebular emission (for instance, $\lambda\lambda5100$-5500). Other small nebulae have been found on objective-prism surveys as objects with bright Hα or [O III] emission lines, but faint continuous spectra.

Much of the physical analysis of nebulae depends on spectrophotometric measurements of emission-line intensities, carried out photographically in earlier years, or almost exclusively with electronic detectors today. The photoelectric scanners used a few years ago had higher intrinsic accuracy for measuring a single line, because of the higher quantum efficiency of the photoelectric effect and the linear response of a photoelectric system, while the photographic plate had the advantage of "multiplexing," or recording very many picture elements (many spectral lines in this example) simultaneously. However, many-channel electronic systems, such as image-dissector systems, Reticons, and CCDs (charge-coupled devices), which retain the advantages of photoelectric systems and add the multiplexing property formerly available only photographically, are now in general use. A fast nebular spectrophotometer can be matched to any telescope, but the larger the aperture of the telescope, the smaller the size of the nebular features that can be accurately measured, or the fainter the small nebulae (those with angular size smaller than the entrance slit or diaphragm).

Radial velocities in nebulae are measured on slit spectrograms or spectral scans. Here again, a fast spectrograph is essential to reach low-surface-brightness objects and a large telescope is required to observe nebular features with small angular size. The electronic detectors mentioned previously have almost completely replaced photography for this work.

Nebular infrared continuum measurements can be made with broad-band photometers and radiometers, using filters to isolate various spectral regions. Spectrophotometry of individual infrared lines requires better wavelength resolution than can be achieved with ordinary filters, and Fabry-Perot interferometers and Fourier-transform spectrographs have just begun to be used for these measurements. In the radio region, filters are used for high wavelength resolution. In all spectral regions, large telescopes are required to measure features with small angular size; in particular, in the radio region long-baseline interferometers are required. Far-infrared measurements must be made from above most of the Earth's atmosphere to minimize strong water-vapor absorption, and ultraviolet measurements must be made from above essentially all the atmosphere, to eliminate ozone absorption. Generally, high-altitude airplanes have been used to carry the far-infrared telescopes, and artificial satellites for the ultraviolet observations.

1.4 Physical Ideas

The source of energy that enables emission nebulae to radiate is, almost always, ultraviolet radiation from stars within or near the nebula. There are one or more hot stars, with surface temperature $T_* \gtrsim 3 \times 10^4$ ° K, near or in almost every nebula; and the ultraviolet photons these stars emit transfer energy to the nebula by photoionization. In nebulae and in practically all astronomical objects, H is by far the most abundant element, and photoionization of H is thus the main energy-input mechanism. Photons with energy greater than 13.6 eV, the ionization potential of H, are absorbed in this process, and the excess energy of each absorbed photon over the ionization potential appears as kinetic energy of a newly liberated photoelectron. Collisions between electrons, and between electrons and ions, distribute this energy and maintain a Maxwellian velocity distribution with temperature T in the range 5,000° K $< T <$ 20,000° K in typical nebulae. Collisions between thermal electrons and ions excite the low-lying energy levels of the ions. Downward radiation transitions from these excited levels have very small transition probabilities, but at the low densities ($N_e \lesssim 10^4$ cm^{-3}) of typical nebulae, collisional deexcitation is even less probable; so almost every excitation leads to emission of a photon, and the nebula thus emits a forbidden-line spectrum that is quite difficult to excite under terrestrial laboratory conditions.

Thermal electrons are recaptured by the ions, and the degree of ionization at each point in the nebula is fixed by the equilibrium between photoionization and recapture. In nebulae in which the central star has an especially high temperature, T_*, the radiation field has a correspondingly high number of high-energy photons, and the nebular ionization is therefore high. In such nebulae collisionally excited lines up to [Ne V] and [Fe VII] may be observed, but the high ionization results from the high energy of the *photons* emitted by the star, and does *not* necessarily indicate a high nebular temperature T, defined by the kinetic energy of the free electrons.

In the recombination process, recaptures occur to excited levels, and the excited atoms thus formed then decay to lower and lower levels by radiative transitions, eventually ending in the ground level. In this process, line photons are emitted, and this is the origin of the H I Balmer- and Paschen-line spectra observed in all gaseous nebulae. Note that the recombination of H^+ gives rise to excited atoms of H^0 and thus leads to the emission of the H I spectrum. Likewise, He^+ recombines and emits the He I spectrum, and in the most highly ionized regions, He^{++} recombines and emits the He II spectrum, the strongest line in the ordinary observed region being $\lambda 4686$. Much weaker recombination lines of C II, C III, C IV, and so on, are also emitted; however, the main excitation process responsible for the observed strengths of such lines with the same spin or multiplicity as the ground term is resonance fluorescence by photons, which is much less effective for H and He lines because the resonance lines of these more abundant elements have greater optical depth.

In addition to the bright-line and continuous spectra emitted by atomic processes, many nebulae also have an infrared continuous spectrum emitted by dust particles heated to a temperature of order 100° K by radiation derived originally from the central star.

Gaseous nebulae may be classified into two main types, ***diffuse nebulae*** or *H II regions,* and ***planetary nebulae.*** Though the physical processes in both types are quite similar, the two groups differ greatly in origin, mass, evolution, and age of typical members; so for some purposes it is convenient to discuss them separately. Nova shells are rare but interesting objects: tiny, rapidly expanding, cool photoionized nebulae. In addition, an even rarer class of objects, supernova remnants, differs greatly from both diffuse and planetary nebulae. We will briefly examine each of these types of objects, and then discuss Seyfert galaxies and other active galactic nuclei, in which much the same physical processes occur, although with differences in detail because considerably higher-energy photons are involved.

1.5 Diffuse Nebulae

Diffuse nebulae or H II regions are regions of interstellar gas in which the exciting star or stars are O- or early B-type stars of Population I. Often there are several exciting stars, a multiple star, or a galactic cluster whose hottest two or three stars are the main sources of ionizing radiation. These hot, luminous stars undoubtedly formed fairly recently from interstellar matter that would otherwise be part of the same nebula they now ionize and thus illuminate. The effective temperatures of the stars are in the range $3 \times 10^4\ ^\circ\mathrm{K} < T_* < 5 \times 10^4\ ^\circ\mathrm{K}$; throughout the nebula, H is ionized, He is singly ionized, and other elements are mostly singly or doubly ionized. Typical densities in the ionized part of the nebula are of order 10 or 10^2 cm^{-3}, ranging to as high as 10^4 cm^{-3}, although undoubtedly small denser regions exist close to or even below the limit of resolvability. In many nebulae dense neutral condensations are scattered throughout the ionized volume. Internal motions occur in the gas with velocities of order 10 km sec^{-1}, approximately the isothermal sound speed. Bright rims, knots, condensations, and so on, are apparent to the limit

FIGURE 1.1
NGC 6611, a bright H II region in which the exciting O stars are members of a star cluster. Note the dark condensations, bright rims, and other fine structure. The area shown is approximately one-half degree in diameter, or 20 pc at the distance of the nebula. Original plate taken in Hα and [N II] $\lambda\lambda$6548, 6583 with the 5-m Hale telescope. *(Palomar Observatory photograph.)*

of resolution. The hot, ionized gas tends to expand into the cooler surrounding neutral gas, thus decreasing the density within the nebula and increasing the ionized volume. The outer edge of the nebula is surrounded by ionization fronts running out into the neutral gas.

The spectra of these "H II regions," as they are often called (because they contain mostly H^+), are strong in H I recombination lines and [N II] and [O II] collisionally excited lines, but the strengths of [O III] and [N III] lines may differ greatly, being stronger in the nebulae with higher central-star temperatures.

These H II regions are observed not only in our Galaxy but also in other nearby galaxies. The brightest H II regions can easily be seen on almost any large-scale photograph, but plates taken in a narrow wavelength band in the red, including Hα and the [N II] lines, are especially effective in showing faint and often heavily reddened H II regions in other galaxies. The H II regions are strongly concentrated to the spiral arms, and indeed are the best objects for tracing the structure of spiral arms in distant galaxies. Radial-velocity measurements of H II regions then give information on the kinematics of Population I objects in our own and other galaxies. Typical masses of observed H II regions are of order 10^2 to $10^4 M_{\odot}$, with the lower limit depending largely on the sensitivity of the observational method used.

1.6 Planetary Nebulae

Planetary nebulae are isolated nebulae, often (but not always) possessing a fair degree of bilateral symmetry, that are actually shells of gas that have been lost in the fairly recent past by their central stars. The name "planetary" is purely historical and refers to the fact that some of the bright planetaries appear as small, disk-like, greenish objects in small telescopes. The central stars of planetary nebulae are old stars, typically with $T_* \approx 5 \times 10^4$ ° K, much hotter than galactic O stars, and often less luminous ($M_V = -3$ to $+5$). The stars are in fact rapidly evolving toward the white-dwarf stage, and the shells are expanding with velocities of order of several times the velocity of sound (25 km sec^{-1} is a typical expansion velocity). However, because they are decreasing in density, their emission is decreasing, and on a cosmic time scale they rapidly become unobservable, with mean lifetimes as planetary nebulae of a few times 10^4 years.

As a consequence of the higher stellar temperatures of their exciting stars, typical planetary nebulae are relatively more highly ionized than H II regions, often including large amounts of He^{++}. Their spectra thus include not only the H I and He I recombination lines, but often also the He II lines; the collisionally excited lines of [O III] and [Ne III] in their spectra are characteristically stronger than those in diffuse nebulae, and [Ne V] is often strong. There is a wide range in the temperatures of planetary-nebula central stars, however,

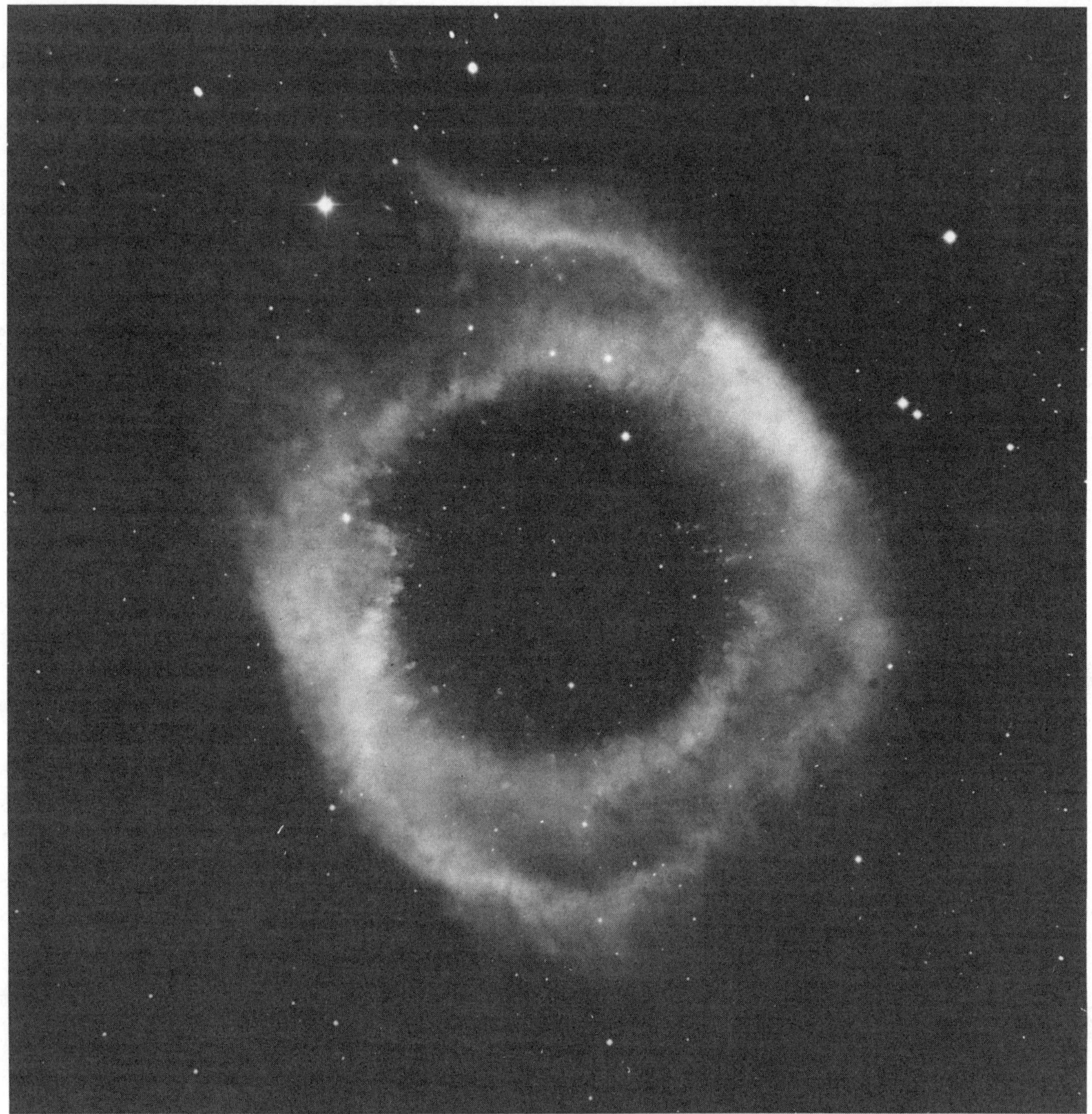

FIGURE 1.2
NGC 7293, a nearby, large, low-surface-brightness planetary nebula. Note the fine structure, including the long narrow radial filaments in the central "hole" of the nebula, which point to the exciting star. The diameter of the nebula is approximately one quarter of a degree, or 0.5 pc at the distance of the nebula. Original plate taken in Hα and [N II] $\lambda\lambda$6548, 6583 with the 5-m Hale telescope. *(Palomar Observatory photograph.)*

and the lower-ionization planetaries have spectra that are quite similar to those of H II regions.

The space distribution and kinematic properties of planetary nebulae indicate that, on the cosmic time scale, they are fairly old objects, usually classified as old Disk Population or old Population I objects. This indicates that the bulk of the planetaries we now see, though relatively young as planetary nebulae, are actually near-terminal stages in the evolution of quite old stars.

Typical densities in observed planetary nebulae range from 10^4 cm^{-3} down to 10^2 cm^{-3}, and typical masses are of order 0.1 $M_\odot$ to 1.0 $M_\odot$. Many planetaries have been observed in other nearby galaxies, especially the Magellanic Clouds and M 31, but their luminosities are so much smaller than the luminosities of the brightest H II regions, that they are difficult to study in great detail. However, spectroscopic measurements of these planetaries give good information on velocities, abundances of the elements, and stellar evolution in these galaxies.

1.7 Nova and Supernova Remnants

Many recent novae are surrounded by small, faint shells with emission-line spectra. As we shall see, they are tiny photoionized nebulae. A few emission nebulae are known to be supernova remnants. The Crab Nebula (NGC 1952), the remnant of the supernova of A.D. 1054, is the best-known example, and small bits of scattered nebulosity are the observable remnants of the much more heavily reddened objects, Tycho's supernova of 1572 and Kepler's supernova of 1604. All three of these supernova remnants have strong nonthermal radio spectra, and several other filamentary nebulae with appearances quite unlike typical diffuse or planetary nebulae have been identified as older supernova remnants by the fact that they have similar nonthermal radio spectra. Two of the best-known examples are the Cygnus Loop (NGC 6960-6992-6995) and IC 443. In the Crab Nebula, the nonthermal synchrotron spectrum observed in the radio-frequency region extends into the optical region, and extrapolation to the ultraviolet region indicates that this synchrotron radiation is probably the source of the photons that ionize the nebula. However, in the other supernova remnants no photoionization source is seen, and much of the energy is instead provided by the conversion of kinetic energy of motion into heat. In other words, the fast-moving filaments collide with ambient interstellar gas, and the energy thus released provides ionization and thermal energy, which later is partly radiated as recombination- and collisional-line radiation. Thus these supernova remnants are objects in which *collisional ionization* occurs, rather than photoionization. However, note that in all the nebulae, *collisional excitation* is caused by the thermal electrons that are energized either by photoionization or by collisional ionization.

1.8 Active Galactic Nuclei

Many galaxies, in addition to having H II regions and planetary nebulae, show in the spectra of their nuclei characteristic nebular emission lines. In most of these objects the gas is evidently photoionized by hot stars in the nucleus, which is thus much like a giant H II region, or perhaps a cluster of many H II regions. The galactic nuclei with the strongest emission lines of this type are often called "extragalactic H II regions" or "starburst galaxies."

Besides these objects, however, a small fraction of spiral galaxies have ionized gas in their nuclei that emits an emission-line spectrum with a wider range of ionization than any H II region. Usually the emission-line profiles show a significantly greater range of velocities than in starburst galaxies. These galaxies, totaling a few percent of all spiral galaxies, are called Seyfert galaxies. Many of the most luminous radio galaxies, typically N, cD, D or E galaxies in form, have nuclei with very similar emission-line spectra. Quasars (quasistellar radio sources) and QSOs (quasistellar objects) are radio-loud and radio-quiet analogues of radio galaxies and Seyfert galaxies; they have similar optical spectra and even greater optical luminosities, but are much rarer in space. All these objects together are called active galactic nuclei. Among them are the most luminous objects in the universe, quasars and QSOs with redshifts up to $z \approx 4$, corresponding to recession velocities of more than $0.9c$.

Much if not all of the ionized gas in active galactic nuclei appears to be photoionized. However, the source of the ionizing radiation is not a hot star or stars. Instead, it is probably an extension to high energies of the blue "featureless continuum" observed in these objects in the optical spectral region. This is probably emitted by an accretion disk around a black hole, or by relativistic particles and perhaps a magnetic field associated with the immediate environs of the black hole. The spectrum of the ionizing radiation, whatever its source, certainly extends to much higher energies than the spectra of the hot stars that ionize H II regions and planetary nebulae. Also, the particle and energy densities are much larger in some ionized regions in active galactic nuclei than in nebulae.

The basic physical principles that govern the structures and emitted spectra of active galactic nuclei are largely the same as those that apply in H II regions and planetary nebulae. However, because of the large proportion of high-energy photons in the ionizing flux, some new physical processes become important in active galactic nuclei, and these cause their structures to differ in important details from those of classical nebulae. These differences can best be analyzed after H II regions and planetary nebulae are well understood. Hence we treat nova shells, supernova remnants, and active galactic nuclei in the final chapters of this book.

References

Every introductory textbook on astronomy contains an elementary general description of gaseous nebulae. The following books and papers are good general references for the entire subject.

Middlehurst, B. M., and Aller, L. H., eds. 1960. *Nebulae and Interstellar Matter.* Chicago: University of Chicago Press.

Terzian, Y. ed. 1968. *Interstellar Ionized Hydrogen.* New York: Benjamin.

Aller, L. H. 1984. *Physics of Thermal Gaseous Nebulae. (Physical Processes in Gaseous Nebulae).* Dordrecht: Reidel.

Spitzer, L. 1978. *Physical Processes in the Interstellar Medium.* New York: Wiley.

Seaton, M. J. 1960. *Reports Progress Phys.* **23,** 313.

Seaton, M. J. 1980. *Q.J.R.A.S.* **21,** 229.

Kaler, J. B. 1985. *Ann. Rev. Astr. Ap.* **23,** 89.

Osterbrock, D. E. 1967. *P.A.S.P.* **79,** 523.

Lynds, B. T. 1965. *Ap. J. Supp.* **12,** 163.

Parker, R. A. R., Gull, T. R., and Kirschner, R. P. 1979. *An Emission-Line Survey of the Milky Way.* NASA SP-434.

Osterbrock, D. E. 1979. *A. J.* **84,** 901.

Osterbrock, D. E. 1984. *Q. J. R. A. S.* **25,** 1.

Weedman, D. W. 1986. *Quasar Astronomy.* Cambridge: Cambridge University Press.

Lynds is a catalogue of bright nebulae, including emission and reflection nebulae, identified on the National Geographic-Palomar Observatory Sky Survey, taken with the 48-inch Schmidt, and gives references to several earlier catalogues. The last three references here deal with active galactic nuclei; all the others deal with H II regions and planetary nebulae.

2

Photoionization Equilibrium

2.1 Introduction

Emission nebulae result from the photoionization of a diffuse gas cloud by ultraviolet photons from a hot "exciting" star or from a cluster of exciting stars. The ionization equilibrium at each point in the nebula is fixed by the balance between photoionizations and recombinations of electrons with the ions. Since hydrogen is the most abundant element, we can get a first idealized approximation to the structure of a nebula by considering a pure H cloud surrounding a single hot star. The ionization equilibrium equation is:

$$N_{\mathrm{H}^0} \int_{\nu_0}^{\infty} \frac{4\pi J_\nu}{h\nu}\, a_\nu(\mathrm{H}^0) d\nu \;=\; N_e N_p \alpha(\mathrm{H}^0,\, T), \tag{2.1}$$

where J_ν is the mean intensity of radiation (in energy units per unit area per unit time per unit solid angle per unit frequency interval) at the point. Thus $4\pi J_\nu / h\nu$ is the number of incident photons per unit area per unit time per unit frequency interval, and $a_\nu(\mathrm{H}^0)$ is the ionization cross section for H by photons with energy $h\nu$ (above the threshold $h\nu_0$); the integral therefore represents the number of photoionizations per H atom per unit time. N_{H^0}, N_e, and N_p are the neutral atom, electron, and proton densities per unit volume, and $\alpha(\mathrm{H}^0, T)$ is the recombination coefficient; so the right-hand side of the equation gives the number of recombinations per unit volume per unit time.

To a first approximation, the mean intensity is simply the radiation emitted by the star reduced by the inverse-square effect of geometrical dilution. Thus

$$4\pi J_\nu \;=\; \frac{R^2}{r^2}\pi F_\nu(0) \;=\; \frac{L_\nu}{4\pi r^2}, \tag{2.2}$$

where R is the radius of the star, $\pi F_\nu(0)$ is the flux at the surface of the star, r is the distance from the star to the point in question, and L_ν is the luminosity of the star per unit frequency interval.

At a typical point in a nebula, the ultraviolet radiation field is so intense that the H is almost completely ionized. Consider, for example, a point in an H II region, with density 10 H atoms and ions per cm^3, 5 pc from a central O6 star with $T_* = 40{,}000°$ K. We will examine the numerical values of all the other variables later, but for the moment we can adopt the following very rough values:

$$\int_{\nu_0}^{\infty} \frac{L_\nu}{h\nu} d\nu \approx 5 \times 10^{48} \text{ photons sec}^{-1};$$

$$a_\nu(\mathrm{H}^0) \approx 6 \times 10^{-18} \text{ cm}^2;$$

$$\int_{\nu_0}^{\infty} \frac{4\pi J_\nu}{h\nu} a_\nu(\mathrm{H}^0)\, d\nu \approx 10^{-8} \text{ sec}^{-1};$$

$$\alpha(\mathrm{H}^0,\, T) \approx 4 \times 10^{-13} \text{ cm}^3 \text{ sec}^{-1}.$$

Substituting these values and taking ξ as the fraction of neutral H, that is, $N_e = N_p = (1 - \xi)N_{\mathrm{H}}$ and $N_{\mathrm{H}^0} = \xi N_{\mathrm{H}}$, where $N_{\mathrm{H}} = 10$ cm^{-3} is the density of H, we find $\xi \approx 4 \times 10^{-4}$, that is, H is very nearly completely ionized.

On the other hand, a finite source of ultraviolet photons cannot ionize an infinite volume, and therefore, if the star is in a sufficiently large gas cloud, there must be an outer edge to the ionized material. The thickness of this transition zone between ionized and neutral gas, since it is due to absorption, is approximately one mean free path of an ionizing photon. Using the same parameters as before, and taking $\xi = 0.5$, we find the thickness

$$d \approx \frac{1}{N_{\mathrm{H}^0} a_\nu} \approx 0.01 \text{ pc},$$

or much smaller than the radius of the ionized nebula. Thus we have the picture of a nearly completely ionized "Strömgren sphere" or H II region, separated by a thin transition region from an outer neutral gas cloud or H I region. In the rest of this chapter we will explore this ionization structure in detail.

First we will examine the photoionization cross section and the recombination coefficients for H, and then use this information to calculate the structure of hypothetical pure H regions. Next we will consider the photoionization cross section and recombination coefficients for He, the second most abundant element, and then calculate more realistic models of H II regions, that take both H and He into account. Finally, we will extend our analysis to other, less-abundant heavy elements; these often do not strongly affect the ionization structure of the nebula, but are always quite important in the thermal balance to be discussed in the next chapter.

2.2 Photoionization and Recombination of Hydrogen

Figure 2.1 is an energy-level diagram of H; the levels are marked with their quantum numbers n (principal quantum number) and L (angular momentum quantum number), and with $S, P, D, F, \ldots$ standing for $L = 0, 1, 2, 3, \ldots$ in the conventional notation. Permitted transitions (which, for one-electron systems, must satisfy the selection rule $\Delta L = \pm 1$) are marked by solid lines in the figure. The transition probabilities $A_{nL,n'L'}$ of these lines are of order 10^4 to 10^8 sec^{-1}, and the corresponding mean lifetimes of the excited levels,

$$\tau_{nL} = \frac{1}{\sum_{n'<n} \sum_{L'=L\pm1} A_{nL,n'L'}}, \tag{2.3}$$

are therefore of order 10^{-4} to 10^{-8} sec. The only exception is the $2\,^2S$ level, from which there are no allowed one-photon downward transitions. However, the transition $2\,^2S \rightarrow 1\,^2S$ does occur with the emission of two photons, and the probability of this process is $A_{2\,^2S,1\,^2S} = 8.23$ sec^{-1}, corresponding to a mean lifetime for the $2\,^2S$ level of 0.12 sec. Even this lifetime is quite short compared with the mean lifetime of an H atom against photoionization, which has been estimated previously as 10^8 sec for the $1\,^2S$ level, and is of the same order of magnitude for the excited levels. Thus, to a very good approximation, we may consider that very nearly all the H^0 is in the $1\,^2S$ level, and that photoionization from this level is balanced by recombination to all levels, and that recombination to any excited level is followed very quickly by radiative transitions downward, leading ultimately to the ground level. This basic approximation greatly simplifies calculations of physical conditions in gaseous nebulae.

The photoionization cross section for the $1\,^2S$ level of H^0, or, in general, of a hydrogenic ion with nuclear charge Z, may be written in the form

$$a_\nu(Z) = \frac{A_0}{Z^2}\left(\frac{\nu_1}{\nu}\right)^4 \frac{e^{4-[(4\tan^{-1}\epsilon)/\epsilon]}}{1 - e^{-2\pi/\epsilon}} \quad \text{for } \nu \geq \nu_1 , \tag{2.4}$$

where

$$A_0 = \frac{2^8\pi}{3e^4}\left(\frac{1}{137.0}\right)\pi a_0^2 = 6.30 \times 10^{-18}\text{cm}^2,$$

$$\epsilon = \sqrt{\frac{\nu}{\nu_1} - 1},$$

and

$$h\nu_1 = Z^2\, h\nu_0 = 13.60\, Z^2 \text{ eV}$$

is the threshold energy. This cross section is plotted in Figure 2.2, which shows that it drops off rapidly with energy, approximately as ν^{-3} not too far above the threshold, which, for H, is at $\nu_0 = 3.29 \times 10^{15}$ sec^{-1} or $\lambda_0 = 912$ Å, so that the higher-energy photons, on the average, penetrate further into neutral gas before they are absorbed.

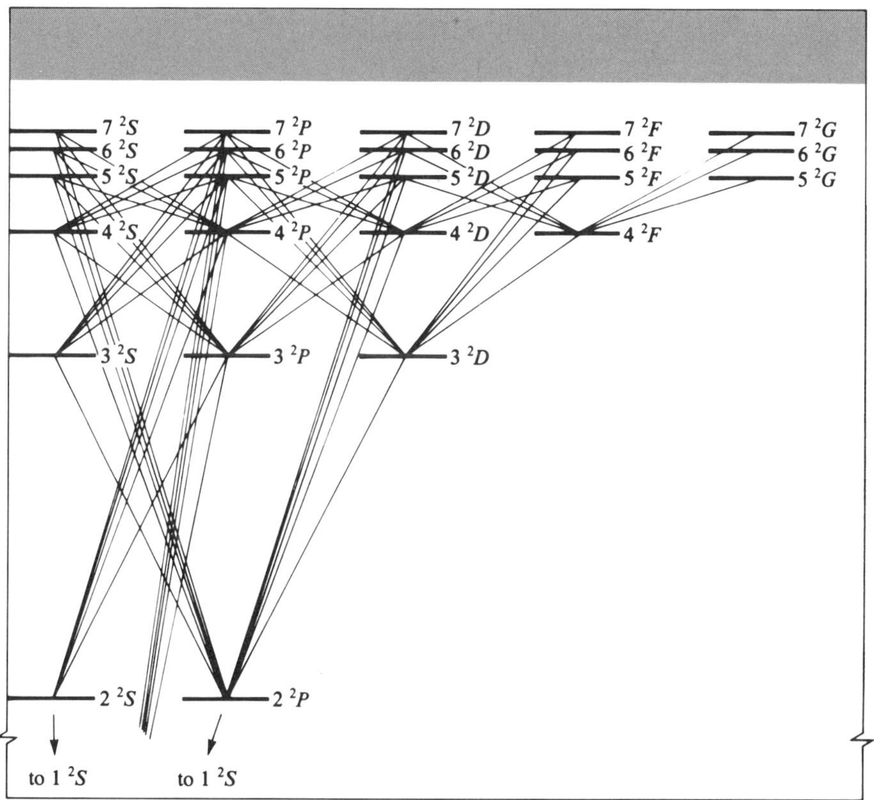

FIGURE 2.1
Partial energy-level diagram of H I, limited to $n \leq 7$ and $L \leq$ G. Permitted radiative transitions to levels $n \leq 4$ are indicated by solid lines.

The electrons produced by photoionization have an initial distribution of energies that depends on $J_\nu a_\nu / h\nu$. However, the cross section for elastic-scattering collisions between electrons is quite large, of order $4\pi(e^2/mv^2)^2 \approx 10^{-13}$ cm^2, and these collisions tend to set up a Maxwell-Boltzmann energy distribution. The recombination cross section, and all the other cross sections involved in the nebulae, are so much smaller that, to a very good approximation, the electron-distribution function is Maxwellian, and therefore all collisional processes occur at rates fixed by the local temperature defined by this Maxwellian. Therefore, the recombination coefficient to a specified level

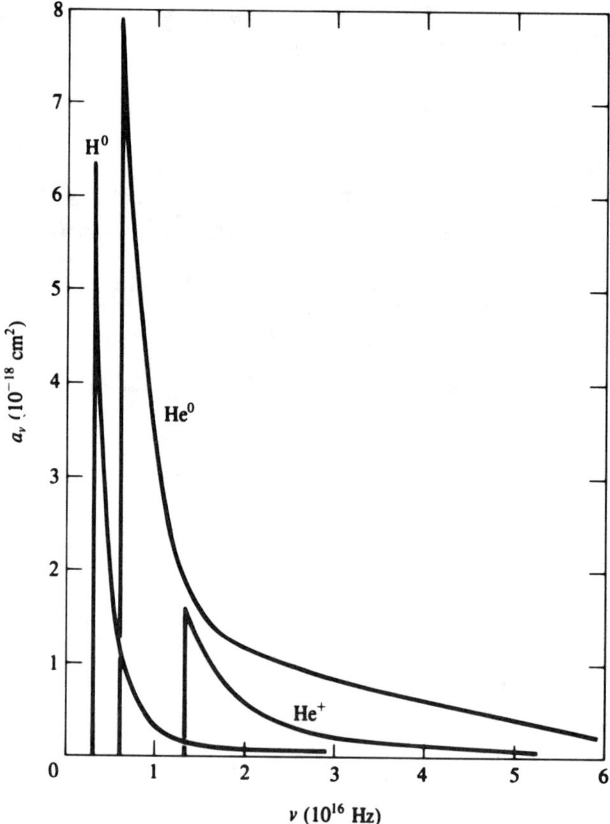

FIGURE 2.2
Photoionization absorption cross sections of H^0, He^0, and He^+.

n^2L may be written

$$\alpha_{n^2L}(H^0, T) = \int_0^\infty v\sigma_{nL}(H^0, v)f(v)dv, \tag{2.5}$$

where

$$f(v) = \frac{4}{\sqrt{\pi}}\left(\frac{m}{2kT}\right)^{3/2} v^2 e^{-mv^2/2kT} \tag{2.6}$$

is the Maxwell-Boltzmann distribution function for the electrons, and $\sigma_{nL}(H^0, v)$ is the recombination cross section to the term n^2L in H^0 for electrons with velocity v. These cross sections vary approximately as v^{-2}, and the recombination coefficients, which are proportional to $v\sigma$, therefore vary approximately as $T^{-1/2}$. A selection of numerical values of α_{n^2L} is given in Table 2.1. Since the mean electron velocities at the temperatures listed are of order 5×10^7 cm sec^{-1}, it can be seen that the recombination cross sections are of order 10^{-20} cm^2 or 10^{-21} cm^2, much smaller than the geometrical cross section of an H atom.

In the nebular approximation discussed previously, recombination to any level n^2L quickly leads through downward radiative transitions to $1\ ^2S$, and the total recombination coefficient is the sum over captures to all levels, ordinarily written

$$\begin{aligned} \alpha_A &= \sum_{n,L} \alpha_{n^2L}(\mathrm{H}^0, T) \\ &= \sum_{n} \sum_{L=0}^{n-1} \alpha_{nL}(\mathrm{H}^0, T) \\ &= \sum_{n} \alpha_n(\mathrm{H}^0, T) \end{aligned} \tag{2.7}$$

where α_n is thus the recombination coefficient to all the levels with principal quantum number n. Numerical values of α_A are also listed in Table 2.1. A typical recombination time is $\tau_r = 1/N_e\alpha_A \approx 3 \times 10^{12}/N_e$ sec $\approx 10^5/N_e$ yr, and deviations from ionization equilibrium are ordinarily damped out in times of this order of magnitude.

2.3 Photoionization of a Pure Hydrogen Nebula

Consider the simple idealized problem of a single star that is a source of ionizing photons in a homogeneous static cloud of H. Only radiation with frequency $\nu \geq \nu_0$ is effective in the photoionization of H from the ground level, and the ionization equilibrium equation at each point can be written

$$N_{\mathrm{H}^0} \int_{\nu_0}^{\infty} \frac{4\pi J_\nu}{h\nu} a_\nu d\nu = N_p N_e \alpha_A(\mathrm{H}^0, T). \tag{2.8}$$

The equation of transfer for radiation with $\nu \geq \nu_0$ can be written in the form

$$\frac{dI_\nu}{ds} = -N_{\mathrm{H}^0} a_\nu I_\nu + j_\nu, \tag{2.9}$$

where I_ν is the specific intensity of radiation and j_ν is the local emission coefficient (in energy units per unit volume per unit time per unit solid angle per unit frequency) for ionizing radiation.

It is convenient to divide the radiation field into two parts, a "stellar" part, resulting directly from the input radiation from the star, and a "diffuse" part, resulting from the emission of the ionized gas,

$$I_\nu = I_{\nu s} + I_{\nu d}. \tag{2.10}$$

The stellar radiation decreases outward because of geometrical dilution and absorption, and since its only source is the star, it can be written

$$4\pi J_{\nu s} = \pi F_{\nu s}(r) = \pi F_{\nu s}(R)\frac{R^2 e^{-\tau_\nu}}{r^2}, \tag{2.11}$$

where $\pi F_{\nu s}(r)$ is the standard astronomical notation for the flux of stellar radiation (per unit area per unit time per unit frequency interval) at r, $\pi F_{\nu s}(R)$ is the flux at the radius of the star R, and τ_ν is the radial optical depth at r,

$$\tau_\nu(r) = \int_0^r N_{\mathrm{H^0}}(r')a_\nu dr', \tag{2.12}$$

which can also be written

$$\tau_\nu(r) = \frac{a_\nu}{a_{\nu_0}}\tau_0(r)$$

in terms of τ_0, the optical depth at the threshold.

The equation of transfer for the diffuse radiation $I_{\nu d}$ is

$$\frac{dI_{\nu d}}{ds} = -N_{\mathrm{H^0}}a_\nu I_{\nu d} + j_\nu, \tag{2.13}$$

and for $kT \ll h\nu_0$ the only source of ionizing radiation is recaptures of electrons from the continuum to the ground $1\,^2S$ level. The emission coefficient for this radiation is

$$j_\nu(T) = \frac{2h\nu^3}{c^2}\left(\frac{h^2}{2\pi mkT}\right)^{3/2} a_\nu\, e^{-h(\nu-\nu_0)/kT} N_p N_e \qquad (\nu > \nu_0), \tag{2.14}$$

which is strongly peaked to $\nu = \nu_0$, the threshold. The total number of

TABLE 2.1
Recombination coefficients[a] $\alpha_{n\,^2L}$ *for H*

$\alpha_{n\,^2L}$	T		
	5000° K	10,000° K	20,000° K
$\alpha_{1\,^2S}$	2.28×10^{-13}	1.58×10^{-13}	1.08×10^{-13}
$\alpha_{2\,^2S}$	3.37×10^{-14}	2.34×10^{-14}	1.60×10^{-14}
$\alpha_{2\,^2P}$	8.33×10^{-14}	5.35×10^{-14}	3.24×10^{-14}
$\alpha_{3\,^2S}$	1.13×10^{-14}	7.81×10^{-15}	5.29×10^{-15}
$\alpha_{3\,^2P}$	3.17×10^{-14}	2.04×10^{-14}	1.23×10^{-14}
$\alpha_{3\,^2D}$	3.03×10^{-14}	1.73×10^{-14}	9.09×10^{-15}
$\alpha_{4\,^2S}$	5.23×10^{-15}	3.59×10^{-15}	2.40×10^{-15}
$\alpha_{4\,^2P}$	1.51×10^{-14}	9.66×10^{-15}	5.81×10^{-15}
$\alpha_{4\,^2D}$	1.90×10^{-14}	1.08×10^{-14}	5.68×10^{-15}
$\alpha_{4\,^2F}$	1.09×10^{-14}	5.54×10^{-15}	2.56×10^{-15}
$\alpha_{10\,^2S}$	4.33×10^{-16}	2.84×10^{-16}	1.80×10^{-16}
$\alpha_{10\,^2G}$	2.02×10^{-15}	9.28×10^{-16}	3.91×10^{-16}
$\alpha_{10\,^2M}$	2.7×10^{-17}	1.0×10^{-17}	$4. \times 10^{-18}$
α_A	6.82×10^{-13}	4.18×10^{-13}	2.51×10^{-13}
α_B	4.54×10^{-13}	2.59×10^{-13}	2.52×10^{-13}

[a] In $cm^3\ sec^{-1}$.

photons generated by recombinations to the ground level is given by the recombination coefficient

$$4\pi \int_{\nu_0}^{\infty} \frac{j_\nu}{h\nu}\, d\nu \;=\; N_p N_e \alpha_1(\mathrm{H}^0, T), \tag{2.15}$$

and since $\alpha_1 = \alpha_{1^2S} < \alpha_A$, the diffuse field $J_{\nu d}$ is smaller than $J_{\nu s}$ on the average, and may be calculated by an iterative procedure. For an optically thin nebula, a good first approximation is to take $J_{\nu d} \approx 0$.

On the other hand, for an optically thick nebula, a good first approximation is based on the fact that no ionizing photons can escape, so that every diffuse radiation-field photon generated in such a nebula is absorbed elsewhere in the nebula

$$4\pi \int \frac{j_\nu}{h\nu} dV = 4\pi \int N_{\mathrm{H}^0} \frac{a_\nu J_{\nu d}}{h\nu} dV, \tag{2.16}$$

where the integration is over the entire volume of the nebula. The so-called on-the-spot approximation amounts to assuming that a similar relation holds locally:

$$J_{\nu d} = \frac{j_\nu}{N_{\mathrm{H}^0} a_\nu}. \tag{2.17}$$

This, of course, automatically satisfies (2.16), and would be exact if all photons were absorbed very close to the point at which they are generated ("on the spot"). This is not a bad approximation because the diffuse radiation-field photons have $\nu \approx \nu_0$, and therefore have large a_ν and correspondingly small mean free paths before absorption.

Making this on-the-spot approximation and using (2.11) and (2.15), we find that the ionization equation (2.8) becomes

$$\frac{N_{\mathrm{H}^0} R^2}{r^2} \int_{\nu_0}^{\infty} \frac{\pi F_\nu(R)}{h\nu} a_\nu e^{-\tau_\nu} d\nu = N_p N_e \alpha_B(\mathrm{H}^0, T), \tag{2.18}$$

where

$$\alpha_B(\mathrm{H}^0, T) = \alpha_A(\mathrm{H}^0, T) - \alpha_1(\mathrm{H}^0, T)$$

$$= \sum_2^{\infty} \alpha_n(\mathrm{H}^0, T).$$

The physical meaning is that in optically thick nebulae, the ionizations caused by stellar radiation-field photons are balanced by recombinations to excited levels of H, while recombinations to the ground level generate ionizing photons that are absorbed elsewhere in the nebula but have no net effect on the overall ionization balance.

For any stellar input spectrum $\pi F_\nu(R)$, the integral on the left-hand side of (2.18) can be tabulated as a known function of τ_0, since a_ν and τ_ν are known functions of ν. Thus, for any assumed density distribution

$$N_{\mathrm{H}}(r) = N_{\mathrm{H}^0}(r) + N_p(r)$$

and temperature distribution $T(r)$, equations (2.18) and (2.12) can be integrated outward to find $N_{\mathrm{H}^0}(r)$ and $N_p(r) = N_e(r)$. Two calculated models for homogeneous nebulae with constant density $N_H = 10$ H atoms plus ions cm^{-3} and constant temperature $T = 7,500°$ K are listed in Table 2.2 and graphed in Figure 2.3. For one of these ionization models, the assumed $\pi F_\nu(R)$ is a blackbody spectrum with $T_* = 40{,}000°$ K, chosen to represent approximately an O6 main-sequence star, while for the other, the $\pi F_\nu(R)$ is a computed model stellar atmosphere with $T_* = 37{,}450°$ K. The table and graph clearly show the expected nearly complete ionization out to a critical radius r_1, at which the ionization drops off abruptly to nearly zero. The central ionized zone is often referred to as an "H II region" ("H^+ region" is a better name), and it is surrounded by an outer neutral H^0 region.

The radius r_1 can be found from (2.18), substituting from (2.12)

$$\frac{d\tau_\nu}{dr} = N_{\mathrm{H}^0} a_\nu$$

and integrating over r:

$$R^2 \int_{\nu_0}^{\infty} \frac{\pi F_\nu(R)}{h\nu}\, d\nu \int_0^{\infty} d(-e^{-\tau_\nu}) = \int_0^{\infty} N_p N_e \alpha_B r^2\, dr$$
$$= R^2 \int_{\nu_0}^{\infty} \frac{\pi F_\nu(R)}{h\nu}\, d\nu.$$

Using the result that the ionization is nearly complete ($N_p = N_e \approx N_{\mathrm{H}}$) within r_1, and nearly zero ($N_p = N_e \approx 0$) outside r_1, this becomes

$$4\pi R^2 \int_{\nu_0}^{\infty} \frac{\pi F_\nu}{h\nu}\, d\nu = \int_{\nu_0}^{\infty} \frac{L_\nu}{h\nu}\, d\nu$$
$$= Q(\mathrm{H}^0) = \frac{4\pi}{3}\, r_1^3\, N_{\mathrm{H}}^2 \alpha_B. \qquad (2.19)$$

Here $4\pi R^2 \pi F_\nu(R) = L_\nu$ is the luminosity of the star at frequency ν (in energy units per time per unit frequency interval), and the physical meaning of (2.19) is that the total number of ionizing photons emitted by the star just balances the total number of recombinations to excited levels within the ionized volume $4\pi r_1^3/3$, often called the Strömgren sphere. Numerical values of radii calculated by using the model stellar atmospheres discussed in Chapter 5 are given in Table 2.3.

TABLE 2.2
Calculated ionization distributions for model H II regions

r(pc)	$T_* = 4 \times 10^4$ ° K Blackbody model		$T_* = 3.74 \times 10^4$ ° K Model stellar atmosphere	
	$\frac{N_p}{N_p + N_{H^0}}$	$\frac{N_{H^0}}{N_p + N_{H^0}}$	$\frac{N_p}{N_p + N_{H^0}}$	$\frac{N_{H^0}}{N_p + N_{H^0}}$
0.1	1.0	4.5×10^{-7}	1.0	4.5×10^{-7}
1.2	1.0	2.8×10^{-5}	1.0	2.9×10^{-5}
2.2	0.9999	1.0×10^{-4}	0.9999	1.0×10^{-4}
3.3	0.9997	2.5×10^{-4}	0.9997	2.5×10^{-4}
4.4	0.9995	4.4×10^{-4}	0.9994	4.5×10^{-4}
5.5	0.9992	8.0×10^{-4}	0.9992	8.1×10^{-4}
6.7	0.9985	1.5×10^{-3}	0.9985	1.5×10^{-3}
7.7	0.9973	2.7×10^{-3}	0.9973	2.7×10^{-3}
8.8	0.9921	7.9×10^{-3}	0.9924	7.6×10^{-3}
9.4	0.977	2.3×10^{-2}	0.979	2.1×10^{-2}
9.7	0.935	6.5×10^{-2}	0.940	6.0×10^{-2}
9.9	0.838	1.6×10^{-1}	0.842	1.6×10^{-1}
10.0	0.000	1.0	0.000	1.0

TABLE 2.3
Calculated radii of Strömgren spheres

Spectral type	M_v	$T_*(° K)$	Log $Q(H^0)$ (photons/sec)	Log $N_e N_p r_1^3$ (N in cm^{-3}; r_1 in pc)	r_1 (pc) ($N_e = N_p$ = 1 cm^{-3})
O5	– 5.6	48,000	49.67	6.07	108
O6	– 5.5	40,000	49.23	5.63	74
O7	– 5.4	35,000	48.84	5.24	56
O8	– 5.2	33,500	48.60	5.00	51
O9	– 4.8	32,000	48.24	4.64	34
O9.5	– 4.6	31,000	47.95	4.35	29
B0	– 4.4	30,000	47.67	4.07	23
B0.5	– 4.2	26,200	46.83	3.23	12

NOTE: $T = 7500°$ K assumed for calculating α_B.

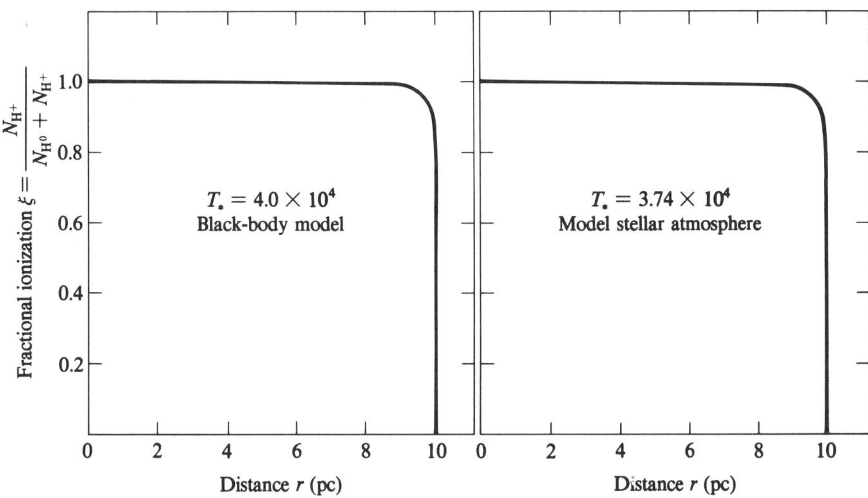

FIGURE 2.3
Ionization structure of two homogeneous pure-H model H II regions.

2.4 Photoionization of a Nebula Containing Hydrogen and Helium

The next most abundant element after H is He, whose relative abundance (by number) is of order 10 percent, and a much better approximation to the ionization structure of an actual nebula is provided by taking both these elements into account. The ionization potential of He is $h\nu_2 = 24.6$ eV, somewhat higher than H; the ionization potential of He^+ is 54.4 eV, but since even the hottest O stars emit practically no photons with $h\nu > 54.4$ eV, second ionization of He does not occur in ordinary H II regions. (The situation is quite different in planetary nebulae, as we shall see later in this chapter.) Thus photons with energy 13.6 eV $< h\nu <$ 24.6 eV can ionize H only, but photons with energy $h\nu > 24.6$ eV can ionize both H and He. As a result, two different types of ionization structure are possible, depending on the spectrum of ionizing radiation and the abundance of He. At one extreme, if the spectrum is concentrated to frequencies just above 13.6 eV and contains only a few photons with $h\nu > 24.6$ eV, then the photons with energy 13.6 eV $< h\nu <$ 24.6 eV keep the H ionized, and the photons with $h\nu > 24.6$ eV are all absorbed by He. The ionization structure thus consists of a small central H^+, He^+ zone surrounded by a larger H^+, He^0 region. At the other extreme, if the input spectrum contains a large fraction of photons with $h\nu > 24.6$ eV, then these photons dominate the ionization of both H and He, the outer boundaries of both ionized zones coincide, and there is a single H^+, He^+ region.

The He^0 photoionization cross section $a_\nu(He^0)$ is plotted in Figure 2.2, along with $a_\nu(H^0)$ and $a_\nu(He^+)$ calculated from equation (2.4). The total recombination coefficients for He to configurations $L \geq 2$ are, to a good approximation, the same as for H, since these levels are hydrogenlike, but because He is a two-electron system, it has separate singlet and triplet levels and

$$\left.\begin{aligned} \alpha_{n\,{}^1L}(He^0, T) &\approx \frac{1}{4}\,\alpha_{n\,{}^2L}(H^0, T) \\ \alpha_{n\,{}^3L}(He^0, T) &\approx \frac{3}{4}\,\alpha_{n\,{}^2L}(H^0, T) \end{aligned}\right\} \quad L \geq 2. \tag{2.20}$$

For the P and particularly the S terms there are sizeable differences between the He and H recombination coefficients. Representative numerical values of the recombination coefficients are included in Table 2.4.

The ionization equations for H and He are coupled by the radiation field with $h\nu > 24.6$ eV, and are straightforward to write down in the on-the-spot approximation, though complicated in detail. First, the photons emitted in recombinations to the ground level of He can ionize either H or He, since these photons are emitted with energies just above $h\nu_2 = 24.6$ eV. The fraction absorbed by H is

$$y = \frac{N_{H^0} a_{\nu_2}(H^0)}{N_{H^0} a_{\nu_2}(H^0) + N_{He^0} a_{\nu_2}(He^0)}, \tag{2.21}$$

and the remaining fraction $1 - y$ is absorbed by He. Second, following recombination to excited levels of He, various photons are emitted that ionize H. Of the recombinations to excited levels of He, approximately three-fourths are to the triplet levels and approximately one-fourth are to the singlet levels. All the captures to triplets lead ultimately through downward radiative transitions to $2\,{}^3S$, which is highly metastable, but which can decay by a one-photon forbidden line at 19.8 eV to $1\,{}^1S$, with transition probability $A_{2\,{}^3S,1\,{}^1S} = 1.27 \times 10^{-4}\ \text{sec}^{-1}$. Competing with this mode of depopulation of $2\,{}^3S$, collisional excitation to the singlet levels $2\,{}^1S$, and $2\,{}^1P$ can also occur with fairly high probability, while collisional transitions to $1\,{}^1S$ or to the continuum are less probable. Since the collisions leading to the singlet levels involve a spin change, only electrons are effective in causing these excitations, and the transition rate per atom in the $2\,{}^3S$ level is

$$N_e q_{2\,{}^3S,2\,{}^1L} = N_e \int_{\frac{1}{2}mv^2=\chi}^{\infty} v\sigma_{2\,{}^3S,\,2\,{}^1L}(v) f(v)\,dv, \tag{2.22}$$

where the $\sigma_{2^3S,2^1L}(v)$ are the electron collision cross sections for these excita-

TABLE 2.4
Recombination coefficients[a] *for* He

	T		
	$5000°K$	$10{,}000°K$	$20{,}000°K$
$\alpha(\mathrm{He}^0, 1\ ^1S)$	2.23×10^{-13}	1.59×10^{-13}	1.14×10^{-13}
$\alpha(\mathrm{He}^0, 2\ ^1S)$	7.64×10^{-15}	5.55×10^{-15}	4.06×10^{-15}
$\alpha(\mathrm{He}^0, 2\ ^1P)$	2.11×10^{-14}	1.35×10^{-14}	8.16×10^{-15}
$\alpha(\mathrm{He}^0, 3\ ^1S)$	2.23×10^{-15}	1.63×10^{-15}	1.19×10^{-15}
$\alpha(\mathrm{He}^0, 3\ ^1P)$	8.92×10^{-15}	5.65×10^{-15}	3.34×10^{-15}
$\alpha(\mathrm{He}^0, 3\ ^1D)$	9.23×10^{-15}	5.28×10^{-15}	2.70×10^{-15}
$\alpha_B(\mathrm{He}^0, \Sigma\ n\ ^1L)$	1.08×10^{-13}	6.27×10^{-14}	3.46×10^{-14}
$\alpha(\mathrm{He}^0, 2\ ^3S)$	1.98×10^{-14}	1.48×10^{-14}	1.13×10^{-14}
$\alpha(\mathrm{He}^0, 2\ ^3P)$	8.78×10^{-14}	5.75×10^{-14}	3.59×10^{-14}
$\alpha(\mathrm{He}^0, 3\ ^3S)$	4.88×10^{-15}	3.77×10^{-15}	2.97×10^{-15}
$\alpha(\mathrm{He}^0, 3\ ^3P)$	3.20×10^{-14}	2.09×10^{-14}	1.30×10^{-14}
$\alpha(\mathrm{He}^0, 3\ ^3D)$	2.84×10^{-14}	1.64×10^{-14}	8.46×10^{-15}
$\alpha_B(\mathrm{He}^0, \Sigma\ n\ ^3L)$	3.26×10^{-13}	2.10×10^{-13}	1.20×10^{-13}
$\alpha_B(\mathrm{He}^0)$	4.34×10^{-13}	2.73×10^{-13}	1.55×10^{-13}

[a] In cm^3 sec^{-1}.

tion processes, and the χ are their energy thresholds. These rate coefficients are listed in Table 2.5, along with the critical electron density $N_c(2\ ^3S)$, defined by

$$N_c(2\ ^3S) = \frac{A_{2\ ^3S,1\ ^1S}}{q_{2\ ^3S,2\ ^1S} + q_{2\ ^3S,2\ ^1P}}, \tag{2.23}$$

at which collisional transitions are equally probable with radiative transitions. In typical H II regions, the electron density $N_e \lesssim 10^2$ cm^{-3}, considerably smaller than N_c, so practically all the atoms leave 2 ^3S by emission of a 19.8 eV-line photon. In contrast, in typical bright planetary nebulae, $N_e \approx 10^4$ cm^{-3}, somewhat larger than N_c, and therefore many of the atoms are transferred to 2 ^1S or 2 ^1P before emitting a line photon. From the ratio of excitation rates, it can be seen that, for instance, at $T = 10^4$ ° K, a fraction 0.83 of the transitions lead to 2 ^1S, and 0.17 to 2 ^1P. If the less probable collisional deexcitating collisions to 1 ^1S are also included, these fractions become 0.78 and 0.16, respectively.

TABLE 2.5
Collisional excitation coefficients from $He^0(2\ ^3S)$

$T(°K)$	$q_{2\ ^3S,2\ ^1S}$ ($cm^3\ sec^{-1}$)	$q_{2\ ^3S,2\ ^1P}$ ($cm^3\ sec^{-1}$)	N_c (cm^{-3})
6,000	1.83×10^{-8}	2.02×10^{-9}	6.2×10^3
8,000	2.40×10^{-8}	3.83×10^{-9}	4.6×10^3
10,000	2.74×10^{-8}	5.51×10^{-9}	3.9×10^3
15,000	2.95×10^{-8}	8.43×10^{-9}	3.3×10^3
20,000	2.83×10^{-8}	9.84×10^{-9}	3.3×10^3
25,000	2.63×10^{-8}	1.05×10^{-8}	3.4×10^3

Of the captures to the singlet-excited levels in He, approximately two-thirds lead ultimately to population of 2 ^1P, while approximately one-third lead to population of 2 ^1S. Atoms in 2 ^1P decay mostly to 1 ^1S with emission of a resonance-line photon at 21.2 eV, but some also decay to 2 ^1S (with emission of 2 ^1S - 2 ^1P at 2.06 μ) with a relative probability of approximately 10^{-3}. The resonance-line photons are scattered by He^0, and therefore, after approximately 10^3 scatterings, a typical photon would, on the average, be converted to a 2.06 μ line photon and thus populate 2 ^1S. However, it is more likely that before a resonance-line photon is scattered this many times, it will photoionize an H atom and be absorbed. He atoms in 2 ^1S decay by two-photon emission (with the sum of the energies 20.6 eV and transition probability 51 sec^{-1}) to 1 ^1S. From the distribution of photons in this continuous spectrum, the probability that a photon is produced that can ionize H is 0.56 per radiative decay from He^0 2 ^1S.

All these He bound-bound transitions produce photons that ionize H but not He, and they can easily be included in the H ionization equation in the on-the-spot approximation. The total number of recombinations to excited levels of He per unit volume per unit time is $N_{He^+} N_e \alpha_B(He^0, T)$, and of these a fraction p generate ionizing photons that are absorbed on the spot. As shown by the preceding discussion, in the low-density limit $N_e \ll N_c$,

$$p \approx \frac{3}{4} + \frac{1}{4}\left[\frac{2}{3} + \frac{1}{3}(0.56)\right] = 0.96,$$

but in the high-density limit $N_e \gg N_c$,

$$p \approx \left[\frac{3}{4}\,(0.78) + \frac{1}{4}\cdot\frac{1}{3}\right](0.56) + \left[\frac{3}{4}(0.16) + \frac{1}{4}\cdot\frac{2}{3}\right] = 0.66.$$

Thus, in the on-the-spot approximation, the ionization equations become

$$\frac{N_{\mathrm{H^0}}R^2}{r^2}\int_{\nu_0}^{\infty}\frac{\pi F_\nu(R)}{h\nu}a_\nu(\mathrm{H^0})e^{-\tau_\nu}d\nu + yN_{\mathrm{He^+}}N_e\alpha_1(\mathrm{He^0},T)$$
$$+\,pN_{\mathrm{He^+}}N_e\alpha_B(\mathrm{He^0},T) = N_pN_e\alpha_B(\mathrm{H^0},T); \qquad (2.24)$$

$$\frac{N_{\mathrm{He^0}}R^2}{r^2}\int_{\nu_2}^{\infty}\frac{\pi F_\nu(R)}{h\nu}a_\nu(\mathrm{He^0})e^{-\tau_\nu}d\nu + (1-y)N_{\mathrm{He^+}}N_e\alpha_1(\mathrm{He^0},T)$$
$$= N_{\mathrm{He^+}}N_e\alpha_A(\mathrm{He^0},T), \qquad (2.25)$$

with

$$\frac{d\tau_\nu}{dr} = N_{\mathrm{H^0}}a_\nu(\mathrm{H^0}) \text{ for } \nu_0 < \nu < \nu_2$$

and (2.26)

$$\frac{d\tau_\nu}{dr} = N_{\mathrm{H^0}}a_\nu(\mathrm{H^0}) + N_{\mathrm{He^0}}a_\nu(\mathrm{He^0}) \text{ for } \nu_2 < \nu,$$

and

$$N_e = N_p + N_{\mathrm{He^+}}.$$

These equations again can be integrated outward step-by-step, and sample models for a diffuse nebula (with $N_{\mathrm{H}} = 10\ \mathrm{cm^{-3}}$, $N_{\mathrm{He}}/N_{\mathrm{H}} = 0.15$) excited by an O6 star and by a B0 star are plotted in Figure 2.4. It can be seen that the hotter O6 star excites a coincident $\mathrm{H^+}$, $\mathrm{He^+}$ zone, while the cooler B0 star has an inner $\mathrm{H^+}$, $\mathrm{He^+}$ zone and an outer $\mathrm{H^+}$, $\mathrm{He^0}$ zone.

Although the exact size of the $\mathrm{He^+}$ zone can be found only from the integration because of the coupling between the H and He ionization by the radiation with $\nu > \nu_2$, the approximate size can easily be found by ignoring the absorption by H in the $\mathrm{He^+}$ zone. This corresponds to setting $y = 0$ in equation (2.25) and $N_{\mathrm{H^0}} = 0$ in the second of equations (2.26), and we then immediately find, in analogy to equation (2.19), that

$$\int_{\nu_2}^{\infty}\frac{L_\nu}{h\nu} = Q(\mathrm{He^0}) = \frac{4\pi}{3}r_2^3N_{\mathrm{He^+}}N_e\alpha_B(\mathrm{He^0}), \qquad (2.27)$$

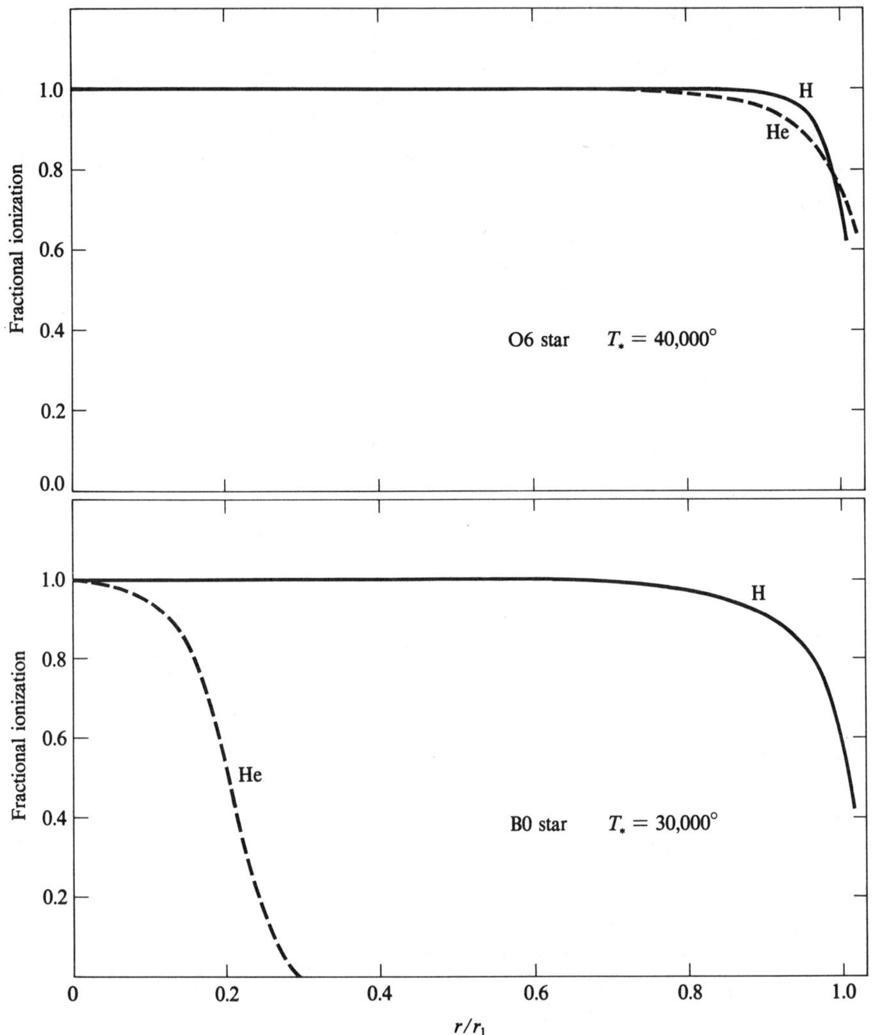

FIGURE 2.4
Ionization structure of two homogeneous H + He model H II regions.

where r_2 is the radius of the He$^+$ zone. Furthermore, since, according to the preceding discussions, $p \approx 1$, the absorptions by He do not greatly reduce the number of photons available for ionizing H; therefore, to a fair approximation,

$$\int_{\nu_0}^{\infty} \frac{L_\nu}{h\nu} d\nu = Q(\mathrm{H}^0) = \frac{4\pi}{3} r_1^3 N_{\mathrm{H}^+} N_e \alpha_B(\mathrm{H}^0). \tag{2.19}$$

If we suppose that the He$^+$ zone is much smaller than the H$^+$ zone, then throughout most of the H$^+$ zone the electrons come only from ionization of H, but in the He$^+$ zone, the electrons come from ionization of both H and He. With this simplification,

$$\left(\frac{r_1}{r_2}\right)^3 = \frac{Q(\mathrm{H}^0)}{Q(\mathrm{He}^0)}\frac{N_{\mathrm{He}}}{N_{\mathrm{H}}}\left(1 + \frac{N_{\mathrm{He}}}{N_{\mathrm{H}}}\right)\frac{\alpha_B(\mathrm{He}^0)}{\alpha_B(\mathrm{H}^0)} \quad \text{if } r_2 < r_1. \tag{2.28}$$

A plot of r_2/r_1, calculated according to this equation for $N_{\mathrm{He}}/N_H = 0.15$ and $T = 7{,}500°$ K, is shown in Figure 2.5, and it can be seen that, for $T_* \geq 40{,}000°$ K, the He$^+$ and H$^+$ zones are coincident, while at significantly lower temperatures, the He$^+$ zone is much smaller. The details of the curve, including the precise effective temperature at which $r_2/r_1 = 1$, are not significant, because of the simplifications made, but the general trends it indicates are correct. For a smaller relative helium abundance, for instance $N_{\mathrm{He}}/N_{\mathrm{H}} = 0.10$ instead of 0.15, r_2/r_1 is larger by approximately 16% at corresponding values of T_*, up to $r_2/r_1 = 1$.

2.5 Photoionization of He$^+$ to He^{++}

Although ordinary O stars of Population I do not radiate any appreciable number of photons with $h\nu > 54.4$ eV (hence galactic H II regions do not have a He^{++} zone), the situation is quite different for the central stars of planetary nebulae. Many of these stars are much hotter than even the hottest O5 stars, and do radiate high-energy photons that produce central He^{++} zones, which are observed by the He II recombination spectra they emit.

The structure of these central He^{++} zones is governed by equations that are very similar to those for pure H$^+$ zones discussed previously, with the threshold, absorption cross section, and recombination coefficient changed from H^0 ($Z = 1$) to He$^+$ ($Z = 2$). This He^{++} zone is, of course, also an H$^+$ zone, and the ionization equations of H^0 and He$^+$ are therefore, in principle, coupled: but in practice they can be fairly well separated. The coupling results from the fact that, in the recombination of He^{++} to form He$^+$, photons are emitted that ionize H^0. Three different mechanisms are involved, namely: recombinations that populate $2\,^2P$, resulting in He II $L\alpha$ emission with $h\nu = 40.8$ eV; recombinations that populate $1\,^2S$, resulting in He II $2\,^2S \to 1\,^2S$ two-photon emission for which $h\nu' + h\nu'' = 40.8$ eV (the spectrum peaks at 20.4 eV, and, on the average, 1.42 ionizing photons are emitted per decay); and recombinations directly to $2\,^2S$ and to $2\,^2P$, resulting in He II Balmer continuum emission, which has the same threshold as the Lyman limit of H and therefore emits a continuous spectrum concentrated just above $h\nu_0$. The He II $L\alpha$ photons are scattered by resonance scattering, and therefore diffuse only slowly away from their point of origin before they are absorbed, while the

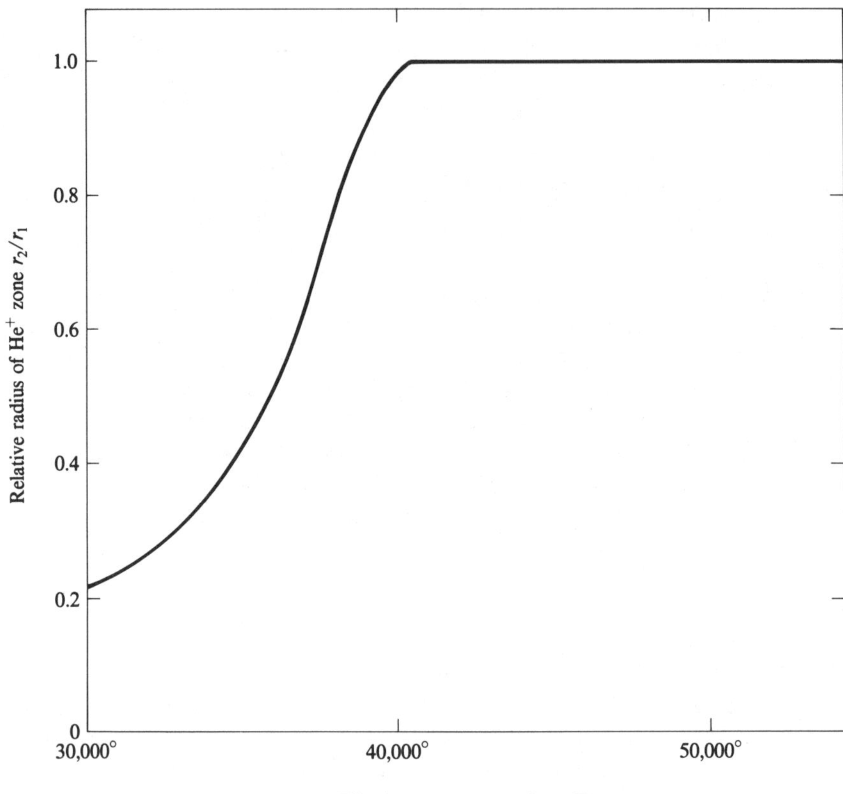

FIGURE 2.5
Relative radius of He$^+$ zone as a function of effective temperature of exciting star.

He II Balmer-continuum photons are concentrated close to the H^0 ionization threshold and therefore have a short mean free path. Both these sources tend to ionize H^0 in the He^{++} zone, and at a "normal" abundance of He, He/H ≈ 0.15, the number of ionizing photons generated in the He^{++} zone by these two processes is nearly sufficient to balance the recombinations of H^+ in this zone and thus to maintain the ionization of H^0. This is shown in Table 2.6, which lists the ionizing-photon generation rates relative to the recombination rates for two temperatures. Thus, to a good approximation, the He II $L\alpha$ and Balmer continuum photons are absorbed by and maintain the ionization of H^0 in the He^{++} zone, but the stellar radiation with 13.6 eV $< h\nu <$ 54.4 eV is not significantly absorbed by the H^0 in the He^{++}, H^+ zone, and that with $h\nu >$ 54.4 eV is absorbed only by the He$^+$. The He II two-photon continuum

TABLE 2.6
Generation of H ionizing photons in the He^{++} *zone*

Number generated per H recombination	$T = 1 \times 10^4$ °K	2×10^4 °K
$N_{He^{++}}\, q(He^+\, L\alpha)/N_{H^+}\alpha_B(H^0)$	0.64	0.66
$N_{He^{++}}\, q(He^+$ 2 photon$)/N_{H^+}\alpha_B(H^0)$	0.36	0.42
$N_{He^{++}}\, q(He^+\, Ba\ c)/N_{H^+}\alpha_B(H^0)$	0.20	0.25

NOTE: Numerical values are calculated assuming that $N(He^{++})/N(H^+) = 0.15$.

is an additional source of ionizing photons for H; most of these photons escape from the He^{++} zone and therefore must be added to the stellar radiation field with $h\nu > 54.4$ eV in the He^+ zone. Of course, a more accurate calculation may be made, taking into account the detailed frequency dependence of each of the emission processes, but since normally the helium abundance is small, only an approximation to its effects is usually required.

Some sample calculations of the ionization structure of a model planetary nebula, with the radiation source a black body at $T_* = 10^5$ ° K are shown in Figure 2.6. The sharp outer edge of the He^{++} zone, as well as the even sharper outer edges of the H^+ and He^+ zones, can be seen in these graphs. There is, of course, an equation that is exactly analogous to (2.19) and (2.27) for the "Strömgren radius" r_3 of the He^{++} zone:

$$\int_{4\nu_0}^{\infty} \frac{L_\nu}{h\nu} d\nu = \frac{4\pi}{3} r_3^3 N_{He^{++}} N_e \alpha_B(He^+, T). \tag{2.29}$$

Thus stellar temperatures $T_* \gtrsim 10^5$ ° K are required for $r_3/r_1 \approx 1$.

2.6 Further Iterations of the Ionization Structure

As described previously, the on-the-spot approximation may be regarded as the first approximation to the ionization and, as will be described in Chapter 3, to the temperature distribution in the nebula. From these a first approximation to the emission coefficient j_ν may be found throughout the model: from j_ν a first approximation to $I_{\nu d}$ and hence J_ν at each point, and from J_ν a better approximation to the ionization and temperature at each point. This iteration procedure can be repeated as many times as desired (given sufficient computing time) and actually converges quite rapidly, but except where high accuracy is required, the first (on-the-spot) approximation is usually sufficient. The higher approximations show that the degree of ionization nearest the star is so high that ionizing photons emitted there are not really absorbed

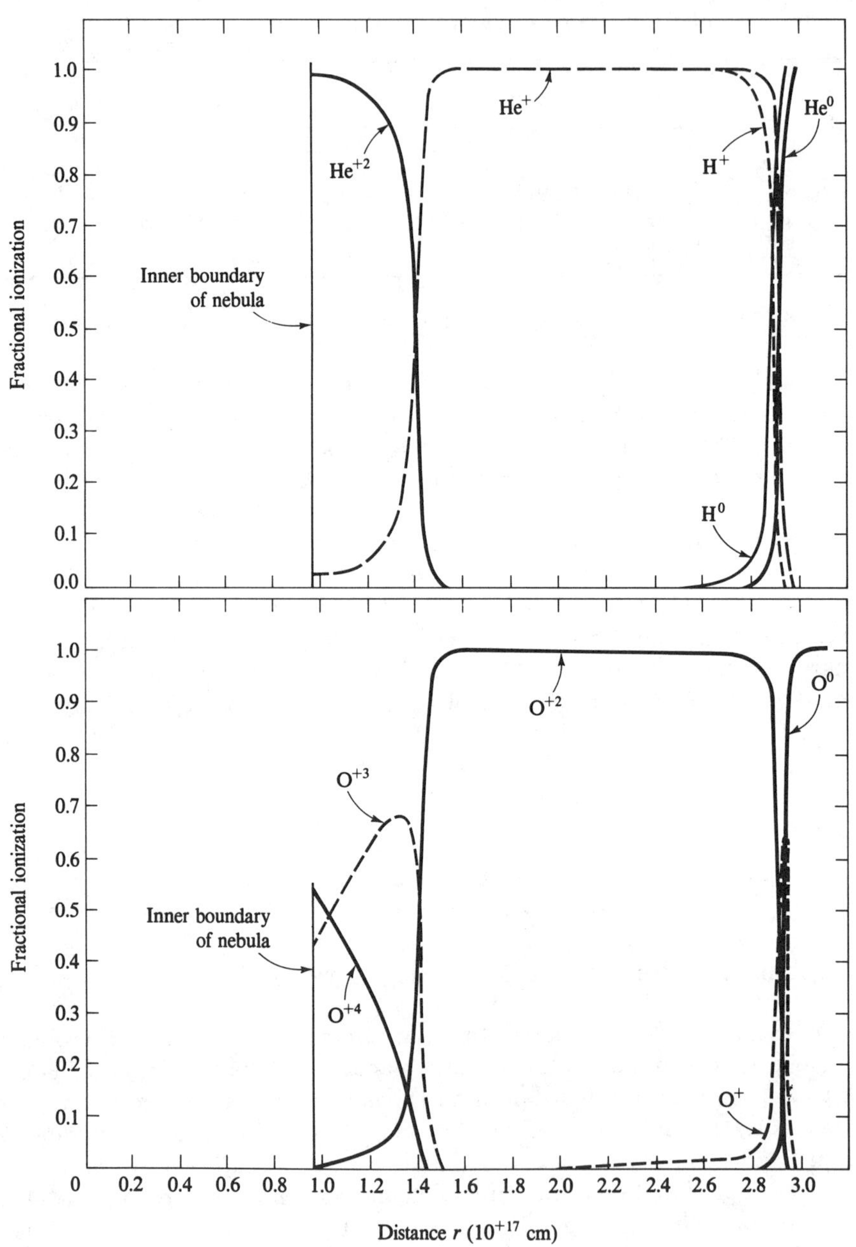

FIGURE 2.6
Ionization structure of H, He (*top*), and O (*bottom*) for a model planetary nebula.

on the spot. There is thus a net flow of these photons from the center to the outer regions. However, the ionization calculated from the on-the-spot approximation is reasonably close to the final exact result.

2.7 Photoionization of Heavy Elements

Finally, let us examine the ionization of the heavy elements, of which O, Ne, C, N, Fe, Si, with abundances (by number) of order 10^{-3} to 10^{-4} that of H, are the most abundant. The ionization-equilibrium equation for any two successive stages of ionization i and $i+1$ of any element X may be written

$$N(X^{+i}) \int_{\nu_i}^{\infty} \frac{4\pi J_\nu}{h\nu} a_\nu(X^{+i}) d\nu = N(X^{+i+1}) N_e \alpha_G(X^{+i}, T), \qquad (2.30)$$

where $N(X^{+i})$ and $N(X^{+i+1})$ are the number densities of the two successive stages of ionization; $a_\nu(X^{+i})$ is the photoionization cross section from the ground level of X^i with the threshold ν_i; and $\alpha_G(X^{+i}, T)$ is the recombination coefficient of the ground level of X^{+i+1} to all levels of X^{+i}. These equations, together with the total number of ions of all stages of ionization,

$$N(X^0) + N(X^{+1}) + N(X^{+2}) + \cdots + N(X^{+n}) = N(X)$$

(presumably known from the abundance of X), completely determine the ionization equilibrium at each point. The mean intensity J_ν, of course, includes both the stellar and diffuse contributions, but the abundances of the heavy elements are so small that their contributions to the diffuse field are negligible, and only the emission by H, He, and He^+ mentioned previously needs to be taken into account.

The required data of numerical values of a_ν and α_G are less readily available for heavy elements, which are many-electron systems, than for H and He. However, approximate calculations, mostly based on Hartree-Fock or close-coupling wave functions, are available for many common ions. For simple ions, photoionization, which is the removal of one outer electron, can lead to only a single level, the ground level of the resulting ion. However, in more complicated ions, photoionization can often lead to any of several levels of the ground configuration of the resulting ion, and correspondingly the photoionization cross section has several thresholds, instead of a single threshold as for simpler ions. An example is neutral O, which can be photoionized by the following schemes:

$$\begin{aligned}
O^0(2p^4\ {}^3P) + h\nu \rightarrow\ & \left.\begin{array}{l} O^+(2p^3\ {}^4S) + ks \\ O^+(2p^3\ {}^4S) + kd \end{array}\right\} \quad h\nu > 13.6 \text{ eV}, \\
& \left.\begin{array}{l} O^+(2p^3\ {}^2D) + ks \\ O^+(2p^3\ {}^2D) + kd \end{array}\right\} \quad h\nu > 16.9 \text{ eV}, \\
& \left.\begin{array}{l} O^+(2p^3\ {}^2P) + ks \\ O^+(2p^3\ {}^2P) + kd \end{array}\right\} \quad h\nu > 18.6 \text{ eV}.
\end{aligned}$$

The calculated photoionization cross section is plotted in Figure 2.7. Note that inner-shell photoionization can also occur; for example, in neutral O,

$$\begin{aligned}
O^0(2s^2\ 2p^4\ {}^3P) + h\nu \rightarrow\ & O^+(2s\ 2p^4\ {}^4P) + kp\} \quad h\nu > 28.4 \text{ eV} \\
& O^+(2s\ 2p^4\ {}^2D) + kp\} \quad h\nu > 34.0 \text{ eV}
\end{aligned}$$

but the thresholds are generally so high that there is little available radiation and they make no contribution to the photoionization in most nebulae. However, they cannot be ignored if the source has a spectrum that does not drop off rapidly at high energies — as is the situation in quasars and other active galactic nuclei.

Actual photoionization cross sections of many-electron atoms and ions show complicated resonance structures, with large variations in quite small frequency intervals. These are smoothed out in a representation like Figure 2.7, but such smoothed cross sections are quite adequate for most applications. A good interpolation formula that fits the smoothed contribution of each threshold ν_T to the photoionization cross section is

$$a_\nu = a_T\left[\beta\left(\frac{\nu}{\nu_T}\right)^{-s} + (1-\beta)\left(\frac{\nu}{\nu_T}\right)^{-s-1}\right] \quad \nu > \nu_T, \tag{2.31}$$

and the total cross section is then the sum of the contributions of the individual thresholds. A list of numerical values of ν_T, a_T, β, and s for common atoms and ions (including H^0, He^0, and He^+) is given in Table 2.7. These interpolation formulae are good approximations for energies within a few times the threshold energy; at higher energies numerically computed values must be used.

The heavy elements do not usually make an appreciable contribution to the optical depth, but in some situations they can, particularly at frequencies just below the He^0 threshold. They can always be included by simple

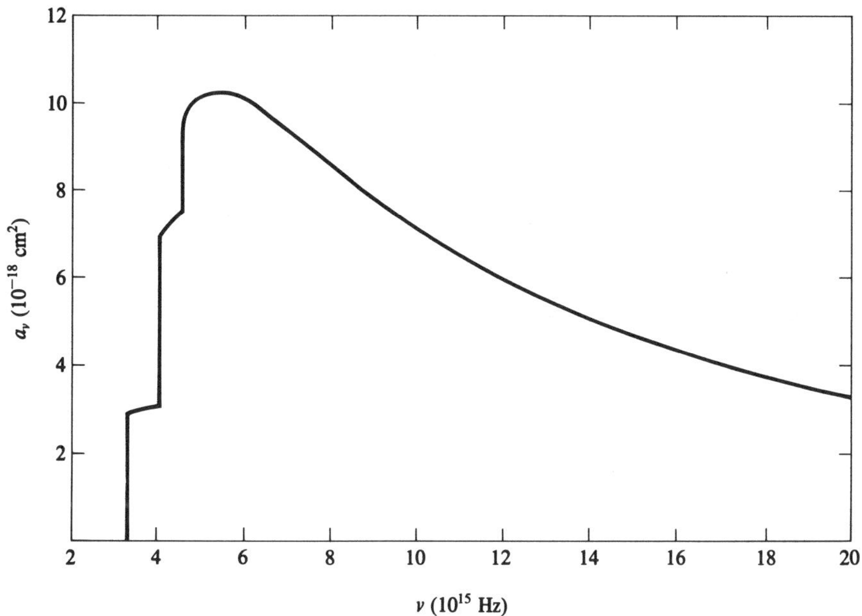

FIGURE 2.7
Absorption cross section of O^0, showing several thresholds resulting from photoionization to several levels of O^+.

generalizations of equations (2.26),

$$\frac{d\tau_\nu}{dr} = N^0_{\rm H}\, a_\nu({\rm H}^0) + \sum_{X,i} N(X^{+i})\, a_\nu(X^{+i}) \text{ for } \nu_0 < \nu < \nu_2$$

and (2.32)

$$\frac{d\tau_\nu}{dr} = N_{\rm H^0}\, a_\nu({\rm H}^0) + N_{\rm He^0}\, a_\nu({\rm He}^0) + \sum_{X,i} N(X^{+i})\, a_\nu(X^{+i}) \text{ for } \nu_2 < \nu$$

In both these equations all stages of ionization of all heavy elements that have thresholds below the frequency for which the optical depth is being calculated are included in the sums over X and i.

TABLE 2.7
Photoionization cross-section parameters

Parent	Resulting ion	$\nu_T(\mathrm{cm}^{-1})$	$a_T(10^{-18}\ \mathrm{cm}^2)$	β	s
$H^0(^2S)$	$H^+(^1S)$	1.097×10^5	6.30	1.34	2.99
$He^0(^1S)$	$He^+(^2S)$	1.983×10^5	7.83	1.66	2.05
$He^+(^2S)$	$He^{+2}(^1S)$	4.389×10^5	1.58	1.34	2.99
$C^0(^3P)$	$C^+(^2P)$	9.09×10^4	12.20	3.32	2.00
$C^+(^2P)$	$C^{+2}(^1S)$	1.97×10^5	4.60	1.95	3.00
$C^{+2}(^1S)$	$C^{+3}(^2S)$	3.86×10^5	1.60	2.60	3.00
$C^{+3}(^2S)$	$C^{+4}(^1S)$	5.21×10^5	0.68	1.00	2.00
$N^0(^4S)$	$N^+(^3P)$	1.17×10^5	11.40	4.29	2.00
$N^+(^3P)$	$N^{+2}(^2P)$	2.39×10^5	6.65	2.86	3.00
$N^{+2}(^2P)$	$N^{+3}(^1S)$	3.83×10^5	2.06	1.63	3.00
$N^{+3}(^1S)$	$N^{+4}(^2S)$	6.25×10^5	1.08	2.60	3.00
$N^{+4}(^2S)$	$N^{+5}(^1S)$	7.90×10^5	0.48	1.00	2.00
$O^0(^3P)$	$O^+(^4S)$	1.098×10^5	2.94	2.66	1.00
$O^0(^3P)$	$O^+(^2D)$	1.363×10^5	3.85	4.38	1.50
$O^0(^3P)$	$O^+(^2P)$	1.500×10^5	2.26	4.31	1.50
$O^+(^4S)$	$O^{+2}(^3P)$	2.836×10^5	7.32	3.84	2.50
$O^{+2}(^3P)$	$O^{+3}(^2P)$	4.432×10^5	3.65	2.01	3.00
$O^{+3}(^2P)$	$O^{+4}(^1S)$	6.244×10^5	1.27	0.83	3.00
$O^{+4}(^1S)$	$O^{+5}(^2S)$	9.187×10^5	0.78	2.60	3.00
$O^{+5}(^2S)$	$O^{+6}(^1S)$	1.114×10^6	0.36	1.00	2.10
$Ne^0(^1S)$	$Ne^+(^2P)$	1.739×10^5	5.35	3.77	1.00
$Ne^+(^2P)$	$Ne^{+2}(^3P)$	3.314×10^5	4.16	2.72	1.50
$Ne^+(^2P)$	$Ne^{+2}(^1D)$	3.572×10^5	2.71	2.15	1.50
$Ne^+(^2P)$	$Ne^{+2}(^1S)$	3.871×10^5	0.52	2.13	1.50
$Ne^{+2}(^3P)$	$Ne^{+3}(^4S)$	5.141×10^5	1.80	2.28	2.00
$Ne^{+2}(^3P)$	$Ne^{+3}(^2D)$	5.551×10^5	2.50	2.35	2.50
$Ne^{+2}(^3P)$	$Ne^{+3}(^2P)$	5.763×10^5	1.48	2.23	2.50
$Ne^{+3}(^4S)$	$Ne^{+4}(^3P)$	7.839×10^5	3.11	1.96	3.00
$Ne^{+4}(^3P)$	$Ne^{+5}(^2P)$	1.020×10^6	1.40	1.47	3.00
$Ne^{+5}(^2P)$	$Ne^{+6}(^1S)$	1.274×10^6	0.49	1.15	3.00

The recombination coefficients for complex ions may be divided into two parts, a radiative and a dielectronic part,

$$\alpha_G(X^{+i},T) \;=\; \alpha_R(X^{+i},T) \;+\; \alpha_d(X^{+i}T). \tag{2.33}$$

The radiative part represents simple bound-free recaptures. Just as in H and He, captures to any level are followed by downward radiative transitions, leading ultimately to the ground level. Thus the radiative recombination coefficient is a sum over all levels, and is dominated by the excited levels, which to a good approximation are hydrogen-like. These summed recombination coefficients are listed for reference in Table 2.8.

In the best calculations, the deviations due to departures of the energy levels from the exact hydrogen-like energies are taken into account through "effective" nonintegral principal quantum numbers. The recombinations to the terms resulting from recaptures of an electron to the innermost unfilled shell, for instance,

$$\left.\begin{array}{l} \mathrm{O}^{++}(2p^2\,{}^3P) \;+\; ks \\ \mathrm{O}^{++}(2p^2\,{}^3P) \;+\; kd \end{array}\right\} \;\;\longrightarrow\;\; \mathrm{O}^{+}(2p^3\,{}^4S) \;+\; h\nu,$$

$$\left.\begin{array}{l} \mathrm{O}^{++}(2p^2\,{}^3P) \;+\; ks \\ \mathrm{O}^{++}(2p^2\,{}^3P) \;+\; kd \end{array}\right\} \;\;\longrightarrow\;\; \mathrm{O}^{+}(2p^3\,{}^2D) \;+\; h\nu,$$

$$\left.\begin{array}{l} \mathrm{O}^{++}(2p^2\,{}^3P) \;+\; ks \\ \mathrm{O}^{++}(2p^2\,{}^3P) \;+\; kd \end{array}\right\} \;\;\longrightarrow\;\; \mathrm{O}^{+}(2p^3\,{}^2S) \;+\; h\nu,$$

are strongly affected by the presence of the other bound electrons, through the Pauli principle, and are far from hydrogen-like. These recombination coefficients can be found from the computed photoionization cross sections, using the Milne relation as explained in Appendix 1.

Since the radiative recombination coefficients result mainly from the sum of captures to excited levels, their temperature dependences are similar to those of H and He. To a good approximation they can be represented by

$$\alpha_R(X^{+i},T) \;=\; \alpha_R(X^{+i},10,000^\circ\,K)\left(\frac{10,000^\circ\,K}{T}\right)^{1/2}[f \;+\; (1-f)\phi(T)], \tag{2.34}$$

and some representative values of $\alpha_R(X^{+i},10,000^\circ\ \mathrm{K})$ and f are listed in Table 2.9. Values of $\phi\ (T)$ for these ions (all have ground levels in which the outermost bound electron is in the shell $n = 2$) are given in Table 2.10. An even simpler, but somewhat coarser approximation is that the temperature dependence for these ions is given by $\alpha_R \propto T^{-0.8}$, which corresponds to a mean value $f = 0.3$.

TABLE 2.8
Recombination coefficients[a] *for* H-*like ions*

	T				
	1250° K	2500° K	5000° K	10,000° K	20,000° K
$\alpha_A = \sum_1^\infty \alpha_n$	1.74×10^{-12}	1.10×10^{-12}	6.82×10^{-13}	4.18×10^{-13}	2.51×10^{-13}
$\alpha_B = \sum_2^\infty \alpha_n$	1.28×10^{-12}	7.72×10^{-13}	4.54×10^{-13}	2.59×10^{-13}	1.43×10^{-13}
$\alpha_C = \sum_3^\infty \alpha_n$	1.03×10^{-12}	5.99×10^{-13}	3.37×10^{-13}	1.83×10^{-13}	9.50×10^{-14}
$\alpha_D = \sum_4^\infty \alpha_n$	8.65×10^{-13}	4.86×10^{-13}	2.64×10^{-13}	1.37×10^{-13}	6.83×10^{-14}

NOTE: In this table, $Z = 1$; for other values of $Z, \alpha(Z,T) = Z\alpha(1, T/Z^2)$.
[a]In $cm^3\ sec^{-1}$.

The additional dielectronic part of the recombination coefficient is larger than the radiative part for many, but not all, heavy ions at nebular temperatures. It results from resonances (at specific energies) in the total recombination cross section, which are related to resonances (at related specific frequencies) in the corresponding photoionization cross section. Physically, these occur at energies at which the incoming free electron can give up nearly all its kinetic energy to exciting a bound level of the ion, thus creating a short-lived doubly excited level of the next lower stage of ionization. This level can then often decay to a singly excited bound level, and then by further successive radiative transitions downward to the ground level.

TABLE 2.9
Radiative recombination coefficients[a] *for heavy ions*

X^{+i}	$\alpha_R(X^{+i}, 10{,}000^\circ\ \mathrm{K})$	f
C^0	4.66×10^{-13}	0.500
C^+	2.45×10^{-12}	0.474
C^{+2}	5.05×10^{-12}	0.325
C^{+3}	8.45×10^{-12}	0.193
N^0	3.92×10^{-13}	0.383
N^+	2.28×10^{-12}	0.425
N^{+2}	5.44×10^{-12}	0.360
N^{+3}	9.55×10^{-12}	0.272
O^0	3.31×10^{-13}	0.260
O^+	2.05×10^{-12}	0.346
O^{+2}	5.43×10^{-12}	0.350
O^{+3}	1.03×10^{-11}	0.314
Ne^0	2.83×10^{-13}	0.104
Ne^+	1.71×10^{-12}	0.193
Ne^{+2}	4.44×10^{-12}	0.245
Ne^{+3}	9.81×10^{-12}	0.274

[a]In $\mathrm{cm^3\ sec^{-1}}$.

An example is the recombination of C^{++} to form C^+. A free electron with kinetic energy 0.41 eV colliding with a C^{++} ion in the ground $2s^2\ {}^1S$ level makes up a five-electron system with the same total energy as a C^+ ion in the $2s2p3d\ {}^2F$ term. There is thus a strong resonance centered at this energy, representing the high probability of collisional excitation of one of the bound $2s$ electrons to $2p$ together with capture of the free electron into the $3d$ shell. This term can then emit a photon decaying with non-zero transition probability to $2s2p^2\ ^2D$, a bound term, which decays ultimately to the ground $2s^22p\ ^2P_{1/2}$ level. This dielectronic transition process may be written

$$\begin{aligned} C^{++}(2s^2\ ^1S)\ +\ \kappa f\ \rightarrow\ & C^+\ [2s2p(^3P)3d]\ ^2F \\ & \downarrow \\ & C^+(2s2p^2\ ^2D)\ +\ h\nu \\ & \downarrow \\ & C^+(2s^22p\ ^2P)\ +\ h\nu. \end{aligned}$$

This is actually the main recombination process for C^{++} at nebular temperatures. Note that spin, orbital angular momentum and parity must all be conserved in the first radiationless transition; for this reason only free f ($l = 3$) electrons are involved in this specific dielectronic recombination process.

TABLE 2.10
Temperature dependence function for recombination to atoms and ions with outermost electron in $n = 2$ shell

T(° K)	$\phi(T)$
5,000	1.317
7,500	1.130
10,000	1.000
15,000	0.839
20,000	0.732

Selected calculated values of $\alpha_d(X^{+i}, T)$ are listed in Table 2.11. The temperature dependence is not simple, because the resonance effects introduce an exponential behavior; so the values are given at three representative temperatures. Comparison with Table 2.9 shows that at $T = 10{,}000°$ K, dielectronic recombination is more important in some situations, but radiative recombination is more important at others. For most but not all of these ions the importance of dielectronic recombination grows with increasing temperature. Both recombination processes must be taken into account in all calculations.

TABLE 2.11
Dielectronic recombination coefficients[a] *for heavy ions*

X^{+i}		$\alpha(X^{+i}, T)$	
	7,500° K	10,000° K	15,000° K
C^0	1.91×10^{-13}	1.84×10^{-13}	1.63×10^{-13}
C^+	7.16×10^{-12}	6.06×10^{-12}	4.79×10^{-12}
C^{+2}	1.68×10^{-11}	1.31×10^{-11}	9.19×10^{-12}
N^0	6.59×10^{-13}	5.22×10^{-13}	3.59×10^{-13}
N^+	1.82×10^{-12}	2.04×10^{-12}	2.14×10^{-12}
N^{+2}	2.21×10^{-11}	2.16×10^{-11}	1.96×10^{-11}
N^{+3}	1.58×10^{-11}	1.54×10^{-11}	1.37×10^{-11}
O^0	7.35×10^{-14}	7.62×10^{-14}	7.74×10^{-14}
O^+	1.96×10^{-12}	1.66×10^{-12}	1.29×10^{-12}
O^{+2}	1.06×10^{-11}	1.14×10^{-11}	1.10×10^{-11}
O^{+3}	3.97×10^{-11}	3.45×10^{-11}	3.03×10^{-11}
O^{+4}	3.67×10^{-12}	5.92×10^{-12}	8.62×10^{-12}

[a]In $cm^3\ sec^{-1}$.

In addition, there is a third contribution, the integrated effect of many higher-energy resonances, which is ordinarily small at nebular temperatures. It can have a non-zero effect, and should be included in the most accurate calculations.

Calculations have been made of the ionization of heavy elements in several model H II regions and planetary nebulae. In H II regions, the common elements, such as O^+ and N^+, tend to be mostly singly ionized in the outer parts of the nebulae, although near the central stars there are often fairly large amounts of O^{++}, N^{++}, and Ne^{++}. Most planetary nebulae have hotter central stars, and the degree of ionization is correspondingly higher. This is shown in Figure 2.6, where the ionization of O is plotted for a calculated model planetary nebula. Note that in this figure the outer edge of the He^{++} zone is also the outer edge of the O^{+3} zone and the inner edge of the O^{++} zone, since O^{++} has an ionization potential 54.9 eV, nearly the same as He^+.

Again, in actual nebulae, density condensations play an important role in complicating the ionization structure; these simplified models do, however, give an overall picture of the ionization.

2.8 Charge-Exchange Reactions

One other atomic process is important in determining the ionization equilibrium of particular light elements, especially near the outer boundaries of radiation-bounded nebulae. This process is charge exchange in two-body reactions with hydrogen. As an example, consider neutral oxygen, which has the charge-exchange reaction with a proton

$$\mathrm{O}^0(^3P) + \mathrm{H}^+ \rightarrow \mathrm{O}^+(^4S) + \mathrm{H}^0(^2S). \tag{2.35}$$

This reaction converts an originally neutral O atom into an O^+ ion, and thus is an ionization process for O. There is an attractive polarization force between O^0 and H^+; in addition, the ionization potentials of O and H are very nearly the same, so that the reaction is very nearly a resonance process. For both these reasons the cross section for this charge-exchange reaction is relatively large. The reaction rate per unit volume per unit time for the reaction can be written

$$N_O N_p \delta(T), \tag{2.36}$$

where $\delta(T)$ is expressed in terms of the reaction cross section $\sigma(v)$ by an integral analogous to equation (2.5),

$$\delta(T) = \int_0^\infty v\, \sigma(v)\, f(v)\, dv. \tag{2.37}$$

Here it should be noted that $f(v)$ is the Maxwell-Boltzmann distribution function for the relative velocity v in the OH^+ center-of-mass system, and thus involves their reduced mass. A selection of computed values of $\delta(T)$ is given in Table 2.12; for instance, in an H II region with $N_p = 10\ \mathrm{cm}^{-3}$, the ionization rate per O atom per unit time is about $10^{-8}\ \mathrm{sec}^{-1}$, comparable with the photoionization rate for the typical conditions adopted in Section 2.1. Likewise, the rate for the inverse reaction

$$\mathrm{O}^+(^4S) + \mathrm{H}^0(^2S) \rightarrow \mathrm{O}^0(^3P) + \mathrm{H}^+ \tag{2.38}$$

can be written

$$N_{\mathrm{O}^+} N_{\mathrm{H}^0}\, \delta'(T). \tag{2.39}$$

TABLE 2.12
Charge-exchange reaction rate coefficients[a]

	O		N	
$T(^\circ$ K)	δ (10^{-9})	δ' (10^{-9})	δ (10^{-12})	δ' (10^{-12})
1,000	1.39	1.96	0.00	1.21
3,000	1.63	1.98	0.15	1.31
5,000	1.70	2.00	0.60	1.21
10,000	1.74	2.00	1.70	1.04

[a]In cm^3 sec^{-1}.

Numerical values of $\delta'(T)$, which of course is related to $\delta(T)$ through an integral form of the Milne relation

$$\frac{\delta'}{\delta} = \frac{9}{8} e^{\Delta E/kT}, \tag{2.40}$$

where $\Delta E = 0.19$ eV is the difference in the ionization potentials of O^0 and H^0, are also listed in Table 2.12. Comparison of Table 2.12 with Tables 2.9 and 2.11 shows that charge exchange has a rate comparable with recombination in converting O^+ to O^0 at the typical H II region conditions, but at the outer edge of the nebula, charge exchange dominates because of the higher density of H^0. At the higher radiation densities that occur in planetary nebulae, charge exchange is not important in the ionization balance of O except near the outer edge of the ionized region. The charge-exchange reactions (2.35) and (2.38) do not appreciably affect the ionization equilibrium of H, because of the low O and O^+ densities.

At temperatures that are high compared with the difference in ionization potentials between O^0 and H^0, the charge-exchange reactions (2.35) and (2.38) tend to set up an equilibrium in which the ratio of species depends only on the statistical weights, since $\delta'/\delta \to 9/8$. It can be seen from Table 2.12 that this situation is closely realized at $T = 10{,}000^\circ$ K. Thus in a nebula, wherever charge-exchange processes dominate the ionization balance of O, its degree of ionization is locked to that of H by the equation

$$\frac{N_{O^0}}{N_{O^+}} = \frac{9}{8}\frac{N_{H^0}}{N_p}. \tag{2.41}$$

The ionization potential of N (14.5 eV) is also close to that of H (13.60 eV), although it is not so close as that of O. Charge exchange was once believed (on the basis of an estimated rate constant) to also be important in the ionization balance of N in the outer boundaries of ionization-bounded nebulae. The charge-exchange reactions equivalent to (2.35) and (2.38) are

$$\mathrm{N}^0(^4S) + \mathrm{H}^+ \rightleftarrows \mathrm{N}^+(^3P) + \mathrm{H}^0(^2S), \tag{2.42}$$

and the calculated reaction rate constants δ (for the reaction proceeding from left to right) and δ' (from right to left) are also listed in Table 2.12. It can be seen that they are much smaller than for the corresponding oxygen reactions, and so in fact charge-transfer is not important for nitrogen except in very high-density nebulae. Because of the different statistical weights,

$$\frac{N_{\mathrm{N}^0}}{N_{\mathrm{N}^+}} \rightarrow \frac{\delta'}{\delta}\frac{N_{\mathrm{H}^0}}{N_p} \rightarrow \frac{2}{9}\frac{N_{\mathrm{H}^0}}{N_p} \tag{2.43}$$

at high temperatures, but because of the higher ionization potential of N, the second part of this equation is not closely approached at $T = 10{,}000°$ K.

Another type of charge-exchange reaction that can be important in the ionization balance of heavy ions is exemplified by

$$\mathrm{O}^{+2}(2s^2 2p^2\ ^3P) + \mathrm{H}^0(1s\ ^2S) \rightarrow \mathrm{O}^+(2s2p^4\ ^4P) + \mathrm{H}^+. \tag{2.44}$$

Note that the O^+ is left in an excited state. This reaction is strongly exothermic, with an energy difference $\Delta E = 6.7$ eV. In addition, there are almost no $\mathrm{O}^+(2s2p^4\ ^4P)$ ions (with excitation energy 14.9 eV) present in the nebula. Hence the inverse reaction, which has this threshold, essentially does not proceed at all at nebular temperatures. Reaction (2.44) and some others like it, have large cross sections because it is a two-body process for which the strong Coulomb repulsion of the products speeds up the process. For the general reaction

$$X^{+i} + \mathrm{H}^0 \rightarrow X^{+(i-1)} + \mathrm{H}^+, \tag{2.45}$$

the rate per unit volume per unit time may be written

$$N_{X^{+i}} N_{\mathrm{H}^0}\, \delta'(T),$$

and calculated values of $\delta'(T)$ are listed for three temperatures in Table 2.13. As the table shows, several of these rate coefficients are quite large. Thus even though the density of H^0 is small in comparison with the electron density, these charge-exchange reactions can be as important as recombination, or even more so, in determining the ionization equilibrium for these heavy ions. Similar charge reactions with He^0 can also make appreciable contributions to the ionization balance.

TABLE 2.13

Charge-exchange reaction coefficients[a] $\delta'(T)$

	T		
X^{+i}	5,000° K	10,000° K	20,000° K
C^{+2}	1.00×10^{-12}	1.00×10^{-12}	1.35×10^{-12}
C^{+3}	3.09×10^{-9}	3.58×10^{-9}	4.22×10^{-9}
N^{+2}	0.78×10^{-9}	0.86×10^{-9}	0.97×10^{-9}
N^{+3}	1.54×10^{-9}	2.93×10^{-9}	5.14×10^{-9}
O^{+2}	0.60×10^{-9}	0.77×10^{-9}	1.03×10^{-9}
O^{+3}	6.34×10^{-9}	8.63×10^{-9}	1.18×10^{-8}
Ne^{+3}	4.00×10^{-9}	5.68×10^{-9}	8.28×10^{-9}

[a] In $cm^3\ sec^{-1}$.

References

Much of the early work on gaseous nebulae was done by H. Zanstra, D. H. Menzel, L. H. Aller, and others. The very important series of papers on physical processes in gaseous nebulae by D. H. Menzel and his collaborators is collected in *Selected Papers on Physical Processes in Ionized Nebulae* (New York: Dover, 1962). The treatment in this chapter is based on ideas that were, in many cases, given in these pioneering papers. The specific formulation and the numerical values used in this chapter are largely based on the references listed here.

Basic papers on ionization structure:

Strömgren, B. 1939, *Ap. J.* **89,** 529.
Hummer, D. G., and Seaton, M. J. 1963. *M.N.R.A.S.* **125,** 437.
Hummer, D. G., and Seaton, M. J. 1964. *M.N.R.A.S.* **127,** 217.

The name "Strömgren sphere" comes from the first of these papers, the pioneering basic treatment that began the whole subject.

Numerical values of H recombination coefficient:

Seaton, M. J. 1959. *M.N.R.A.S.* **119**, 81

Burgess, A. 1964. *Mem. R.A.S.* **69**, 1.

Pengelly, R.M. 1964. *M.N.R.A.S.* **127**, 145.

Hummer, D. G., and Storey, P. J. 1987. *M.N.R.A.S.* **224,** 80.

(Tables 2.1, 2.6, and 2.8 are based on these references.)

Numerical values of He recombination coefficient:

Burgess, A., and Seaton, M. J. 1960. *M.N.R.A.S.*, **121**, 471.

Robbins, R. R. 1968. *Ap. J.* **151,** 497.

Robbins, R. R. 1970. *Ap. J.* **160**, 519.

Brown, R. L., and Mathews, W. G. 1970. *Ap. J.* **160**, 939.

(Table 2.4 is based on these references.)

Numerical values of He ($2\ ^3S \rightarrow 2^1S$ and 2^1P) collisional cross sections:

Berrington, K. A., and Kingston, A. E. 1987. *J. Phys. B.* **20**, 6631.

Table 2.5 is based on this reference. It also gives collisional excitation rates to the higher $n = 3$ singlets, which are negligible at low temperatures and make at most a 10 percent correction at the highest temperature in Table 2.5. It also gives the $1^1S \rightarrow 2\ ^3S$ collisional excitation rate from which the inverse collisional deexcitation rate can be computed by equation (3.21).

Numerical values of the H and He^+ photoionization cross sections:

Hummer, D. G., and Seaton, M. J. 1963. *M.N.R.A.S.* **125,** 437.

(Figure 2.2 is based on this reference and the two following ones.)

Numerical values of the He photoionization cross section:

Bell, K. L., and Kingston, A. E. 1967. *Proc. Phys. Soc.* **90,** 31.

Brown, R. L. 1971. *Ap. J.* **164,** 387.

Numerical values of He ($2\ ^1S \rightarrow 1\ ^1S$) and He ($2\ ^3S \rightarrow 1\ ^1S$) transition probabilities, including the frequency distribution in the $2\ ^1S \rightarrow 1\ ^1$ S two-photon spectrum:

Drake, G. W. F., Victor, G. A., and Dalgarno, A. 1969. *Phys Rev.* **180,** 25.

Drake, G. W . 1971. *Phys. Rev. A* **3,** 908.

Jacobs, V. 1971. *Phys. Rev. A* **4,** 939.

Numerical values of heavy-element photoionization cross sections:

Flower, D. R. 1968. *Planetary Nebulae* (IAU Symposium No. 34), ed. D. E. Osterbrock and C. R. O'Dell. Dordrecht: Reidel, p. 205.

Henry, R. J. W. 1970. *Ap. J.* **161,** 1153.

(Table 2.7 is based on the preceding two references.)

Reilman, R. F. and Manson, S. T. 1979. *Ap. J. Supp.* **40,** 815.

Actual resonance calculations are complicated, and the results are difficult to reproduce either graphically or in tabular form, because of the very rapid variations with energy. An example is O^0:

Pradhan, A. K. 1978. *J. Phys. B.* **11,** L729.

Interpolation formulae for representing the actual resonance structure of photoionization cross sections:

Butler, K., Mendoza, C., and Zeippen, C. J. 1984. *M.N.R.A.S.* **213,** 345.

Radiative recombination of heavy elements:

Gould, R. J. 1978. *Ap. J.* **219,** 250.

(Tables 2.9 and 2.10 are based on this reference.)

Dielectronic recombination:

Storey, P. J. 1983. *Planetary Nebulae* (IAU Symposium No. 103), ed D. R. Flower. Dordrecht: Reidel, p. 199.

Nussbaumer, H. and Storey, P. J. 1985. *Astr. Ap.* **126,** 75.

Burgess, A. 1964. *Ap. J.* **139,** 776.

(Table 2.11 is based on the second of these references.) The last reference discusses the integrated contribution of many higher-order resonances to the recombination coefficient that is briefly mentioned in the text. It is ordinarily negligible at nebular temperatures.

Charge-exchange reactions of O^+ and N^+:

Chamberlain, J. W. 1956. *Ap. J.* **124,** 390.

Dalgarno, A. 1978. *Planetary Nebulae, Observations and Theory* (IAU Symposium No. 76), ed. Y. Terzian. Dordrecht: Reidel, p. 139.

Butler, K. and Dalgarno, A. 1979. *Ap. J.* **234,** 765.

(Table 2.12 is based on the last two of these references.)

Charge-exchange reactions of higher stages of ionization:

Pequignot, D., Aldrovandi, S. M. V., and Stasinska, G. 1978. *Astr. Ap.* **63,** 313.

Butler, S. E., Heil, T. G., and Dalgarno, A. 1980. *Ap. J.* **241,** 442.

Butler, S. E., and Dalgarno, A. 1980. *Ap. J.* **241,** 838.

Shields, G. A., Aller, L. H., Keyes, C. D., and Czyzak, S. J. 1981. *Ap. J.* **248,** 569.

Table 2.13 is based on the second of these references.

Calculations of model H II regions:

Hjellming, R. M. 1966. *Ap. J.* **143,** 420.

Rubin, R. H. 1968. *Ap. J.* **153,** 761.

Rubin, R. H. 1983. *Ap. J.* **274,** 671.

(Figure 2.4 is based on the second of these references. The last emphasizes the importance of heavy-element opacity in some situations.)

Calculations of model planetary nebulae:

Harrington, J. P. 1969. *Ap. J.* **156,** 903.

Flower, D. R. 1969. *M.N.R.A.S.* **146,** 171.

Harrington, J. P. 1979. *Planetary Nebulae, Observations and Theory* (IAU Symposium No. 76) ed. Y. Terzian. Dordrecht: Reidel, p. 15.

(Figure 2.6 is based on the second of these references.)

References to other later, still more detailed models of H II regions and planetary nebulae are given in Chapter 5.

Model atmospheres for early-type stars are discussed in more detail in Chapter 5. In the present chapter, it is easiest to say that the simplest models are black bodies; a much better approximation is provided by models in which the continuous spectrum is calculated; and the best models are those that also include the effects of the absorption lines, since they are strong and numerous in the ultraviolet.

Models for continuous spectrum:

Hummer, D. G., and Mihalas, D. 1970. *M.N.R.A.S.* **147,** 339.

Models with line blanketing:

Kurucz, R. L. 1979. *Ap. J. Supp.* **40,** 1.

Temperature scale and models with line blanketing:

Morton, D. C. 1969. *Ap. J.* **158,** 629.

3

Thermal Equilibrium

3.1 Introduction

The temperature in a static nebula is fixed by the equilibrium between heating by photoionization and cooling by recombination and by radiation from the nebula. When a photon of energy $h\nu$ is absorbed and causes an ionization of H, the photoelectron produced has an initial energy $\frac{1}{2}mv^2 = h(\nu - \nu_0)$, and we may think of an electron being "created" with this energy. The electrons thus produced are rapidly thermalized, as indicated in Chapter 2, and in ionization equilibrium these photoionizations are balanced by an equal number of recombinations. In each recombination, a thermal electron with energy $\frac{1}{2}\, mv^2$ disappears, and an average of this quantity over all recombinations represents the mean energy that "disappears" per recombination. The difference between the mean energy of a newly created photoelectron and the mean energy of a recombining electron represents the net gain in energy by the electron gas per ionization process. In equilibrium this net energy gain is balanced by the energy lost by radiation, chiefly by electron collisional excitation of bound levels of abundant ions, followed by emission of photons that can escape from the nebula. Free-free emission, or bremsstrahlung, is another, less-important radiative energy-loss mechanism.

3.2 Energy Input by Photoionization

Let us first examine the energy input by photoionization. As in Chapter 2, it is simplest to begin by considering a pure H nebula. At any specific point in the nebula, the energy input (per unit volume per unit time) is

$$G(\mathrm{H}) = N_{\mathrm{H}^0} \int_{\nu_0}^{\infty} \frac{4\pi J_\nu}{h\nu} h(\nu - \nu_0) a_\nu(\mathrm{H}^0) d\nu. \qquad (3.1)$$

Furthermore, since the nebula is in ionization equilibrium, we may eliminate N_{H^0} by substituting equation (2.8), giving

$$\begin{aligned} G(\mathrm{H}) &= N_e N_p \alpha_A(\mathrm{H}^0,\, T) \frac{\displaystyle\int_{\nu_0}^{\infty} \frac{J_\nu}{h\nu} h(\nu - \nu_0) a_\nu(\mathrm{H}^0) d\nu}{\displaystyle\int_{\nu_0}^{\infty} \frac{J_\nu}{h\nu} a_\nu(\mathrm{H}^0) d\nu} \\ &= N_e N_p \alpha_A(H^0,\, T)\, \tfrac{3}{2}\, kT_i. \end{aligned} \tag{3.2}$$

From this equation it can be seen that the mean energy of a newly created photoelectron depends on the form of the ionizing radiation field, but not on the absolute strength of the radiation. The rate of creation of photoelectrons depends on the strength of the radiation field, or, as equation (3.2) shows, on the recombination rate. The quantity $\frac{3}{2}\, kT_i$ represents the initial temperature of the newly created photoelectrons. For assumed blackbody spectra with $J_\nu = B_\nu(T_*)$, it is easy to show that $T_i \approx T_*$ so long as $kT_* < h\nu_0$. For any known J_ν (for instance, the emergent spectrum from a model atmosphere), the integration can be carried out numerically; a short list of representative values of T_i is given in Table 3.1. Note that the second column in the table, $\tau_0 = 0$, corresponds to photoionization by the emergent model-atmosphere spectrum. At larger distances from the star, the spectrum of the ionizing radiation is modified by absorption in the nebula, the radiation nearest the series limit being most strongly attenuated because of the frequency dependence of the absorption coefficient. Therefore, the higher-energy photons penetrate further into the gas, and the mean energy of the photoelectrons produced at larger optical depths from the star is higher. This effect is shown for a pure H nebula in the columns labeled with values of τ_0, the optical depth at the ionization limit.

3.3 Energy Loss by Recombination

The kinetic energy lost by the electron gas (per unit volume per unit time) in recombination can be written

$$L_R(\mathrm{H}) = N_e N_p kT \beta_A(\mathrm{H}^0,\, T), \tag{3.3}$$

where

$$\beta_A(\mathrm{H}^0,\, T) = \sum_{n=1}^{\infty} \beta_n(\mathrm{H}^0,\, T) = \sum_{n=1}^{\infty} \sum_{L=0}^{n-1} \beta_{nL}(\mathrm{H}^0,\, T), \tag{3.4}$$

with

$$\beta_{nL}(\mathrm{H}^0,\, T) = \frac{1}{kT} \int_0^{\infty} v \sigma_{nL}(\mathrm{H}^0, T)\, \tfrac{1}{2}\, mv^2 f(v) dv. \tag{3.5}$$

TABLE 3.1
Mean input energy of photoelectrons

Model stellar atmosphere T_*(° K)	T_i(° K)			
	$\tau_0 = 0$	$\tau_0 = 1$	$\tau_0 = 5$	$\tau_0 = 10$
3.0×10^4	1.46×10^4	1.81×10^4	3.61×10^4	5.45×10^4
3.5×10^4	2.15×10^4	2.65×10^4	4.67×10^4	6.31×10^4
4.0×10^4	2.67×10^4	3.38×10^4	6.52×10^4	9.57×10^4
5.0×10^4	3.50×10^4	4.47×10^4	8.47×10^4	11.87×10^4

The left-hand side of equation (3.5) is thus effectively a kinetic-energy-averaged recombination coefficient. Note that since the recombination cross sections are approximately proportional to v^{-2}, the electrons of lower kinetic energy are preferentially captured, and the mean energy of the captured electrons is somewhat less than $\frac{3}{2}\ kT$. Calculated values of β_1 and β_A are listed in Table 3.2.

TABLE 3.2
Recombination cooling coefficient[a]

T(° K)	β_A	β_1	β_B
2,500	8.93×10^{-13}	3.13×10^{-13}	5.80×10^{-13}
5,000	5.42×10^{-13}	2.20×10^{-13}	3.22×10^{-13}
10,000	3.23×10^{-13}	1.50×10^{-13}	1.73×10^{-13}
20,000	1.88×10^{-13}	9.58×10^{-14}	9.17×10^{-14}

[a] In $cm^3\ sec^{-1}$.

In a pure H nebula that had no radiation losses, the thermal equilibrium equation would be

$$G(\mathrm{H}) = L_R(\mathrm{H}), \tag{3.6}$$

and the solution for the nebular temperature would give a $T > T_i$ because of the "heating" due to the preferential capture of the slower electrons.

The radiation field J_ν in equation (3.1) should, of course, include the diffuse radiation as well as the stellar radiation modified by absorption. This can easily be included in the on-the-spot approximation, since, according to it, every emission of an ionizing photon during a recombination to the level $n = 1$ is balanced by absorption of the same photon at a nearby spot in the nebula. Thus production of photons by the diffuse radiation field and recombinations to the ground level can simply be omitted from the gain and loss rates, leading to the equations

$$G_{OTS}(\mathrm{H}) = N_{\mathrm{H}^0} \int_{\nu_0}^{\infty} \frac{4\pi J_{\nu s}}{h\nu} h(\nu - \nu_0) a_\nu(\mathrm{H}^0) d\nu$$

$$= N_e N_p \alpha_B(H^0, T) \frac{\displaystyle\int_{\nu_0}^{\infty} \frac{J_{\nu s}}{h\nu} h(\nu - \nu_0) a_\nu(\mathrm{H}^0) d\nu}{\displaystyle\int_{\nu_0}^{\infty} \frac{J_{\nu s}}{h\nu} a_\nu(\mathrm{H}^0) d\nu} \qquad (3.7)$$

and

$$L_{OTS}(\mathrm{H}) = N_e N_p kT \beta_B(\mathrm{H}^0, T), \qquad (3.8)$$

with

$$\beta_B(\mathrm{H}^0, T) = \sum_{n=2}^{\infty} \beta_n(\mathrm{H}^0, T). \qquad (3.9)$$

The on-the-spot approximation is not as accurate for the thermal equilibrium as it is in the ionization equation, because of the fairly large difference in $h(\nu - \nu_0)$ between the ionizing photons in the stellar and diffuse radiation fields, but it may be improved by further iterations if necessary.

The generalization to include He in the heating and recombination cooling rates is straightforward to write, namely,

$$G = G(\mathrm{H}) + G(\mathrm{He}), \qquad (3.10)$$

where

$$G(\mathrm{He}) = N_e N_{\mathrm{He}^+} \alpha_A(\mathrm{He}^0, T) \frac{\displaystyle\int_{\nu_2}^{\infty} \frac{J\nu}{h\nu} h(\nu - \nu_2) a_\nu(\mathrm{He}^0) d\nu}{\displaystyle\int_{\nu_2}^{\infty} \frac{J_\nu}{h\nu} a_\nu(\mathrm{He}^0) d\nu} \qquad (3.11)$$

and

$$L_R = L_R(\mathrm{H}) + L_R(\mathrm{He}), \qquad (3.12)$$

with

$$L_R(\mathrm{He}) = N_e N_{\mathrm{He}^+} kT \beta_A(\mathrm{He}^0, T), \qquad (3.13)$$

and so on.

It can be seen that the heating and recombination cooling rates are proportional to the densities of the ions involved; so the contributions of the heavy elements, which are much less abundant than H and He, can, to a good approximation, be omitted from these rates.

3.4 Energy Loss by Free-Free Radiation

Next we will examine cooling by radiation processes that do not involve recombination. In most circumstances such cooling is far more important than the recombination cooling, and therefore dominates the thermal equilibrium. A minor contributor to the cooling rate, which nevertheless is important because it can occur even in a pure H nebula, is free-free radiation or bremsstrahlung, in which a continuous spectrum is emitted. The rate of cooling by this process by ions of charge Z, integrated over all frequencies, is, to a fair approximation,

$$\begin{aligned} L_{\rm FF}(Z) &= 4\pi j_{ff} \\ &= \frac{2^5 \pi e^6 Z^2}{3^{3/2} hmc^3} \left(\frac{2\pi kT}{m} \right)^{1/2} g_{ff} \; N_e N_+ \\ &= 1.42 \times 10^{-27} Z^2 T^{1/2} g_{ff} N_e N_+ \end{aligned} \tag{3.14}$$

in ergs cm^{-3} sec^{-1}, where N_+ is the number density of the ions. Again H^+ dominates the free-free cooling, because of its abundance, and He^+ can be included with H^+ (since both have $Z = 1$) by writing $N_+ = N_p + N_{\rm He^+}$. The numerical factor g_{ff} is called the mean Gaunt factor for free-free emission; it is a slowly varying function of N_e and T, generally for nebular conditions in the range $1.0 < g_{ff} < 1.5$, and a good average value to adopt is $g_{ff} \approx 1.3$.

3.5 Energy Loss by Collisionally Excited Line Radiation

A far more important source of radiative cooling is collisional excitation of low-lying energy levels of common ions, such as O^+, O^{++}, and N^+. These ions make a significant contribution in spite of their low abundance because they have energy levels with excitation potentials of the order of kT, but all the levels of H and He have much higher excitation potentials, and therefore are usually not important as collisionally excited coolants. Let us therefore examine how an ion is excited to level 2 by electron collisions with ions in the lower level 1. The cross section for excitation $\sigma_{12}(v)$ is a function of electron velocity and is zero below the threshold $\chi = h\nu_{21}$. Not too far above the threshold, the main dependence of the excitation cross section is $\sigma \propto v^{-2}$

(because of the focusing effect of the Coulomb force); so it is convenient to express the collision cross sections in terms of the collision strength $\Omega(1,2)$ defined by

$$\sigma_{12}(v) = \frac{\pi \hbar^2}{m^2 v^2} \frac{\Omega(1,2)}{\omega_1} \quad \text{for} \quad \tfrac{1}{2}\, mv^2 > \chi, \tag{3.15}$$

where $\Omega(1,2)$ is a function of electron velocity (or energy) but is often approximately constant near the threshold, and ω_1 is the statistical weight of the lower level.

There is a relation between the cross section for de-excitation, $\sigma_{21}(v)$, and the cross section for excitation, namely

$$\omega_1 v_1^2 \sigma_{12}(v_1) = \omega_2 v_2^2 \sigma_{21}(v_2), \tag{3.16}$$

where v_1 and v_2 are related by

$$\tfrac{1}{2}\, mv_1^2 = \tfrac{1}{2}\, mv_2^2 + \chi. \tag{3.17}$$

Equation (3.16) can easily be derived from the principle of detailed balancing, which states that in thermodynamic equilibrium each microscopic process is balanced by its inverse. Thus in this particular case, the number of excitations caused by collisions with electrons in the velocity range v_1 to $v_1 + dv_1$ is just balanced by the de-excitations caused by collisions that produce electrons in the same velocity range. Thus

$$N_e N_1 v_1 \sigma_{12}(v_1) f(v_1) dv_1 = N_e N_2 v_2 \sigma_{21}(v_2) f(v_2) dv_2,$$

and using the Boltzmann equation of thermodynamic equilibrium,

$$\frac{N_2}{N_1} = \frac{\omega_2}{\omega_1} e^{-\chi/kT},$$

we derive the relation (3.16). Combining equations (3.15) and (3.16), so that the de-excitation cross section can be expressed in terms of the collision strength $\Omega(1,2)$,

$$\sigma_{21}(v_2) = \frac{\pi \hbar^2}{m^2 v_2^2} \frac{\Omega(1,2)}{\omega_2}; \tag{3.18}$$

that is, the collision strengths are symmetrical in 1 and 2.

The total collisional de-excitation rate per unit volume per unit time is

$$N_e N_2 q_{21} = N_e N_2 \int_0^\infty v \sigma_{21} f(v) dv$$

$$= N_e N_2 \left(\frac{2\pi}{kT}\right)^{1/2} \frac{\hbar^2}{m^{3/2}} \frac{\Omega(1,2)}{\omega_2}$$

$$= N_e N_2 \frac{8.629 \times 10^{-6}}{T^{1/2}} \frac{\Omega(1,2)}{\omega_2} \tag{3.19}$$

(in $\text{cm}^{-3}\ \text{sec}^{-1}$) if $\Omega(1,2)$ is a constant. In general, the mean value

$$\Omega(1,2) = \int_0^\infty \Omega(1,2;\ E) e^{-E/kT} d\left(\frac{E}{kT}\right) \tag{3.20}$$

should be used in equation (3.19), where $E = \frac{1}{2}mv_2^2$. Likewise, the collisional excitation rate per unit volume per unit time is $N_e N_1 q_{12}$, where

$$q_{12} = \frac{\omega_2}{\omega_1} q_{21} e^{-\chi/kT}. \tag{3.21}$$

The collision strengths must be calculated quantum-mechanically, and some of the most important numerical values are listed in Tables 3.3 through 3.7. Each collision strength in general consists of a part that varies slowly with energy, on which, in many cases, there are superimposed resonance contributions that vary rapidly with energy; but when the cross sections are integrated over a Maxwellian distribution, as in almost all astrophysical applications, the effect of the exact positions of the resonances tends to be averaged out. Therefore, it is usually sufficient to use the collision strength averaged over resonances, and this simpler quantity is given in Tables 3.3 to 3.7, evaluated at $T = 10{,}000°$ K, a representative nebular temperature. It is convenient to remember that, for an electron with the mean energy at a typical nebular temperature, $T \approx 7{,}500°$ K, the cross sections for excitation and de-excitation are $\sigma \approx 10^{-15}\Omega/\omega\ \text{cm}^2$.

Note that there is a simple relation for the collision strengths between a term consisting of a single level and a term consisting of various levels, namely,

$$\Omega(SLJ,\ S'L'J') = \frac{(2J' + 1)}{(2S' + 1)(2L' + 1)}\ \Omega(SL,\ S'L') \tag{3.22}$$

if either $S = 0$ or $L = 0$. The factors $(2J' + 1)$ and $(2S' + 1)(2L' + 1)$ are the statistical weights of the level and of the term, respectively. On account of this relation, the rate of collisional excitation in p^2 or p^4 ions (such as O^{++}) from the ground ^3P term to the excited (singlet) ^1D and ^1S levels is very nearly independent of the distribution of ions among $^3P_0$, $^3P_1$, and $^3P_2$.

For all the low-lying levels of the ions listed in Tables 3.3 through 3.5, the excited levels arise from the same electron configurations as the ground level. Radiative transitions between these excited levels and the ground level are therefore forbidden by the electric-dipole selection roles, but can occur by magnetic-dipole and/or electric-quadrupole transitions. These are the well-known forbidden lines, many of which are observed in nebular spectra, the best known in the optical region, and others, thanks to recent advances in technology, in the infrared or ultraviolet. Transition probabilities, as well as wavelengths for the observable lines, are listed in Tables 3.8 through 3.10.

TABLE 3.3
Collision strengths for p and p^5 ions

Ion	$\Omega(^2P_{1/2}, {}^2P_{3/2})$	Ion	$\Omega(^2P_{1/2}, {}^2P_{3/2})$
C^+	2.90	Si^+	5.58
N^{+2}	1.08	Si^{+5}	0.24
O^{+3}	2.36	S^{+3}	6.42
Ne^{+5}	0.43	Ar^{+5}	0.80
Ne^+	0.30	Ar^+	0.64
Mg^{+3}	0.30	Ca^{+3}	1.06

For an ion with a single excited level, in the limit of very low electron density every collisional excitation is followed by the emission of a photon, and the cooling rate per unit volume is therefore

$$L_C = N_e N_1 q_{12} h\nu_{21}. \tag{3.23}$$

However, if the density is sufficiently high, collisional de-excitation is not negligible and the cooling rate is reduced. The equilibrium equation for the balance between the excitation and de-excitation rates of the excited level is, in general,

$$N_e N_1 q_{12} = N_e N_2 q_{21} + N_2 A_{21}, \tag{3.24}$$

and the solution is

$$\frac{N_2}{N_1} = \frac{N_e q_{12}}{A_{21}} \left[\frac{1}{1 + \frac{N_e q_{21}}{A_{21}}} \right], \tag{3.25}$$

so the cooling rate is

$$L_C = N_2 A_{21} h\nu_{21} = N_e N_1 q_{12} h\nu_{21} \left[\frac{1}{1 + \dfrac{N_e q_{21}}{A_{21}}} \right]. \tag{3.26}$$

It can be seen that as $N_e \to 0$, we recover equation (3.23), but as $N_e \to \infty$,

$$L_C \to N_1 \frac{\omega_2}{\omega_1} e^{-\chi/kT} A_{21} h\nu_{21}, \tag{3.27}$$

the thermodynamic-equilibrium cooling rate.

Some ions have only two low-lying levels and may be treated by this simple formalism, but most ions have more levels, and all ions with ground configurations p^2, p^3, or p^4 have five low-lying levels. Examples are O^{++} and N^+, whose energy-level diagrams are shown in Figure 3.1. For such ions, collisional and radiative transitions can occur between any of the levels, and excitation and de-excitation cross sections and collision strengths exist between all pairs of the levels.

The equilibrium equations for each of the levels $i = 1, 5$ thus become

$$\sum_{j \neq i} N_j N_e q_{ji} + \sum_{j>i} N_j A_{ji} = \sum_{j \neq i} N_i N_e q_{ij} + \sum_{j<i} N_i A_{ij}, \tag{3.28}$$

which, together with the total number of ions

$$\sum_j N_j = N, \tag{3.29}$$

can be solved for the relative population in each level, and then for the collisionally excited radiative cooling rate

$$L_C = \sum_i N_i \sum_{j<i} A_{ij} h\nu_{ij}. \tag{3.30}$$

TABLE 3.4
Collision strengths for p^2 and p^4 ions

Ion	$\Omega(^3P, {}^1D)$	$\Omega(^3P, {}^1S)$	$\Omega(^1D, {}^1S)$	$\Omega(^3P_0, {}^3P_1)$	$\Omega(^3P_0, {}^3P_2)$	$\Omega(^3P_1, {}^3P_2)$	$\Omega(^3P, {}^5S)$
N^+	2.68	0.35	0.41	0.40	0.28	1.13	1.27
O^{+2}	2.17	0.28	0.62	0.54	0.27	1.29	1.18
Ne^{+4}	1.78	0.25	0.52	0.24	0.12	0.58	1.51
Ne^{+2}	1.65	0.17	0.23	0.35	0.31	1.13	–
S^{+2}	8.39	1.19	1.88	2.64	1.11	5.79	–
Ar^{+4}	3.72	1.18	1.25	0.26	0.32	1.04	–
Ar^{+2}	4.74	0.68	0.82	1.18	0.53	2.24	–

TABLE 3.5
Collision strengths for p^3 ions

Ion	$\Omega(^4S,\ ^2D)$	$\Omega(^4S,\ ^2P)$	$\Omega(^2D_{3/2},\ ^2D_{5/2})$	$\Omega(^2D_{3/2},\ ^2P_{1/2})$
O^+	1.34	0.40	1.17	0.28
Ne^{+3}	1.40	0.47	1.36	0.34
S^+	6.98	2.28	7.59	1.52
Ar^{+3}	3.24	0.44	6.13	1.67

Ion	$\Omega(^2D_{3/2},\ ^2P_{3/2})$	$\Omega(^2D_{5/2},\ ^2P_{1/2})$	$\Omega(^2D_{5/2},\ ^2P_{3/2})$	$\Omega(^2P_{1/2},\ ^2P_{3/2})$
O^+	0.41	0.30	0.73	0.29
Ne^{+3}	0.51	0.37	0.90	0.34
S^+	3.38	2.56	4.79	2.38
Ar^{+3}	2.47	1.79	4.44	2.33

TABLE 3.6
Collision strengths for $^2S - {}^2P$ transitions

Ion	$\Omega(2s\ {}^2S, 2p\ {}^2P)$	Ion	$\Omega(3s\ {}^2S, 3p\ {}^2P)$
C^{+3}	8.88	Mg^{+}	16.5
N^{+4}	6.65	Si^{+3}	17.0
O^{+5}	5.00		

TABLE 3.7
Collision strengths for $^1S - {}^3P$ transitions

Ion	$\Omega(2s^2\ {}^1S, 2s2p\ {}^3P)$	Ion	$\Omega(3s^2\ {}^1S, 3s3p\ {}^3P)$
C^{+2}	1.05	Si^{+2}	5.43
N^{+3}	0.85	S^{+4}	0.91
O^{+4}	0.72		
Ne^{+6}	0.17		

In the low-density limit, $N_e \to 0$, this becomes a sum of terms like (3.23), but if

$$N_e q_{ij} > \sum_{k<i} A_{ik}$$

for any i, j, collisional de-excitation is not negligible and the complete solution must be used. In fact, for any level i, a critical density $N_c(i)$ may be defined as

$$N_c(i) = \sum_{j<i} A_{ij} \bigg/ \sum_{j \neq i} q_{ij}, \qquad (3.31)$$

so that for $N_e < N_c(i)$, collisional de-excitation of level i is negligible, but for $N_e > N_c(i)$ it is important. Critical densities for levels that are most important in radiative cooling are listed in Table 3.11.

TABLE 3.8
Transition probabilities of p^2 ions

Transition	[N II]		[O III]		[Ne V]		[S III]		[Ar V]	
	Transition probability (sec^{-1})	Wave-length (Å)	Transition probability (sec^{-1})	Wave-length (Å)	Transition probability (sec^{-1})	Wave-length (Å)	Transition probability (sec^{-1})	Wave-length (Å)	Transition probability (sec^{-1})	Wave-length (Å)
$^1D_2 - {}^1S_0$	1.1	5754.6	1.8	4363.2	2.8	2974.8	2.2	6312.1	3.3	4625.5
$^3P_2 - {}^1S_0$	1.5×10^{-4}	3070.8	7.8×10^{-4}	2331.4	6.7×10^{-3}	1592.7	1.0×10^{-2}	3797.2	5.7×10^{-2}	2786.0
$^3P_1 - {}^1S_0$	3.4×10^{-2}	3062.8	2.2×10^{-1}	2321.0	4.2	1575.2	8.0×10^{-1}	3721.7	6.6	2691.1
$^3P_2 - {}^1D_2$	3.0×10^{-3}	6583.4	2.0×10^{-2}	5006.9	3.6×10^{-1}	3425.9	5.8×10^{-2}	9530.9	4.8×10^{-1}	7005.7
$^3P_1 - {}^1D_2$	1.0×10^{-3}	6548.1	6.7×10^{-3}	4958.9	1.3×10^{-1}	3345.8	2.2×10^{-2}	9068.9	2.0×10^{-1}	6435.1
$^3P_0 - {}^1D_2$	5.4×10^{-7}	6527.1	2.7×10^{-6}	4931.0	2.4×10^{-5}	3300.1	5.8×10^{-6}	8829.5	3.5×10^{-5}	6133.1
$^3P_1 - {}^3P_2$	7.5×10^{-6}	122 μ	9.8×10^{-5}	51.8 μ	4.6×10^{-3}	14.3 μ	2.1×10^{-3}	18.7 μ	2.7×10^{-2}	7.9 μ
$^3P_0 - {}^3P_2$	1.2×10^{-12}	76 μ	3.0×10^{-11}	32.7 μ	5.1×10^{-9}	9.0 μ	4.6×10^{-8}	12.0 μ	1.2×10^{-6}	4.9 μ
$^3P_0 - {}^3P_1$	2.1×10^{-6}	204 μ	2.6×10^{-5}	88.4 μ	1.3×10^{-3}	24.3 μ	4.7×10^{-4}	33.5 μ	8.0×10^{-3}	13.1 μ
$^3P_2 - {}^5S_2$	$1.1 \times 10^{+2}$	2142.8	$5.2 \times 10^{+2}$	1666.2	$6.1 \times 10^{+3}$	1146.1	–	1728.9	–	–
$^3P_1 - {}^5S_1$	$4.8 \times 10^{+1}$	2139.0	$2.1 \times 10^{+2}$	1660.8	$2.4 \times 10^{+3}$	1137.0	–	1713.1	–	–

TABLE 3.9
Transition probabilities of p^3 ions

	[O II]		[Ne IV]		[S II]		[Ar IV]	
Transition	Transition probability (sec^{-1})	Wavelength (Å)	Transition probability (sec^{-1})	Wavelength (Å)	Transition probability (sec^{-1})	Wavelength (Å)	Transition probability (sec^{-1})	Wavelength (Å)
$^2P_{1/2} - ^2P_{3/2}$	1.4×10^{-10}	—	2.7×10^{-9}	—	1.0×10^{-6}	—	4.9×10^{-5}	—
$^2D_{5/2} - ^2P_{3/2}$	1.1×10^{-1}	7319.9	4.0×10^{-1}	4714.2	1.8×10^{-1}	10320.5	6.0×10^{-1}	7237.3
$^2D_{3/2} - ^2P_{3/2}$	5.8×10^{-2}	7330.7	4.4×10^{-1}	4724.2	1.3×10^{-1}	10286.7	7.9×10^{-1}	7170.6
$^2D_{5/2} - ^2P_{1/2}$	5.6×10^{-2}	7318.8	1.1×10^{-1}	4715.6	7.8×10^{-2}	10370.5	1.2×10^{-1}	7331.4
$^2D_{3/2} - ^2P_{1/2}$	9.4×10^{-2}	7329.6	3.9×10^{-1}	4725.6	1.6×10^{-1}	10336.4	6.0×10^{-1}	7262.8
$^4S_{3/2} - ^2P_{3/2}$	5.8×10^{-2}	2470.3	1.3	1602.0	2.2×10^{-1}	4068.6	2.6	2853.6
$^4S_{3/2} - ^2P_{1/2}$	2.4×10^{-2}	2470.2	5.2×10^{-1}	1602.1	9.1×10^{-2}	4076.4	8.6×10^{-1}	2868.2
$^2D_{5/2} - ^2D_{3/2}$	1.3×10^{-7}	—	1.5×10^{-6}	—	3.3×10^{-7}	—	2.3×10^{-5}	—
$^4S_{3/2} - ^2D_{5/2}$	3.6×10^{-5}	3728.8	4.8×10^{-4}	2424.5	2.6×10^{-4}	6716.4	1.8×10^{-3}	4711.3
$^4S_{3/2} - ^2D_{3/2}$	1.8×10^{-4}	3726.0	5.5×10^{-3}	2421.8	8.8×10^{-4}	6730.8	2.2×10^{-2}	4740.2

TABLE 3.10
Transition probabilities of p^4 ions

Transition	[O I] Transition probability (sec^{-1})	[O I] Wavelength (Å)	[Ne III] Transition probability (sec^{-1})	[Ne III] Wavelength (Å)	[Ar III] Transition probability (sec^{-1})	[Ar III] Wavelength (Å)
$^1D_2 - ^1S_0$	1.2	5577.4	2.7	3342.5	2.6	5191.8
$^3P_2 - ^1S_0$	2.9×10^{-4}	2958.4	3.9×10^{-3}	1793.7	4.2×10^{-2}	3005.2
$^3P_1 - ^1S_0$	7.3×10^{-2}	2972.3	2.0	1814.6	3.9	3109.2
$^3P_2 - ^1D_2$	6.3×10^{-3}	6300.3	1.7×10^{-1}	3868.8	3.1×10^{-1}	7135.8
$^3P_1 - ^1D_2$	2.1×10^{-3}	6363.8	5.4×10^{-2}	3967.5	8.2×10^{-2}	7751.1
$^3P_0 - ^1D_2$	7.3×10^{-7}	6391.5	8.5×10^{-6}	4011.6	2.2×10^{-5}	8036.3
$^3P_1 - ^3P_0$	1.7×10^{-5}	146 μ	1.2×10^{-3}	36.0 μ	5.2×10^{-3}	21.8 μ
$^3P_2 - ^3P_0$	1.0×10^{-10}	44 μ	2.2×10^{-8}	10.9 μ	2.4×10^{-6}	6.4 μ
$^3P_2 - ^3P_1$	8.9×10^{-5}	63 μ	6.0×10^{-3}	15.6 μ	3.1×10^{-2}	9.0 μ

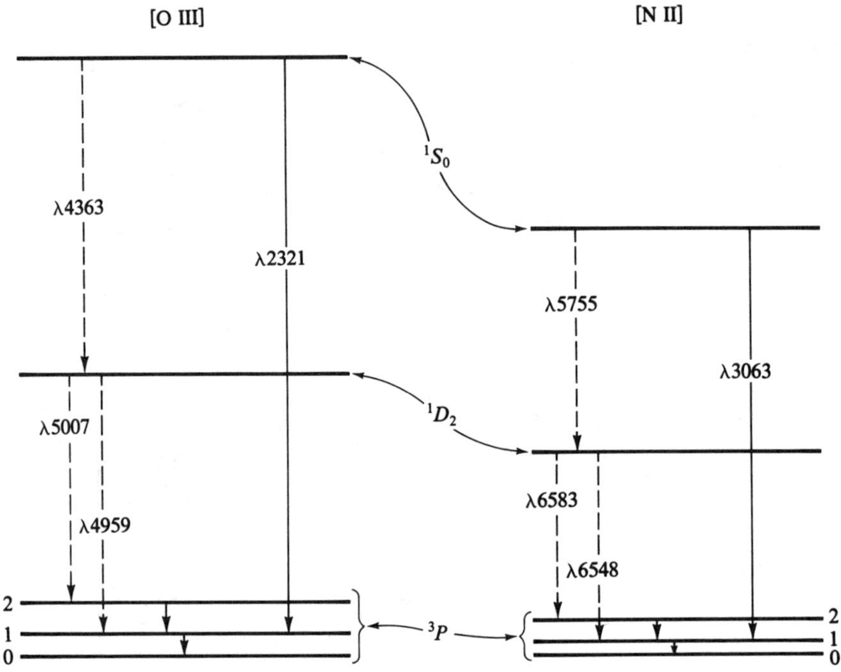

FIGURE 3.1
Energy-level diagram for lowest terms of [O III], all from ground $2p^2$ configuration, and for [N II], of the same isoelectronic sequence. Splitting of the ground ^3P term has been exaggerated for clarity. Emission lines in the optical region are indicated by dashed lines, and by solid lines in the infrared and ultraviolet. Only the strongest transitions are indicated.

TABLE 3.11
Critical densities for collisional de-excitation

Ion	Level	$N_c(\text{cm}^{-3})$	Ion	Level	$N_c(\text{cm}^{-3})$
C II	$^2P_{3/2}$	8.5×10^1	O III	$^1D_2$	7.0×10^5
			O III	$^3P_2$	3.8×10^3
C III	$^3P_2$	5.4×10^5	O III	$^3P_1$	1.7×10^3
N II	$^1D_2$	8.6×10^4	Ne II	$^2P_{1/2}$	6.6×10^5
N II	$^3P_2$	3.1×10^2			
N II	$^3P_1$	1.8×10^2	Ne III	$^1D_2$	7.9×10^6
			Ne III	$^3P_0$	2.0×10^4
N III	$^2P_{3/2}$	3.2×10^3	Ne III	$^3P_1$	1.8×10^5
N IV	$^3P_2$	1.4×10^6	Ne V	$^1D_2$	1.6×10^7
			Ne V	$^3P_2$	3.8×10^5
O II	$^2D_{3/2}$	1.6×10^4	Ne V	$^3P_1$	1.8×10^5
O II	$^2D_{5/2}$	3.1×10^3			

NOTE: All values are calculated for $T = 10,000^\circ$ K.

3.6 Energy Loss by Collisionally Excited Line Radiation of H

H^+, the most abundant ion in nebulae, has no bound levels and no lines, but H^0, although its fractional abundance is low, may affect the radiative cooling in a nebula. The most important excitation processes from the ground $1\,^2S$ term are to $2\,^2P$, followed by emission of an $L\alpha$ photon with $h\nu = 10.2$ eV, and to $2\,^2S$, followed by emission of two photons in the $2\,^2S \rightarrow 1\,^2S$ continuum with $h\nu' + h\nu'' = 10.2$ eV and transition probability $A_{2^2S,1^2S} = 8.23\ \text{sec}^{-1}$. Cross sections for excitation of neutral atoms by electrons do not vary as v^{-2}, but rise from zero at the threshold, peak at energies several times the threshold, and then decline at high energies, often with superimposed resonances. Nevertheless the mean collision strengths, integrated over the Maxwellian velocity distribution of the electrons as defined by (3.19) and (3.20), for these transitions and for $1\,^2S \rightarrow 3\,^2S$, $3\,^2P$, and $3\,^2D$ are quantities that vary fairly slowly, as Table 3.12 shows. Accurate cross sections are not available for $n > 3$, and therefore only rough estimates of the radiative cooling by higher levels of H are available, but they seem to make only a minor contribution, except at very high temperatures.

TABLE 3.12

Effective collision strengths for H I

T (° K)	$\Omega(1\,^2S,\,2\,^2S)$	$\Omega(1\,^2S,\,2\,^2P)$	$\Omega(1\,^2S,\,3\,^2S)$	$\Omega(1\,^2S,\,3\,^2P)$	$\Omega(1\,^2S,\,3\,^2D)$
5,000	0.25	0.40	0.05	0.11	0.05
10,000	0.26	0.40	0.07	0.16	0.06
12,500	0.26	0.42	0.08	0.18	0.06
15,000	0.26	0.44	0.08	0.20	0.07
20,000	0.27	0.49	0.09	0.22	0.08

3.7 Resulting Thermal Equilibrium

The temperature at each point in a static nebula is determined by the equilibrium between heating and cooling rates, namely,

$$G = L_R + L_{FF} + L_C. \tag{3.32}$$

The collisionally excited radiative cooling rate L_C is a sum (over all transitions of all ions) of individual terms like (3.23), (3.26), or (3.30). In the low-density limit, since all the terms in G, L_R, L_{FF}, and L_C are proportional to N_e and to the density of some ion, equation (3.32) and therefore the resulting temperatures are independent of the total density, but do depend on the relative abundances of the various ions. When collisional de-excitation begins to be important, the cooling rate at a given temperature is decreased, and the equilibrium temperature for a given radiation field is therefore somewhat increased.

To understand better the concepts here, let us consider an example, namely, an H II region with "normal" abundances of the elements. We will adopt $N(\text{O})/N(\text{H}) = 7 \times 10^{-4}$, $N(\text{Ne})/N(\text{H}) = 9 \times 10^{-5}$, and $N(\text{N})/N(\text{H}) = 9 \times 10^{-5}$. Let us suppose that O, Ne, and N are each 80 percent singly ionized and 20 percent doubly ionized, that H is 0.1 percent neutral, and that the remainder is ionized. Some of the individual contributions to the radiative cooling (in the low-density limit) and the total radiative cooling $L_C + L_{FF}$ are shown in Figure 3.2. For each level the contribution is small if $kT \ll \chi$, then increases rapidly and peaks at $kT \approx \chi$, and then decreases slowly for $kT > \chi$. The total radiative cooling, composed of the sum of the individual contributions, continues to rise with increasing T as long as there are levels with excitation energy $\chi > kT$. It can be seen that, for the assumed composition and ionization, O^{++} dominates the radiative cooling contribution at low temperatures, and O^{+} at somewhat higher temperatures. At all temperatures shown, the contribution of collisional excitation of H^0 is small.

It is convenient to rewrite equation (3.32) in the form

$$G - L_R = L_{FF} + L_C,$$

where $G - L_R$ is then the "effective heating rate," representing the net energy gained in photoionization processes, with the recombination losses already subtracted. This effective heating rate is also shown in Figure 3.2, for model stellar atmospheres with various temperatures. Notice in the figure that the calculated nebular temperature at which the curves cross and at which equation (3.32) is satisfied is rather insensitive to the input stellar radiation field. Typical nebular temperatures are $T \approx 7{,}000°$ K, according to Figure 3.2, with somewhat higher temperatures for hotter stars or larger optical depths.

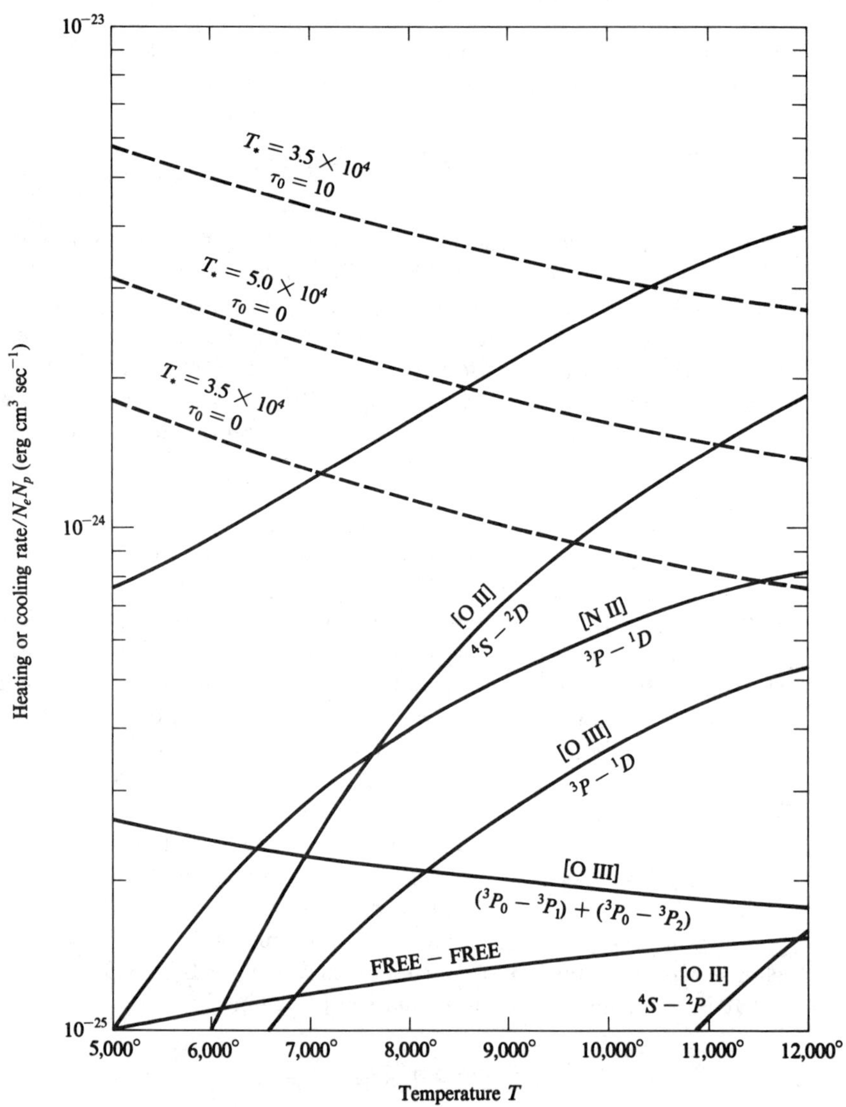

FIGURE 3.2

Net effective heating rates (G-L_R) for various stellar input spectra, shown as dashed curves. Total radiative cooling rate ($L_{FF} + L_C$) for the simple approximation to the H II region described in the text is shown as highest solid black curve, and the most important individual contributors to radiative cooling are shown by labeled solid curves. The equilibrium temperature is given by the intersection of a dashed curve and the highest solid curve. Note how the increased optical depth τ_0 or increased stellar temperature T_* increases T by increasing G.

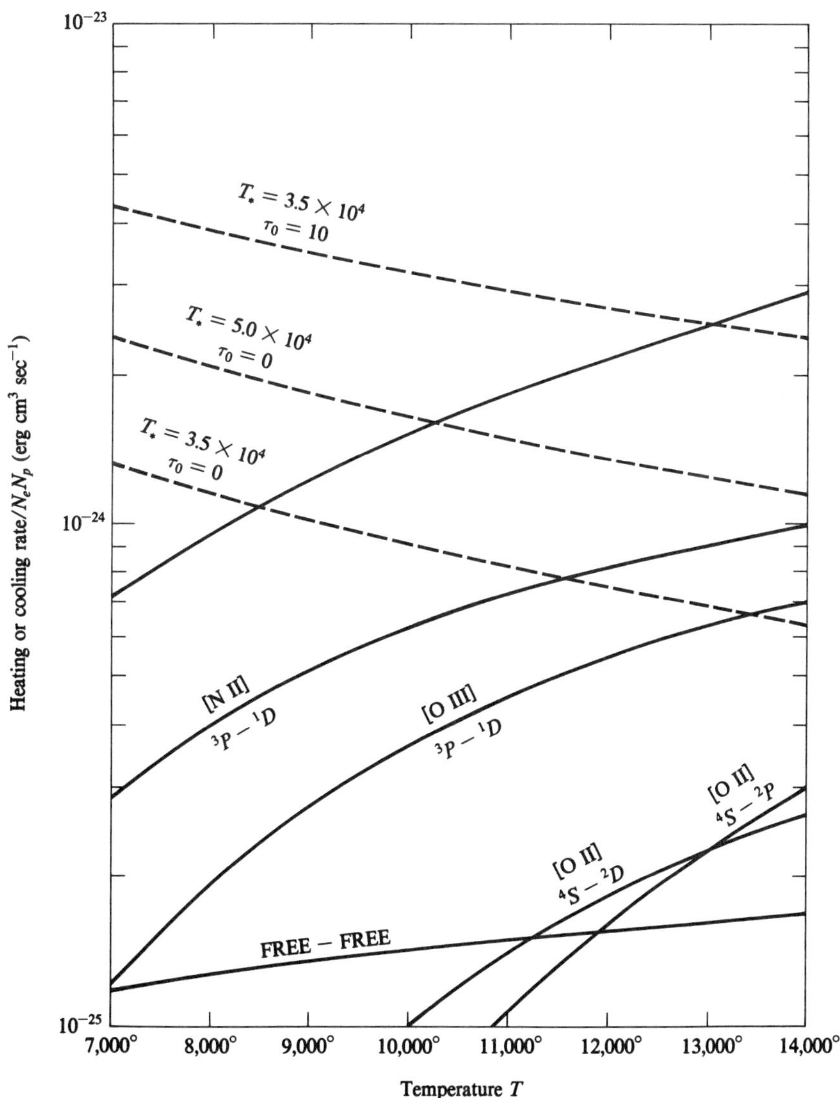

FIGURE 3.3
Same as Figure 3.2, except that collisional de-excitation at $N_e = 10^4$ cm^{-3} has been approximately taken into account in the radiative cooling rates.

At high electron densities, collisional de-excitation can appreciably modify the radiative cooling rate and therefore the resulting nebular temperature. For instance, at $N_e \approx 10^4$ cm^{-3}, a density that occurs in condensations in many H II regions, the [O II] $^4S - ^2D$ and [O III] $^3P_0 - ^3P_1$ and $^3P_0 - ^3P_2$ transitions are only about 10 percent effective, [N II] $^3P_0 - ^3P_1$ and $^3P_0 - ^3P_2$ are only about 1 percent effective, and [N III] $^2P_{1/2} - ^2P_{3/2}$ is about 20 percent effective, as Table 3.11 shows. Figure 3.3 shows the effective cooling rate for this situation, with the abundances and ionization otherwise as previously described, and demonstrates that appreciably higher temperatures occur at high densities. Similarly, lower abundances of the heavy elements tend to decrease the cooling rate and thus to increase the resulting equilibrium temperature.

Under conditions of very high ionization, however, as in the central part of a planetary nebula, the ionization is high enough that there is very little H^0, O^+, or O^{++}, and then the radiative cooling is appreciably decreased. Under these conditions the main coolants are Ne^{+4} and C^{+3}, and the nebular temperature may be $T \lesssim 2 \times 10^4$ ° K. Detailed results obtained from models of both H II regions and planetary nebulae are discussed in Chapter 5.

References

The basic papers on thermal equilibrium are:

Spitzer, L. 1948. *Ap. J.* **107,** 6.

Spitzer, L. 1949. *Ap. J.* **109,** 337.

Spitzer, L., and Savedoff, M. P. 1950. *Ap. J.* **111,** 593.

Some additional work, including the on-the-spot approximation and the effects of collisional de-excitation, is described in

Burbidge, G. R., Gould, R. J., and Pottasch, S. R. 1963. *Ap. J.* **138,** 945.

Osterbrock, D. E. 1965. *Ap. J.* **142,** 1423.

Numerical values of recombination coefficients β are given in

Hummer, D. G., and Seaton, M. J. 1963. *M.N.R.A.S.* **125,** 437.

(Table 3.2 is based on this reference.)

Basic papers on collisional excitation and the methods used to calculate the collision strengths are:

Seaton, M. J. 1968. *Advances in Atomic and Molecular Physics* **4,** 331.

Seaton, M. J. 1975. *Advances in Atomic and Molecular Physics* **11**, 83.

This material is further elucidated in the book

Burke, P. G., Eissner, W. B., Hummer, D. G., and Percival, I. C. eds. 1983. *Atoms in Astrophysics.* New York: Plenum Press.

Numerical values of collision strengths are widely scattered through the physics literature, but most of the best available published values have been collected, tabulated and evaluated by

Mendoza, C. 1983. *Planetary Nebulae* (IAU Symposium No. 103) ed. D. R. Flower. Dordrecht: Reidel, p. 143.

This review paper, on which Tables 3.3 – 3.7 are based, lists all the original references.

Even better, more recently published values also included in these tables, are

Hayes, M. A., and Nussbaumer, H. 1983. *Astr. Ap.* **124,** 279. (O^{+3})

Hayes, M. A., and Nussbaumer, H. 1984. *Astr. Ap.* **134,** 193. (C^{+})

Butler, K., and Storey, P. J. 1988. *M.N.R.A.S.* in preparation. (N^{+2})

Butler, K., and Mendoza, C. 1984. *M.N.R.A.S.* **208,** 17P. (Ne^{+2})

Keenan, F. P., Johnson, C. T., Kingston, A. E., and Dufton, P. L. 1985. *M.N.R.A.S.* **214,** 37P. (Si^{+})

Berrington, K. 1985. *J. Phys. B.* **18,** L395 (C^{+2})

Johnson, C. T., and Kingston, A. E. 1987. *J. Phys. B.* **20**, 5757. (Ne^{+})

Zeippen, C. J., Butler, K., and Le Bourlot, J. 1987. *Astr. Ap.* **188**, 251. (Ar^{+3})

Numerical values of transition probabilities are very conveniently listed in the 1983 review by Mendoza, which also gives the original references. Tables 3.8 – 3.10 are based on it.

The best wavelengths come from nebular measurements made with high-dispersion spectrographs. The fundamental reference, from which nearly all the wavelengths used in the book and listed in Tables 3.8 – 3.10 have been taken, is

Bowen, I. S. 1960. *Ap. J.* **132,** 1.

A few later and better values are from

Lutz, J. H., and Seaton, M. J. 1979. *M.N.R.A.S.* **187,** 1P. [Ne IV]

Hippelein, H., and Münch, G. 1981. *Astr. Ap.* **95,** 100. [S III]

Smith, P. L., Magnusson, C. E., and Zetterberg, P. O. 1984. *Ap. J.* **277,** L79. S III]

DeRobertis, M. M., Osterbrock, D. E., and McKee, C. F. 1985. *Ap. J.* **293,** 459. [O II]

The most accurately calculated numerical values of collisional excitation cross sections for H at the time of writing are in

Callaway, J. 1985. *Phys. Rev. A.* **32,** 775.

Callaway, J., Unnikrishnan, K., and Oza, D. H. 1987. *Phys. Rev. A.,* **36,** 2576.

(Table 3.12 is based on these references.)

4

Calculation of Emitted Spectrum

4.1 Introduction

The radiation emitted by each element of volume in a gaseous nebula depends upon the abundances of the elements, determined by the previous evolutionary history of the gas, and on the local ionization, density and temperature, determined by the radiation field and the abundances as described in the preceding two chapters. The most prominent spectral features are the emission lines, and many of these are the collisionally excited lines described in the preceding chapter on thermal equilibrium. The formalism developed there to calculate the cooling rate, and thus the thermal equilibrium, may be taken over unchanged to calculate the strength of these lines. If we could observe all the lines in the entire spectral region from the extreme ultraviolet to the far infrared, we could measure directly the cooling rate at each observed point in the nebula. Many of the most important lines in the cooling, for instance, [O II] $\lambda\lambda 3726$, 3729 and [O III] $\lambda\lambda 4959$, 5007, are in the optical region and are easily measured. Other lines that are also important in the cooling, such as [O III] $2p^2\ ^3P_0 - 2p^2\ ^3P_1$ $\lambda 88.4\ \mu$, and $^3P_1 - {}^3P_2$ $\lambda 51.8\ \mu$, are in the far infrared region, while still others, such as C IV $\lambda\lambda 1548$, 1551, are in the satellite ultraviolet. These regions have only recently become observable, from high-flying airplanes and artificial satellites, respectively. Therefore far less data for the far-infrared and satellite ultraviolet spectral regions are available, generally with relatively poor angular resolution, but this situation is very rapidly improving.

For historical reasons, astronomers tend to refer to the chief emission lines of gaseous nebulae as *forbidden* lines. Actually, it is better to think of the bulk of the lines as *collisionally excited* lines, which arise from levels within a few volts of the ground level and which therefore can be excited by collisions with thermal electrons. In fact, in the ordinary optical region all these collisionally excited lines are forbidden lines, because in the common ions all the excited levels within a few volts of the ground level arise from the same electron configuration as the ground level itself, and thus radiative transitions are forbidden by the parity selection rule. However, at wavelengths just slightly below the ultraviolet cutoff of the Earth's atmosphere, collisionally excited lines begin to appear that are not forbidden lines; for example, Mg II $3s\,^2S - 3p\,^2P$ $\lambda\lambda 2796$, 2803, C IV $2s\,^2S - 2p\,^2P$ $\lambda\lambda 1548$, 1551, and Si IV $3s\,^2S - 3p\,^2P$ $\lambda\lambda 1394$, 1403. All these lines were calculated to be strong in the spectra of gaseous nebulae before the ultraviolet spectral region could be observed, and these calculations have since been confirmed from satellites in orbit above the Earth's atmosphere.

In addition to the collisionally excited lines, the recombination lines of H I, He I, and He II are characteristic features of the spectra of gaseous nebulae. They are emitted by atoms undergoing radiative transitions in cascading down to the ground level following recombinations to excited levels. In the remainder of this chapter, these recombination emission processes will be discussed in more detail, and then the related topic of resonance-fluorescence excitation of observable lines of other elements will be considered. Finally, the continuum-emission processes, which are the bound-free and free-free analogues of the bound-bound transitions emitted in the recombination-line spectrum, will be examined.

4.2 Optical Recombination Lines

The recombination-line spectrum of H I is emitted by H atoms that have been formed by captures of electrons into excited levels and that are cascading by downward radiative transitions to the ground level. In the limit of very low density, the only processes that need be considered are captures and downward-radiative transitions. Thus the equation of statistical equilibrium for any level nL may be written

$$N_p N_e \alpha_{nL}(T) + \sum_{n'>n}^{\infty} \sum_{L'} N_{n'L'} A_{n'L',nL} = N_{nL} \sum_{n''=1}^{n-1} \sum_{L''} A_{nL,n''L''}. \quad (4.1)$$

Note in general $A_{n'L'}, A_{n''L''} \neq 0$ only if $L' = L'' \pm 1$.

It is convenient to express the population in terms of the dimensionless factors b_{nL} that measure the deviation from thermodynamic equilibrium at the local T, N_e, and N_p. Since in thermodynamic equilibrium, the Saha equation

$$\frac{N_p N_e}{N_{1S}} = \left(\frac{2\pi m k T}{h^2}\right)^{3/2} e^{-h\nu_0/kT}, \quad (4.2)$$

and the Boltzmann equation

$$\frac{N_{nL}}{N_{1S}} = (2L + 1)e^{-\chi_n/kT}, \tag{4.3}$$

apply, the population in the level nL in thermodynamic equilibrium may be written

$$N_{nL} = (2L + 1)\left(\frac{h^2}{2\pi mkT}\right)^{3/2} e^{X_n/kT} N_p N_e, \tag{4.4}$$

where

$$X_n = h\nu_0 - \chi_n = \frac{h\nu_0}{n^2} \tag{4.5}$$

is the ionization potential of the level nL. Therefore, in general, the population may be written

$$N_{nL} = b_{nL}(2L + 1)\left(\frac{h^2}{2\pi mkT}\right)^{3/2} e^{X_n/kT} N_p N_e, \tag{4.6}$$

and $b_{nL} = 1$ in thermodynamic equilibrium.

Substituting this expression in (4.1),

$$\begin{aligned} &\frac{\alpha_{nL}}{(2L+1)}\left(\frac{2\pi mkT}{h^2}\right)^{3/2} e^{-X_n/kT} \\ &+ \sum_{n'>n}^{\infty} \sum_{L'} b_{n'L'} A_{n'L',nL}\left(\frac{2L'+1}{2L+1}\right) e^{(X_{n'}-X_n)/kT} \\ &= b_{nL} \sum_{n''=1}^{n-1} \sum_{L''} A_{nL,n''L''}, \end{aligned} \tag{4.7}$$

it can be seen that the b_{nL} factors are independent of density as long as recombination and downward-radiative transitions are the only relevant processes. Furthermore, it can be seen that the equations (4.7) can be solved by a systematic procedure working downward in n, for if the b_{nL} are known for all $n \geq n_K$, then the n equations (4.7), with $L = 0, 1, \ldots, n-1$ for $n = n_K - 1$, each contain a single unknown b_{nL} and can be solved immediately, and so on successively downward.

It is convenient to express the solutions in terms of the cascade matrix $C_{nL,n'L'}$, which is the probability that population of nL is followed by a transition to $n'L'$ via all possible cascade routes. The cascade matrix can be generated directly from the probability matrix $P_{nL,n'L'}$,, which gives the probability that population of the level nL is followed by a direct radiative transition to $n'L'$,

$$P_{nL,n'L'} = \frac{A_{nL,n'L'}}{\sum_{n''=1}^{n-1} \sum_{L''} A_{nL,n''L''}}, \tag{4.8}$$

which is zero unless $L' = L \pm 1$.

Hence, for $n' = n - 1$,

$$C_{nL,n-1L'} = P_{nL,n-1L'};$$

for $n' = n - 2$,

$$C_{nL,n-2L'} = P_{nL,n-2L'} + \sum_{L''=L'\pm 1} C_{nL,n-1L''} P_{n-1L'',n-2L'};$$

and for $n' = n - 3$,

$$\begin{aligned} C_{nL,n-3L'} &= P_{nL,n-3L'} \\ &+ \sum_{L''=L'\pm 1} (C_{nL,n-1L''} P_{n-1L'',n-3L'} + C_{nL,n-2L''} P_{n-2L'',n-3L'}) \end{aligned}$$

so that if we define

$$C_{nL,nL''} = \delta_{LL''}, \tag{4.9}$$

then in general

$$C_{nL,n'L'} = \sum_{n''>n'}^{n} \sum_{L''=L'\pm 1} C_{nL,n''L''} P_{n''L'',n'L'}. \tag{4.10}$$

The solutions of the equilibrium equations (4.1) may be immediately written down, for the population of any level nL is fixed by the balance between recombinations to all levels $n' \geq n$ that lead by cascades to nL and downward radiative transitions from nL:

$$N_p N_e \sum_{n'=n}^{\infty} \sum_{L'=0}^{n'-1} \alpha_{n'L'}(T) C_{n'L',nL} = N_{nL} \sum_{n''=1}^{n-1} \sum_{L''=L\pm 1} A_{nL,n''L''}. \tag{4.11}$$

It is convenient to express the results in this form because once the cascade matrix has been calculated, it can be used to find the b_{nL} factors or the populations N_{nL} at any temperature, or even for cases in which the population occurs by other nonradiative processes, such as collisional excitation from the ground level or from an excited level. To carry out the solutions, it can be seen from (4.11) that it is necessary to fit series in n, n', L, and L' to $C_{nL,n'L'}$ and $\alpha_{nL}(T)$, and extrapolate these series as $n \rightarrow \infty$. Once the populations N_{nL} have been found, it is simple to calculate the emission coefficient in each line

$$j_{nn'} = \frac{h\nu_{nn'}}{4\pi} \sum_{L=0}^{n-1} \sum_{L'=L\pm1} N_{nL} A_{nL,n'L'}. \tag{4.12}$$

The situation we have been considering is commonly called Case A in the theory of recombination-line radiation, and assumes that all line photons emitted in the nebula escape without absorption and therefore without causing further upward transitions. Case A is thus a good approximation for gaseous nebulae that are optically thin in all H I resonance lines, but in fact such nebulae can contain only a relatively small amount of gas and are mostly too faint to be easily observed.

Nebulae that contain observable amounts of gas generally have quite large optical depths in the Lyman resonance lines of H I. This can be seen from the equation for the central line-absorption cross section,

$$a_0(\mathrm{L}n) = \frac{3\lambda_{n1}^3}{8\pi} \left(\frac{m_{\mathrm{H}}}{2\pi kT}\right)^{1/2} A_{nP,1S}, \tag{4.13}$$

where λ_{n1} is the wavelength of the line. Thus, at a typical temperature $T = 10{,}000°$ K, the optical depth in Lα is about 10^4 times the optical depth at the Lyman limit $\nu = \nu_0$ of the ionizing continuum, and an ionization-bounded nebula with $\tau_0 \approx 1$ therefore has $\tau(\mathrm{L}\alpha) \approx 10^4$, $\tau(\mathrm{L}\beta) \approx 10^3$, $\tau(L8) \approx 10^2$, and $\tau(L18) \approx 10$. In each scattering there is a finite probability that the Lyman-line photon will be converted to a lower-series photon plus a lower member of the Lyman series. Thus, for instance, each time an $L\beta$ photon is absorbed by an H atom, raising it to the 3 ^2P level, the probability that this photon is scattered is $P_{31,10} = 0.882$, while the probability that it is converted to $H\alpha$ is $P_{31,20} = 0.118$, so after nine scatterings an average $L\beta$ photon is converted to $H\alpha$ (plus two photons in the 2 $^2S \rightarrow 1^2S$ continuum) and cannot escape from the nebula. Likewise, an average $L\gamma$ photon is transformed, after a relatively few scatterings, either into a $P\alpha$ photon plus an $H\alpha$ photon plus an Lα photon, or into an $H\beta$ photon plus two photons in the 2 ^2S-1 ^2S continuum. Thus, for these large optical depths, a better approximation than Case A is the opposite assumption that every Lyman-line photon is scattered many times and is converted (if $n \geq 3$) into lower-series photons plus either Lα or two-continuum photons. This large optical depth approximation is called Case B, and is more accurate than Case A for most nebulae. However, it is clear that the real situation is intermediate, and is similar to Case B for the lower Lyman lines, but progresses continuously to a situation nearer Case A as $n \rightarrow \infty$ and $\tau(\mathrm{L}n) \rightarrow 1$.

Under Case B conditions, any photon emitted in an $n\,^2P \rightarrow 1\,^2S$ transition is immediately absorbed nearby in the nebula, thus populating the $n\,^2P$ level in another atom. Hence, in Case B, the downward-radiative transitions to $1\,^2S$ are simply omitted from consideration, and the sums in the equilibrium equations (4.1), (4.7), (4.8), and (4.11) are terminated at $n'' = n_0 = 2$ instead of at $n_0 = 1$ as in Case A. The detailed transition between Cases A and B will be discussed in Section 4.5.

Selected numerical results from the recombination spectrum of H I are listed in Tables 4.1 and 4.2 for Cases A and B, respectively. Note that, in addition to the emission coefficient $j_{42} = j_{\mathrm{H}\beta}$ and the relative intensities of the other lines, it is also sometimes convenient to use the effective recombination coefficient, defined by

$$N_p N_e \alpha_{nn'}^{eff} = \sum_{L=0}^{n-1} \sum_{L'=L\pm 1} N_{nL} A_{nL,n'L'} = \frac{4\pi j_{nn'}}{h\nu_{nn'}}. \qquad (4.14)$$

For hydrogen-like ions of nuclear charge Z, all the transition probabilities $A_{nL,n'L'}$ are proportional to Z^4, so the $P_{nL,n'L'}$ and $C_{nL,n'L'}$ matrices are independent of Z. The recombination coefficients α_{nL} scale as

$$\alpha_{nL}(Z,T) = Z\,\alpha_{nL}(1, T/Z^2);$$

the effective recombination coefficients scale in this same way, and since the energies $h\nu_{nn'}$ scale as

$$\nu_{nn'}(Z) = Z^2 \nu_{nn'}(1),$$

the emission coefficient is

$$j_{nn'}(Z,T) = Z^3 j_{nn'}(1, T/Z^2). \qquad (4.15)$$

Thus the calculations for H I at a temperature T can also be applied to He II at $T' = 4T$. In Table 4.3 some of the main features of the He II recombination-line spectrum are listed for Case B, with the strongest line in the optical spectrum, $\lambda 4686$ ($n = 4 \rightarrow 3$), as the reference line. Note that the Fowler series ($n \rightarrow 3$) except for $\lambda 4686$ and $\lambda 3203$, and the entire "Balmer" series ($n \rightarrow 2$) are in the satellite ultraviolet spectral region.

Next let us return to the H I recombination lines and examine the effects of collisional transitions at finite nebular densities. The largest collisional cross sections involving the excited levels of H are for transitions $nL \rightarrow nL \pm 1$,

TABLE 4.1
H I recombination lines (Case A)

	T			
	2,500° K	5,000° K	10,000° K	20,000° K
$4\pi j_{H\beta}/N_p N_e$ (erg cm^3 sec^{-1})	2.70×10^{-25}	1.54×10^{-25}	8.30×10^{-26}	4.21×10^{-26}
$\alpha^{eff}_{H\beta}$ (cm^3 sec^{-1})	6.61×10^{-14}	3.78×10^{-14}	2.04×10^{-14}	1.03×10^{-14}
Balmer-line intensities relative to Hβ				
$j_{H\alpha}/j_{H\beta}$	3.42	3.10	2.86	2.69
$j_{H\gamma}/j_{H\beta}$	0.439	0.458	0.470	0.485
$j_{H\delta}/j_{H\beta}$	0.237	0.250	0.262	0.271
$j_{H\epsilon}/j_{H\beta}$	0.143	0.153	0.159	0.167
$j_{H8}/j_{H\beta}$	0.0957	0.102	0.107	0.112
$j_{H9}/j_{H\beta}$	0.0671	0.0717	0.0748	0.0785
$j_{H10}/j_{H\beta}$	0.0488	0.0522	0.0544	0.0571
$j_{H15}/j_{H\beta}$	0.0144	0.0155	0.0161	0.0169
$j_{H20}/j_{H\beta}$	0.0061	0.0065	0.0068	0.0071
Lyman-line intensities relative to Hβ				
$j_{L\alpha}/j_{H\beta}$	33.0	32.5	32.7	34.0
Paschen-line intensities relative to corresponding Balmer lines				
$j_{P\alpha}/j_{H\beta}$	0.684	0.562	0.466	0.394
$j_{P\beta}/j_{H\gamma}$	0.609	0.527	0.460	0.404
$j_{P\gamma}/j_{H\delta}$	0.565	0.504	0.450	0.406
j_{P8}/j_{H8}	0.531	0.487	0.443	0.404
j_{P10}/j_{H10}	0.529	0.481	0.439	0.399
j_{P15}/j_{H15}	0.521	0.465	0.429	0.396
j_{P20}/j_{H20}	0.508	0.462	0.426	0.394

TABLE 4.2
He I recombination lines (Case B)

	T			
	2,500° K	5,000° K	10,000° K	20,000° K
$4\pi j_{H\beta}/N_p N_e$ (erg cm^3 sec^{-1})	3.72×10^{-25}	2.20×10^{-25}	1.24×10^{-25}	6.62×10^{-26}
$\alpha_{H\beta}^{eff}$ (cm^3 sec^{-1})	9.07×10^{-14}	5.37×10^{-4}	3.03×10^{-14}	1.62×10^{-14}
Balmer-line intensities relative to Hβ				
$j_{H\alpha}/j_{H\beta}$	3.30	3.05	2.87	2.76
$j_{H\gamma}/j_{H\beta}$	0.444	0.451	0.466	0.474
$j_{H\delta}/j_{H\beta}$	0.241	0.249	0.256	0.262
$j_{H\epsilon}/j_{H\beta}$	0.147	0.153	0.158	0.162
$j_{H8}/j_{H\beta}$	0.0975	0.101	0.105	0.107
$j_{H9}/j_{H\beta}$	0.0679	0.0706	0.0730	0.0744
$j_{H10}/j_{H\beta}$	0.0491	0.0512	0.0529	0.0538
$j_{H15}/j_{H\beta}$	0.0142	0.0149	0.0154	0.0156
$j_{H20}/j_{H\beta}$	0.0059	0.0062	0.0064	0.0065
Paschen-line intensities relative to corresponding Balmer lines				
$j_{P\alpha}/j_{H\beta}$	0.528	0.427	0.352	0.293
$j_{P\beta}/j_{H\gamma}$	0.473	0.415	0.354	0.308
$j_{P\gamma}/j_{H\delta}$	0.440	0.398	0.354	0.313
j_{P8}/j_{H8}	0.421	0.388	0.350	0.321
j_{P10}/j_{H10}	0.422	0.389	0.350	0.320
j_{P15}/j_{H15}	0.415	0.383	0.344	0.321
j_{P20}/j_{H20}	0.407	0.387	0.344	0.323
Brackett-line intensities relative to corresponding Balmer lines				
$j_{Br\alpha}/j_{H\gamma}$	0.326	0.242	0.179	0.135
$j_{Br\beta}/j_{H\delta}$	0.294	0.232	0.184	0.145
$j_{Br\gamma}/j_{H\epsilon}$	0.263	0.217	0.178	0.146
j_{Br8}/j_{H8}	0.254	0.214	0.177	0.147
j_{Br10}/j_{H10}	0.243	0.208	0.174	0.148
j_{Br15}/j_{H15}	0.223	0.198	0.171	0.148
j_{Br20}/j_{H20}	0.215	0.200	0.171	0.149

TABLE 4.3
He II recombination lines (Case B)

	T			
	5,000° K	10,000° K	20,000° K	40,000° K
$4\pi j_{\lambda 4686}/N_{He^{++}}N_e$ (erg cm^3 sec^{-1})	3.14×10^{-24}	1.58×10^{-24}	7.54×10^{-25}	3.48×10^{-25}
$\alpha^{eff}_{\lambda 4686}$ (cm^3 sec^{-1})	7.40×10^{-13}	3.72×10^{-13}	1.77×10^{-13}	8.20×10^{-14}
"Balmer"-line ($n \rightarrow 2$) intensities relative to $\lambda 4686$				
$j_{32}/j_{\lambda 4686}$	0.560	0.625	0.714	8.15
$j_{42}/j_{\lambda 4686}$	0.154	0.189	0.234	2.84
$j_{52}/j_{\lambda 4686}$	0.066	0.084	0.106	1.32
$j_{72}/j_{\lambda 4686}$	0.022	0.028	0.036	0.45
$j_{102}/j_{\lambda 4686}$	0.007	0.009	0.012	0.15
Fowler-line intensities ($n \rightarrow 3$) relative to $\lambda 4686$				
$j_{53}/j_{\lambda 4686}$	0.355	0.398	0.438	0.469
$j_{63}/j_{\lambda 4686}$	0.173	0.201	0.232	0.257
$j_{83}/j_{\lambda 4686}$	0.065	0.078	0.092	0.104
$j_{103}/j_{\lambda 4686}$	0.033	0.039	0.047	0.052
Pickering-line ($n \rightarrow 4$) intensities relative to $\lambda 4686$				
$j_{54}/j_{\lambda 4686}$	0.295	0.274	0.256	0.237
$j_{64}/j_{\lambda 4686}$	0.131	0.134	0.135	0.134
$j_{74}/j_{\lambda 4686}$	0.0678	0.0734	0.0779	0.0799
$j_{84}/j_{\lambda 4686}$	0.0452	0.0469	0.0506	0.0527
$j_{94}/j_{\lambda 4686}$	0.0280	0.0315	0.0345	0.0364
$j_{104}/j_{\lambda 4686}$	0.0198	0.0226	0.0249	0.0262
$j_{124}/j_{\lambda 4686}$	0.0106	0.0124	0.0139	0.0149
$j_{154}/j_{\lambda 4686}$	0.0050	0.0060	0.0069	0.0075
$j_{204}/j_{\lambda 4686}$	0.0020	0.0024	0.0029	0.0031
Pfund-line ($n \rightarrow 5$) intensities relative to corresponding Pickering lines				
j_{65}/j_{64}	0.825	0.713	0.634	0.566
j_{75}/j_{74}	0.807	0.734	0.659	0.593
j_{85}/j_{84}	0.708	0.705	0.646	0.590
j_{105}/j_{104}	0.727	0.690	0.643	0.599
j_{155}/j_{154}	0.640	0.650	0.623	0.600
j_{205}/j_{204}	0.600	0.625	0.586	0.613

which have essentially zero energy difference. Collisions with both electrons and protons can cause these angular-momentum-changing transitions, but because of the small energy difference, protons are more effective than electrons; for instance, representative values of the mean cross sections for thermal protons at $T \approx 10{,}000°$ K are $\sigma_{2^2S\rightarrow 2^2P} \approx 3 \times 10^{-10}$ cm^2, $\sigma_{10^2L\rightarrow 10^2L\pm 1} \approx 4 \times 10^{-7}$cm^2, and $\sigma_{20^2L\rightarrow 20^2L\pm 1} \approx 6 \times 10^{-6}$ cm^2. (Both of the latter are evaluated for $L \approx n/2$.) These collisional transitions must then be included in the equilibrium equations, which are modified from (4.1) to read

$$\begin{aligned} N_p N_e \alpha_{nL}(T) &+ \sum_{n'>n}^{\infty} \sum_{L'=L\pm 1} N_{n'L'} A_{n'L',nL} \\ &+ \sum_{L'=L\pm 1} N_{nL'} N_p q_{nL',nL} \\ &= N_{nL} \left[\sum_{n''=n_0}^{n-1} \sum_{L''=L\pm 1} A_{nL,n''L''} + \sum_{L''=L\pm 1} N_p q_{nL,nL''} \right], \end{aligned} \tag{4.16}$$

where $n_0 = 1$ or 2 for Cases A and B, respectively, and

$$q_{nL,n'L'} \equiv q_{nL,n'L'}(T) = \int_0^{\infty} v\sigma_{nL\rightarrow n'L'} f(v) dv \tag{4.17}$$

is the collisional transition probability per proton per unit volume. For sufficiently large proton densities, the collisional terms dominate, and because of the principle of detailed balancing, they tend to set up a thermodynamic equilibrium distribution of the various L levels within each n; that is, they tend to make

$$\frac{N_{nL}}{N_{nL'}} = \frac{(2L + 1)}{(2L' + 1)}$$

or

$$N_{nL} = \frac{(2L + 1)}{n^2} N_n, \tag{4.18}$$

which is equivalent to $b_{nL} = b_n$, independent of L, where

$$N_n = \sum_{L=0}^{n-1} N_{nL}$$

is the total population in the levels with the same principal quantum number n. Since the cross sections $\sigma_{nL\to nL\pm1}$ increase with increasing n, but the transition probabilities $A_{nL,n'L\pm1}$ decrease, equations (4.18) become increasingly good approximations with increasing n, and there is therefore (for any density and temperature) a level n_{cL} (for coupled angular momentum) above which they apply. For H at $T \approx 10{,}000°$ K, this level is approximately $n_{cL} \approx 15$ at $N_p \approx 10^4$ cm^{-3}, $n_{cL} \approx 30$ at $N_p \approx 10^2$ cm^{-3} and $n_{cL} \approx 45$ at $N_p \approx 1$ cm^{-3}.

Exactly the same type of effect occurs in the He II spectrum, because it also has the property that all the levels nL with the same n are degenerate. The He II lines are emitted in the H^+, He^{++} zone of a nebula, so both protons and He^{++} ions (thermal α-particles) can cause collisional, angular-momentum-changing transitions in excited levels of He^+. The cross sections $\sigma_{nL\to\, nL\pm1}$ actually are larger for the He^{++} ions than for the H^+ ions, and both of them must be taken into account in the He^{++} region. The principal quantum numbers above which (4.18) applies for He II at $T \approx 10{,}000°$ K are approximately $n_{cL} \approx 22$ for $N_p \approx 10^4$ cm^{-3}, and $n_{cL} \approx 32$ for $N_p \approx 10^2$ cm^{-3}.

After the angular-momentum-changing collisions at fixed n, the next largest collisional transition rates occur for collisions in which n changes by $\pm$ 1, and of these the strongest are those for which L also changes by $\pm$ 1. For this type of transition, collisions with electrons are more effective than collisions with protons, and representative cross sections for thermal electrons at $T \approx 10,000°$ K are of order $\sigma_{nL\to n\pm1,L\pm1} \approx 10^{-16}$ cm^2. The effects of these collisions can be incorporated into the equilibrium equations by a straightforward generalization of (4.16). Indeed, since the cross sections for collisions $\sigma_{nL\to n\pm\Delta n,L\pm1}$ decrease with increasing Δn (but not too rapidly), collisions with $\Delta n = 1, 2, 3, \ldots$ must all be included. The computational work required to set up and solve the equilibrium equations numerically becomes increasingly complicated and lengthy, but is straightforward in principle. It is clear that the collisions tend to couple levels with $\Delta L = \pm\ 1$ and small Δn, and that this coupling increases with increasing N_e (and N_p) and with increasing n. With collisions taken into account, the b_{nL} factors and the resulting emission coefficients are no longer independent of density.

Some calculated results for H I, including these collisional effects, are given in Table 4.4, which shows that the density dependence is rather small. Therefore this table, together with Table 4.2, which applies in the limit $N_e \to 0$, enables the H-line emission coefficients to be evaluated over a wide range of densities and temperatures. Similarly, Table 4.5 shows calculated results for the He II recombination spectrum at finite densities and may be used in conjunction with Table 4.3, which applies in the same limit.

Exactly the same formalism can be applied to He I recombination lines, treating the singlet and triplets as completely separate systems, since all transition probabilities between them are quite small. The He I triplets, therefore, always follow Case B, because downward radiative transitions to $1\ ^1S$ essen-

TABLE 4.4

H I recombination lines (Case B)

	T						
	5000° K		10,000° K			20,000° K	
N_e (cm^{-3})	10^2	10^4	10^2	10^4	10^6	10^2	10^4
$4\pi j_{H\beta}/N_p N_e$ (erg cm^3 sec^{-1})	2.20×10^{-25}	2.22×10^{-25}	1.24×10^{-25}	1.24×10^{-25}	1.25×10^{-25}	0.658×10^{-25}	0.659×10^{-25}
$\alpha^{eff}_{H\beta}$ (cm^3 sec^{-1})	5.38×10^{-14}	5.44×10^{-14}	3.02×10^{-14}	3.03×10^{-14}	3.07×10^{-14}	1.61×10^{-14}	1.61×10^{-14}
Balmer-line intensities relative to Hβ							
$j_{H\alpha}/j_{H\beta}$	3.04	3.00	2.86	2.85	2.81	2.75	2.74
$j_{H\gamma}/j_{H\beta}$	0.458	0.460	0.468	0.469	0.471	0.475	0.476
$j_{H\delta}/j_{H\beta}$	0.251	0.253	0.259	0.260	0.262	0.264	0.264
$j_{H\epsilon}/j_{H\beta}$	0.154	0.155	0.159	0.159	0.163	0.163	0.163
$j_{H8}/j_{H\beta}$	0.102	0.102	0.105	0.105	0.110	0.107	0.107
$j_{H9}/j_{H\beta}$	0.0709	0.0714	0.0731	0.0734	0.0786	0.0746	0.0746
$j_{H10}/j_{H\beta}$	0.0515	0.0520	0.0530	0.0533	0.0590	0.0540	0.0541
$j_{H15}/j_{H\beta}$	0.0153	0.0163	0.0156	0.0162	0.0214	0.0158	0.0161
$j_{H20}/j_{H\beta}$	0.0066	0.0082	0.0066	0.0075	0.0105	0.0066	0.0072
Paschen-line intensities relative to corresponding Balmer lines							
$j_{P\alpha}/j_{H\beta}$	0.410	0.396	0.338	0.332	0.317	0.284	0.281
$j_{P\beta}/j_{H\gamma}$	0.402	0.396	0.348	0.345	0.335	0.305	0.305
$j_{P\gamma}/j_{H\delta}$	0.393	0.388	0.349	0.346	0.339	0.312	0.311
j_{P8}/j_{H8}	0.382	0.381	0.348	0.348	0.333	0.317	0.316
j_{P10}/j_{H10}	0.379	0.377	0.347	0.345	0.325	0.318	0.316
j_{P15}/j_{H15}	0.375	0.363	0.347	0.339	0.313	0.319	0.315
j_{P20}/j_{H20}	0.371	0.346	0.346	0.327	0.309	0.320	0.309
Brackett-line intensities relative to corresponding Balmer lines							
$j_{Br\alpha}/j_{H\gamma}$	0.227	0.215	0.171	0.166	0.154	0.132	0.127
$j_{Br\beta}/j_{H\delta}$	0.222	0.214	0.175	0.172	0.163	0.141	0.140
$j_{Br\gamma}/j_{H\epsilon}$	0.214	0.209	0.175	0.173	0.163	0.144	0.143
$j_{Br\delta}/j_{H8}$	0.209	0.206	0.174	0.172	0.160	0.146	0.145
j_{Br10}/j_{H10}	0.204	0.200	0.172	0.170	0.152	0.146	0.146
j_{Br15}/j_{H15}	0.197	0.186	0.170	0.164	0.137	0.147	0.143
j_{Br20}/j_{H20}	0.193	0.169	0.169	0.154	0.133	0.147	0.138

TABLE 4.5
He II recombination lines (Case B)

	T			
	5,000° K	10,000° K		20,000° K
N_e (cm^{-3})	10^4	10^4	10^6	10^4
$4\pi j_{\lambda 4686}/N_{He^{++}}N_e$ (erg cm^3 sec^{-1})	2.96×10^{-24}	1.49×10^{-24}	1.44×10^{-24}	7.21×10^{-25}
$\alpha^{eff}_{\lambda 4686}$ (cm^3 sec^{-1})	6.98×10^{-13}	3.52×10^{-13}	3.40×10^{-13}	1.70×10^{-13}
"Balmer"-line ($n \rightarrow 2$) intensities relative to λ4686				
$j_{32}/j_{\lambda 4686}$	5.90	6.56	6.79	7.42
$j_{42}/j_{\lambda 4686}$	1.66	2.00	2.12	2.44
$j_{52}/j_{\lambda 4686}$	0.72	0.89	0.96	1.12
$j_{72}/j_{\lambda 4686}$	0.24	0.30	0.32	0.38
$j_{102}/j_{\lambda 4686}$	0.08	0.10	0.11	0.13
Fowler-line ($n \rightarrow 3$) intensities relative to λ4686				
$j_{53}/j_{\lambda 4686}$	0.375	0.412	0.427	0.449
$j_{63}/j_{\lambda 4686}$	0.186	0.213	0.224	0.241
$j_{83}/j_{\lambda 4686}$	0.069	0.082	0.087	0.095
$j_{103}/j_{\lambda 4686}$	0.034	0.041	0.043	0.048
Pickering-line ($n \rightarrow 4$) intensities relative to λ4686				
$j_{54}/j_{\lambda 4686}$	0.290	0.272	0.265	0.254
$j_{64}/j_{\lambda 4686}$	0.132	0.135	0.137	0.136
$j_{74}/j_{\lambda 4686}$	0.072	0.077	0.079	0.080
$j_{84}/j_{\lambda 4686}$	0.045	0.049	0.051	0.052
$j_{94}/j_{\lambda 4686}$	0.030	0.033	0.034	0.036
$j_{104}/j_{\lambda 4686}$	0.021	0.024	0.025	0.026
$j_{124}/j_{\lambda 4686}$	0.012	0.013	0.014	0.014
$j_{154}/j_{\lambda 4686}$	0.0060	0.0067	0.0073	0.0074
$j_{204}/j_{\lambda 4686}$	0.0025	0.0028	0.0035	0.0031
Pfund-line ($n \rightarrow 5$) intensities relative to corresponding Pickering lines				
j_{65}/j_{64}	0.795	0.699	0.655	0.620
j_{75}/j_{74}	0.768	0.697	0.670	0.634
j_{85}/j_{84}	0.740	0.682	0.662	0.628
j_{105}/j_{104}	0.706	0.664	0.650	0.620
j_{155}/j_{154}	0.675	0.643	0.623	0.609
j_{205}/j_{204}	0.665	0.634	0.585	0.604

tially do not occur. For the singlets, Case B is ordinarily a better approximation than Case A for observed nebulae, though the optical depths are lower for all lines than for the corresponding lines of H by a factor of approximately the abundance ratio. An extra complication is that He I $1\,^1S - n\,^1P$ line photons can photoionize H^0, and thus may be destroyed before they are converted into lower-energy photons. Calculated (Case B) results for the strongest He I lines are summarized in Table 4.6, with $\lambda4471$ ($2\,^3P - 4\,^3D$) as the reference line. Note that only H itself and the ions of its isoelectronic sequence have energy levels with the same n but different L degenerate, so for He I, Table 4.6 lists the $j_{n\,(2S+1)L,n'\,(2S+1)L'}$ rather than $j_{nn'}$ as for H. The radiative-transfer effects on the He I triplets, discussed in Section 4.6, and the collisional-excitation effects, discussed in Section 4.8, are not included in this table.

4.3 Optical Continuum Radiation

In addition to the line radiation emitted in the bound-bound transition previously described, recombination processes also lead to the emission of rather weak continuum radiation in free-bound and free-free transitions. Because hydrogen is the most abundant element, the H I continuum, emitted in the recombination of protons with electrons, is the strongest, and the He II continuum may also be significant if He is mostly doubly ionized, but the He I continuum is always weaker. In the ordinary optical region the free-bound continua are stronger, but in the infrared and radio regions the free-free continuum dominates. In addition, there is a continuum resulting from the two-photon decay of the $2\,^2S$ level of H, which is populated by recombinations and subsequent downward cascading. In this section we will examine each of these sources of continuous radiation.

The H I free-bound continuum radiation at frequency ν results from recombinations of free electrons with velocity v to levels with principal quantum number $n \geq n_1$, where

$$h\nu = \tfrac{1}{2}mv^2 + X_n \tag{4.19}$$

and

$$h\nu \geq X_{n_1} = \frac{h\nu_0}{n_1^2}; \tag{4.20}$$

its emission coefficient per unit frequency interval per unit solid angle per unit time per unit volume is therefore

$$j_\nu = \frac{1}{4\pi} N_p N_e \sum_{n=n_1}^{\infty} \sum_{L=0}^{n-1} v\sigma_{nL}(\mathrm{H}^0,\ v) f(v) h\nu \frac{dv}{d\nu}. \tag{4.21}$$

TABLE 4.6

He I recombination lines (Case B)

	T						
	5,000° K		10,000° K			20,000° K	
N_e (cm^{-3})	10^2	10^4	10^2	10^4	10^6	10^2	10^4
$4\pi j_{\lambda4471}/N_{He^+}N_e$ (erg cm^3 sec^{-1})	1.16×10^{-25}	1.17×10^{-25}	6.06×10^{-26}	6.08×10^{-26}	6.16×10^{-26}	2.94×10^{-26}	2.95×10^{-26}
$\alpha^{eff}_{\lambda4471}$ (cm^3 sec^{-1})	2.61×10^{-14}	2.64×10^{-14}	1.36×10^{-14}	1.37×10^{-14}	1.38×10^{-14}	0.662×10^{-14}	0.663×10^{-14}
				Triplet lines:			
$j_{\lambda5876}/j_{\lambda4471}$	3.02	3.01	2.75	2.76	2.73	2.58	2.58
$j_{\lambda4026}/j_{\lambda4471}$	0.458	0.459	0.474	0.474	0.476	0.487	0.487
$j_{\lambda3820}j_{\lambda4471}$	0.251	0.251	0.264	0.264	0.265	0.274	0.274
$j_{\lambda7065}/j_{\lambda4471}$	0.244	0.243	0.330	0.328	0.325	0.478	0.477
$j_{\lambda10830}/j_{\lambda4471}$	3.98	3.96	4.42	4.42	4.41	5.02	5.01
$j_{\lambda3889}/j_{\lambda4471}$	1.89	1.90	2.26	2.26	2.27	2.79	2.79
$j_{\lambda3187}/j_{\lambda4471}$	0.748	0.747	0.916	0.917	0.920	1.16	1.16
				Singlet lines:			
$j_{\lambda6678}/j_{\lambda4471}$	—	0.867	—	0.791	0.780	—	0.731
$j_{\lambda4922}/j_{\lambda4471}$	—	0.276	—	0.274	0.274	—	0.271
$j_{\lambda5016}/j_{\lambda4471}$	—	0.512	—	0.588	0.590	—	0.689
$j_{\lambda3965}/j_{\lambda4471}$	—	0.199	—	0.234	0.235	—	0.279

The recombination cross sections $\sigma_{nL}(\mathrm{H}^0, v)$ can be calculated from the photoionization cross sections $a_\nu(\mathrm{H}^0,\ nL)$ by the Milne relation, as shown in Appendix 1.

The free-free (or bremsstrahlung) continuum emitted by free electrons accelerated in Coulomb collisions with positive ions (which are mostly H^+, He^+, or He^{++} in nebulae) of charge Z has an emission coefficient

$$j_\nu = \frac{1}{4\pi} N_+ N_e \frac{32 Z^2 e^4 h}{3 m^2 c^3} \left(\frac{\pi h \nu_0}{3kT}\right)^{1/2} e^{-h\nu/kT} g_{ff}(T, Z, \nu), \tag{4.22}$$

where $g_{ff}(T, Z, \nu)$ is a Gaunt factor. Thus the emission coefficient for the H I recombination continuum, including both bound-free and free-free contributions, may be written

$$j_\nu(\mathrm{H\ I}) = \frac{1}{4\pi} N_p N_e \gamma_\nu(\mathrm{H}^0,\ T). \tag{4.23}$$

Numerical values for γ_ν, as calculated from equations (4.21) and (4.22), are given in Table 4.7. Likewise, the contributions to the continuum-emission coefficient from He I and He II may be written

$$\begin{aligned} j_\nu(\mathrm{He\ I}) &= \frac{1}{4\pi} N_{\mathrm{He}^+} N_e \gamma_\nu(\mathrm{He}^0, T), \\ j_\nu(\mathrm{He\ II}) &= \frac{1}{4\pi} N_{\mathrm{He}^{++}} N_e \gamma_\nu(\mathrm{He}^+,\ T), \end{aligned} \tag{4.24}$$

and numerical values of the γ_ν are listed in Tables 4.8 and 4.9. The calculation for He II is exactly analogous to that for H I, while for He I the only complication is that there is no L degeneracy and equations (4.19), (4.20), and (4.21) must be appropriately generalized. Figure 4.1 shows these calculated values of γ_ν, and also shows the large discontinuities at the ionization potentials of the various excited levels. Note that for a typical He abundance of approximately 10 percent of that of H, if the He is mostly doubly ionized, then the He II contribution to the continuum is roughly comparable to that of H I, but if the He is mostly singly ionized, the He I contribution to the continuum is only about 10 percent of the H I contribution.

TABLE 4.7
H I continuous-emission coefficient[a] $\gamma_\nu(\mathrm{H}^0, T)$

λ (Å)	$\nu(10^{14}\mathrm{Hz})$	T: 5,000° K	10,000° K	15,000° K	20,000° K
10,000	2.998	6.23	5.86	5.35	4.98
8,204+	3.654-	3.30	4.31	4.35	4.25
8,204−	3.654+	25.36	11.54	8.31	6.81
7,000	4.283	13.51	8.67	6.87	5.90
5,696	5.263	5.17	5.52	5.09	4.70
4,500	6.662	1.388	2.88	3.29	3.40
4,000	7.495	0.653	1.946	2.54	2.79
3,646+	8.224-	0.343	1.380	2.017	2.35
3,646−	8.224+	71.7	24.8	14.89	10.65
3,122	9.603	17.83	13.19	9.81	7.84
2,600	11.530	2.857	5.39	5.43	5.07

[a] In 10^{-40} erg cm^3 sec^{-1} Hz^{-1}.

An additional important source of continuum emission in nebulae is the two-photon decay of the 2 ^2S level of H I, which is populated by direct recombinations and by cascades following recombinations to higher levels. The transition probability for this two-photon decay is $A_{2\,^2S,1\,^2S} = 8.23$ sec^{-1}, and the sum of the energies of the two photons is $h\nu' + h\nu'' = h\nu_{12} = h\nu(\mathrm{L}\alpha) = (3/4)h\nu_0$. The probability distribution of the emitted photons is therefore symmetric around the frequency $(1/2)\nu_{12} = 1.23 \times 10^{15}$ sec^{-1}, corresponding to $\lambda = 2431$ A. The emission coefficient in this two-photon continuum may be written

$$j_\nu(2q) = \frac{1}{4\pi} N_{2\,^2S} A_{2\,^2S,\,1\,^2S} 2hyP(y), \tag{4.25}$$

where $P(y)dy$ is the normalized probability per decay that one photon is emitted in the range of frequencies $y\nu_{12}$ to $(y + dy)\nu_{12}$.

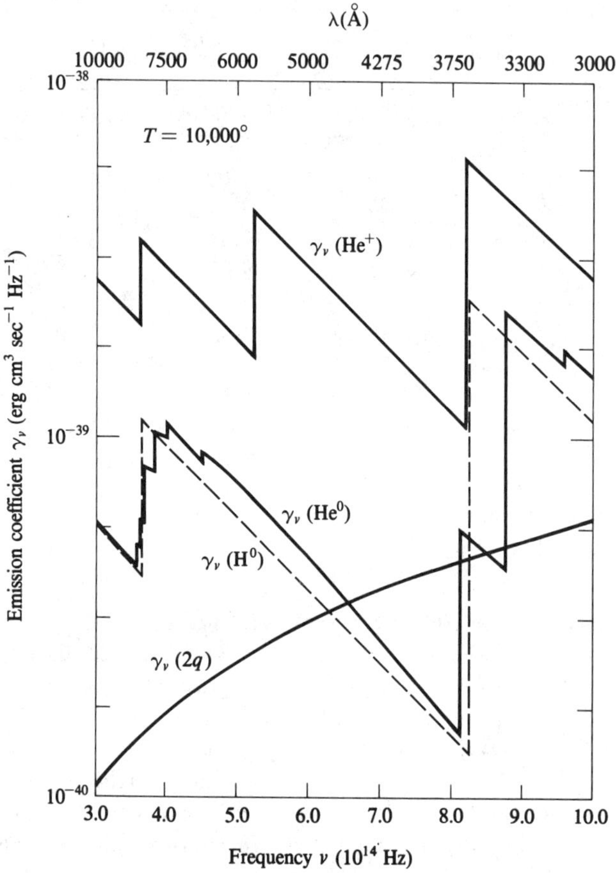

FIGURE 4.1
Frequency variation of continuous-emission coefficient $\gamma_\nu(\mathrm{H}^0)$, $\gamma_\nu(\mathrm{He}^0)$, $\gamma_\nu(\mathrm{He}^+)$, and $\gamma_\nu(2\mathrm{q})$ in the low-density limit $N_e \to 0$, all at $T = 10{,}000°$ K.

TABLE 4.8
He I continuous-emission coefficient[a] $\gamma_\nu(He^0, T)$

λ (Å)	$\nu(10^{14}$Hz)	T: 5,000° K	10,000° K	15,000° K	20,000° K
10,000	2.998	6.23	5.86	5.36	4.98
8268+	3.626−	3.40	4.36	4.38	4.27
8268−	3.626+	5.66	5.10	4.79	4.54
8197+	3.657−	5.52	5.04	4.74	4.50
8197−	3.657+	8.05	5.88	5.20	4.81
8195+	3.658−	8.05	5.88	5.20	4.81
8195−	3.658+	13.06	8.50	6.65	5.73
7849+	3.819−	14.14	8.08	6.44	5.59
7849−	3.819+	21.21	10.61	7.86	6.52
7440+	4.029−	18.04	10.00	7.58	6.35
7440−	4.029+	18.58	10.18	7.68	6.41
6636+	4.518−	12.41	8.66	6.94	5.96
6636−	4.518+	13.71	9.09	7.17	6.11
5696	5.263	7.07	6.76	5.91	5.30
4500	6.662	1.775	3.31	3.63	3.66
4000	7.495	0.759	2.07	2.62	2.85
3680+	8.147−	0.406	1.45	2.06	2.37
3680−	8.147+	13.37	5.69	4.38	3.87
3422+	8.761−	7.22	4.32	3.65	3.37
3422−	8.761+	64.1	23.0	13.88	9.96
3122+	9.603−	27.6	15.87	10.96	8.44
3122−	9.603+	32.5	17.45	11.83	9.00
2600	11.530	5.0	7.14	6.60	5.91

[a] In 10^{-40} erg cm^3 sec^{-1} Hz^{-1}.

TABLE 4.9
He II continuous-emission coefficient[a] γ_ν *(He$^+$, T)**

		T			
λ (Å)	$\nu(10^{14} Hz)$	5,000° K	10,000° K	15,000° K	20,000° K
10,000	2.998	40.9	28.8	24.0	21.2
8200+	3.654−	21.4	21.1	19.48	18.12
8200−	3.654+	67.3	36.2	27.8	23.4
7000	4.283	35.8	27.0	22.8	20.2
5694+	5.263−	13.69	17.05	16.75	16.05
5694−	5.263+	92.3	42.8	29.9	25.2
4500	6.662	22.7	22.3	19.99	18.20
4000	7.495	10.16	15.07	15.39	14.98
3644+	8.224−	5.14	10.69	12.24	12.63
3644−	8.224+	156.5	60.4	39.6	30.2
3122	9.603	38.6	31.7	25.8	22.0
2600	11.530	6.15	12.79	14.07	14.05

[a] In 10^{-40} erg cm^3 sec^{-1} Hz^{-1}.

To express this two-photon continuum-emission coefficient in terms of the proton and electron density, it is necessary to calculate the equilibrium population of N_{2^2S} in terms of these quantities. In sufficiently low-density nebulae, two-photon decay is the only mechanism that depopulates $2\,^2S$, and the equilibrium is given by

$$N_p N_e \alpha^{eff}_{2\,^2S}(\mathrm{H}^0,\, T) \;=\; N_{2\,^2S} A_{2^2S,\,1\,^2S}, \tag{4.26}$$

where $\alpha^{eff}_{2^2S}$ is the effective recombination coefficient for populating $2\,^2S$ by direct recombinations and by recombinations to higher levels followed by cascades to $2\,^2S$. However, at finite densities, angular-momentum-changing collisions of protons and electrons with H atoms in the $2\,^2S$ level shift the atoms to $2\,^2P$ and thus remove them from $2\,^2S$. The protons are more effective than electrons, whose effects, however, are not completely negligible, as can be seen from the values of the collisional transition rates per $2\,^2S$ atom, in Table 4.10. With these collisional processes taken into account, the equilibrium population in $2\,^2S$ is given by

$$N_p N_e \alpha^{eff}_{2\,^2S}(\mathrm{H}^0,\, T) \;=$$

$$N_{2\,^2S}\{A_{2\,^2S,\,1\,^2S} + N_p q^p_{2\,^2S,\,2\,^2P} + N_e q^e_{2\,^2S,\,2\,^2P}\}. \tag{4.27}$$

From Table 4.10, it can be seen that collisional de-excitation of $2\,^2S$ via $2\,^2P$ is more important than two-photon decay for $N_p \geq 10^4$ cm^{-3}; so at densities approaching this value, equation (4.27) must be used instead of equation (4.26). Thus combining equations (4.25) and (4.27), we can write the emission coefficient as

$$j_\nu(2q) = \frac{1}{4\pi} N_p N_e \gamma_\nu(2q), \tag{4.28}$$

where

$$\gamma_\nu(2q) = \frac{\alpha^{eff}_{2\,^2S}(\mathrm{H}^0, T) g_\nu}{1 + \left(\dfrac{N_p q^p_{2\,^2S,\,2\,^2P} + N_e q^e_{2\,^2S,\,2\,^2P}}{A_{2\,^2S,\,1\,^2S}} \right)} \tag{4.29}$$

The quantity $\alpha^{eff}_{2^2S}$ is tabulated in Table 4.11 and g_ν is tabulated in Table 4.12. The two-photon continuum is also plotted in Figure 4.1 for $T = 10{,}000°$ K and in the low-density limit $N_p \approx N_e \ll 10^4$ cm^{-3}. It can be seen that this continuum is quite significant in comparison with the H I continua, particularly just above the Balmer limit at $\lambda 3646$ A. Note that although the two-photon continuum is symmetric about $\nu_{12}/2$ if expressed in *photons* per unit frequency interval, it is not symmetric about $\lambda 2431$ if expressed per unit wavelength interval, nor is either symmetric if expressed in energy units rather than photons. From equation (4.25), if $\nu > \nu_{12}/2$,

$$g_\nu = \frac{\nu}{\nu'} g_{\nu'},$$

where $\nu' = \nu - \nu_{12}/2$.

4.4 Radio-Frequency Continuum and Line Radiation

The line and continuous spectra described in Sections 4.2 and 4.3 extend to arbitrarily low frequency, and in fact give rise to observable features in the radio-frequency spectrum region. Though this "thermal" radio-frequency radiation is a natural extension of the optical line and continuous spectra, it is somewhat different in detail, because in the radio-frequency region $h\nu \ll kT$, and stimulated emission, which is proportional to $e^{-h\nu/kT}$, is therefore much more important in that region than in the ordinary optical region. We will examine the continuous spectrum first, and then the recombination-line spectrum.

TABLE 4.10
Collisional transition rates for H I 2 ^2S, 2 ^2P*

	T	
	10,000° K	20,000° K
	Protons	
$q^p_{2\,^2S,\,2\,^2P_{1/2}}$	2.51×10^{-4}	2.08×10^{-4}
$q^p_{2\,^2S,\,2\,^2P_{3/2}}$	2.23×10^{-4}	2.19×10^{-4}
	Electrons	
$q^e_{2\,^2S,\,2\,^2P_{1/2}}$	0.22×10^{-4}	0.17×10^{-4}
$q^e_{2\,^2S,\,2\,^2P_{3/2}}$	0.35×10^{-4}	0.27×10^{-4}
	Total	
$q_{2\,^2S,\,2\,^2P}$	5.31×10^{-4}	4.71×10^{-4}

[a] In cm^3 sec^{-1}.

TABLE 4.11
Effective recombination coefficient[a] to H (2 ^2S)

T(° K)	$\alpha^{eff}_{2\,^2S}$
5,000	1.38×10^{-13}
10,000	0.838×10^{-13}
15,000	0.625×10^{-13}
20,000	0.506×10^{-13}

[a] In cm^3 sec^{-1}.

TABLE 4.12
Spectral distribution of H I two-photon emission

λ (A)	$\nu(10^{14}$ Hz)	$g_\nu(10^{-27}$ erg Hz^{-1})
∞	0.0	0.0
24,313	1.23	0.303
12,157	2.47	0.978
8104	3.70	1.836
6078	4.93	2.78
4863	6.17	3.78
4052	7.40	4.80
3473	8.64	5.80
3039	9.87	6.78
2701	11.10	7.74
2431	12.34	8.62

In the radio-frequency region, the continuum is due to free-free emission, and the emission coefficient is given by the same equation (4.22) that applies in the optical region. However, in the radio-frequency region, the Gaunt factor $g_{ff}(T, Z, v) \not\approx 1$, as in the optical region, but rather

$$g_{ff}(T,Z,\nu) = \frac{\sqrt{3}}{\pi}\left\{ ln\left(\frac{8k^3\,T^3}{\pi^2 Z^2 e^4 m\nu^2}\right)^{1/2} - \frac{5\gamma}{2}\right\}, \qquad (4.30)$$

where $\gamma = 0.577$ is Euler's constant. Numerically, this is approximately

$$g_{ff}(T,Z,\nu) = \frac{\sqrt{3}}{\pi}\left(ln\frac{T^{3/2}}{Z\nu} + 17.7\right),$$

with T in $^\circ$ K and ν in Hz, and thus at $T \approx 10{,}000^\circ$ K, $\nu \approx 10^3$ MHz, $g_{ff} \approx 10$.

The free-free effective absorption coefficient can then be found from Kirchoff's law, and is

$$\kappa_\nu = N_+ N_e \frac{16\pi^2 Z^2 e^6}{(6\pi mkT)^{3/2}\nu^2 c} g_{ff} \qquad (4.31)$$

per unit length. Note that this effective absorption coefficient is the difference between the true absorption coefficient and the stimulated emission coefficient, since the stimulated emission of a photon is exactly equivalent to a negative absorption process; in the radio-frequency region ($h\nu \ll kT$) the stimulated emissions very nearly balance the true absorptions and the correction for stimulated emission $(1 - e^{-h\nu/kT}) \approx h\nu/kT \ll 1$.

Substituting numerical values and fitting powers to the weak temperature and frequency dependence of g_{ff},

$$\begin{aligned}\tau_\nu &= \int \kappa_\nu \, ds \\ &= 8.24 \times 10^{-2}\, T^{-1.35} \nu^{-2.1} \int N_+ N_e \, ds \\ &= 8.24 \times 10^{-2}\, T^{-1.35} \nu^{-2.1}\, E. \end{aligned} \tag{4.32}$$

In this formula T must be measured in ° K, ν in GHz, and E, the so-called emission measure, in cm^{-6} pc. It can be seen from equations (4.31) or (4.32) that at sufficiently low frequency all nebulae become optically thick; for example, an H II region with $N_e \approx N_p \approx 10^2$ cm^{-3} and a diameter 10 pc has $\tau_\nu \approx 1$ at $\nu \approx 200$ MHz, and a planetary nebula with $N_e \approx 3 \times 10^3$ cm^{-3} and a diameter of 0.1 pc has $\tau_\nu \approx 1$ at $\nu \approx 600$ MHz. Thus, in fact, many nebulae are optically thick at observable low frequencies and optically thin at observable high frequencies. The equation of radiative transfer,

$$\frac{dI_\nu}{ds} = -\kappa_\nu I_\nu + j_\nu \tag{4.33}$$

or

$$\frac{dI_\nu}{d\tau_\nu} = -I_\nu + \frac{j_\nu}{\kappa_\nu} = -I_\nu + B_\nu(T), \tag{4.34}$$

has the solution for no incident radiation

$$I_\nu = \int_0^{\tau_\nu} B_\nu(T) e^{-\tau_\nu} d\tau_\nu. \tag{4.35}$$

In the radio-frequency region,

$$B_\nu(T) = \frac{2h\nu^3}{c^2} \frac{1}{e^{h\nu/kT} - 1} \approx \frac{2\nu^2 kT}{c^2} \tag{4.36}$$

is proportional to T, so it is conventional in radio astronomy to measure intensity in terms of brightness temperature, defined by $T_{b\nu} = c^2 I_\nu / 2\nu^2 k$. Hence (4.35) can be rewritten

$$T_{b\nu} = \int_0^{\tau_\nu} T e^{-\tau_\nu} d\tau_\nu, \tag{4.37}$$

and for an isothermal nebula, this becomes

$$T_{b\nu} = T(1 - e^{-\tau_\nu}) \left\{ \begin{array}{lll} \to T\tau_\nu & \text{as} & \tau_\nu \to 0 \\ \to T & \text{as} & \tau_\nu \to \infty \end{array} \right\}.$$

Thus the radio-frequency continuum has a spectrum in which $T_{b\nu}$ varies approximately as ν^{-2} at high frequency and is independent of ν at low frequency.

The H I recombination lines of very high n also fall in the radio-frequency spectral region and have been observed in many gaseous nebulae. Some specific examples of observed lines are H 109α (the transition with $\Delta n = 1$ from $n = 110$ to $n = 109$) at $\nu = 5008.89$ MHz, $\lambda = 5.99$ cm, H 137β (the transition with $\Delta n = 2$ from $n = 139$ to $n = 137$) at $\nu = 5005.0$ MHz, $\lambda = 6.00$ cm, and so on. The emission coefficients in these radio recombination lines may be calculated from equations similar to those described in Section 4.2 for the shorter wavelength optical recombination lines. For all lines observed in the radio-frequency region, $n > n_{cL}$ defined there, so that at a fixed $n, N_{nL} \propto (2L + 1)$, and only the populations N_n need be considered. One additional process, in addition to those described in Section 4.2, must also be taken into account, namely collisional ionization of levels with large n and its inverse process, three-body recombination,

$$\mathrm{H}^0(n) + e \rightleftarrows \mathrm{H}^+ + e + e.$$

The rate of collisional ionization per unit volume per unit time from level n may be written

$$N_n N_e \quad \overline{v\sigma_{\text{ionization}}(n)} \quad = N_n N_e q_{n,i}(T), \tag{4.38}$$

and the rate of three-body recombination per unit time per unit volume may be written $N_p N_e^2 \phi_n(T)$; so, from the principle of detailed balancing,

$$\phi_n(T) = n^2 \left(\frac{h^2}{2\pi mkT} \right)^{3/2} e^{X_n/kT} q_{n,i}(T). \tag{4.39}$$

Thus the equilibrium equation that is analogous to equation (4.16) becomes, at high n,

$$N_p N_e[\alpha_n(T) + N_e \phi_n(T)] + \sum_{n'>n}^{\infty} N_{n'} A_{n',n} + \sum_{n'=n_0}^{\infty} N_{n'} N_e q_{n',n}$$
$$= N_n \left[\sum_{n'=n_0}^{n-1} A_{n,n'} + \sum_{n'=n_0}^{\infty} N_e q_{n,n'}(T) + N_e q_{n,i}(T) \right], \tag{4.40}$$

where

$$A_{n,n'} = \frac{1}{n^2} \sum_{L,L'} (2L + 1) A_{nL,n'L'} \tag{4.41}$$

is the mean transition probability averaged over all the L levels of the upper principal quantum number. These equations can be expressed in terms of b_n instead of N_n, and the solutions can be found numerically by standard matrix-inversion techniques. Note that since the coefficients b_n have been defined with respect to thermodynamic equilibrium at the local T, N_e, and N_p, the coefficient b for the free electrons is identically unity, and therefore $b_n \to 1$ as $n \to \infty$. Some calculated values of b_n for $T = 10{,}000°$ K and various N_e are plotted in Figure 4.2, which shows that the increasing importance of collisional transitions as N_e increases makes $b_n \approx 1$ at lower and lower n.

To calculate the emission in a specific recombination line, it is again necessary to solve the equation of transfer, taking account of the effects of stimulated emission. In this case, for an $n, \Delta n$ line between the upper level $m = n + \Delta n$ and the lower level n, if $k_{\nu l}$ is the true line-absorption coefficient, then the line-absorption coefficient, corrected for stimulated emission, to be used in the equation of transfer is

$$k_{\nu L} = k_{\nu l}\left(1 - \frac{b_m}{b_n} e^{-h\nu/kT}\right), \tag{4.42}$$

since it is the net difference between the rates of upward absorption processes and of downward-induced emissions. If we expand (4.42) in a power series, it becomes

$$k_{\nu L} = k_{\nu l}\left(\frac{b_m}{b_n}\frac{h\nu}{kT} - \frac{d\, ln\, b_n}{dn}\,\Delta n\right). \tag{4.43}$$

Since $b_m/b_n \approx 1$ and $h\nu \ll kT$, the line-absorption coefficient can become negative, implying positive maser action, if $(d \ln b_n)/dn$ is sufficiently large. Calculated values of this derivative are therefore also shown in Figure 4.2. Since, for typical observed lines $h\nu/kT \approx 10^{-5}$, it can be seen from this figure that the maser effect is in fact often quite important. We will again use these concepts and expressions to calculate the strengths of the radio-frequency recombination lines in Chapter 5.

4.5 Radiative Transfer Effects in H I

For most of the emission lines observed in nebulae there is no radiative-transfer problem; in most lines the nebulae are optically thin, and any line photon emitted simply escapes. However, in some lines, especially the resonance lines of abundant atoms, the optical depths are appreciable, and scattering and absorption must be taken into account in calculating the expected line strengths. Two extreme assumptions, Case A, a nebula with vanishing optical thickness in all the H I Lyman lines, and Case B, a nebula with large optical depths in all the Lyman lines, have already been discussed in Section 4.2; and although these two cases do not require the detailed radiative-transfer solution, in the intermediate cases a more sophisticated treatment is necessary. Other radiative transfer problems arise in connection with the He I triplets, the conversion of He II Lα into observable O III line radiation by the Bowen resonance-fluorescence process, and fluorescence excitation of other lines by stellar continuum radiation. In this section some general concepts about the escape of line photons from nebulae will be discussed in the context of the H I Lyman and Balmer lines, and then in succeeding sections these same concepts will be applied to the other problems mentioned.

In a static nebula the only line-broadening mechanisms are thermal Doppler broadening and radiative damping, and in the cores of the lines the line-absorption coefficient has the Doppler form

$$\kappa_{\nu l} = k_{0l} e^{-(\Delta\nu/\Delta\nu_D)^2} = k_{0l} e^{-x^2}, \tag{4.44}$$

where

$$k_{0l} = \frac{\lambda^2}{8\pi^{3/2}} \frac{\omega_j}{\omega_i} \frac{A_{j,i}}{\Delta\nu_D} = \frac{\sqrt{\pi}\, e^2 f_{ij}}{mc\, \Delta\nu_D} \tag{4.45}$$

is the line-absorption cross section per atom at the center of the line,

$$\Delta\nu_D = \sqrt{\frac{2kT}{Mc^2}}\, \nu_0$$

is the thermal Doppler width, and f_{ij} is the f-value between the lower and upper level i, j. Small-scale turbulence can presumably be taken into account as a further source of broadening of the line-absorption coefficient, and larger-scale turbulence and expansion of the nebula must be treated by considering the frequency shift between the emitting and absorbing volumes.

In a static nebula, a photon emitted at a particular point in a particular direction and with a normalized frequency x from the center of the line has a probability $e^{-\tau_x}$ of escaping from the nebula without further scattering and absorption, where τ_x is the optical depth from the point to the edge of the nebula in this direction and at this frequency. Averaging over all directions

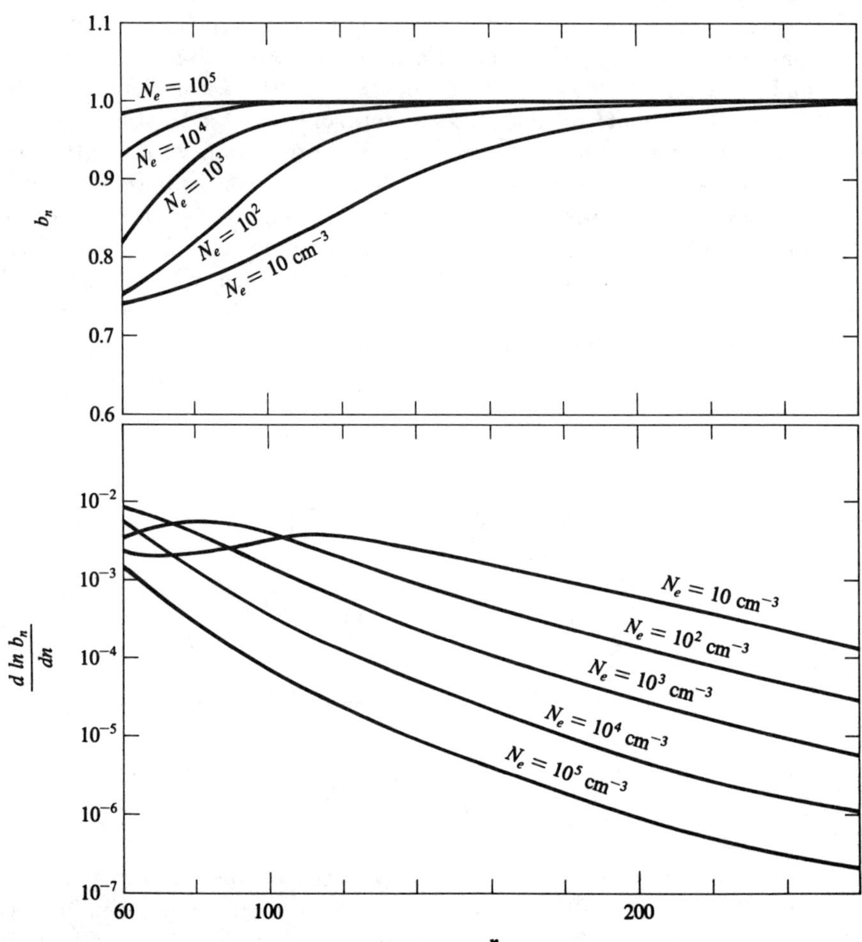

FIGURE 4.2
Dependence of b_n and $\dfrac{d \ln b_n}{dn}$ on n at various densities: $T = 10,000°$ K.

gives the mean escape probability from this point and at this frequency, and further averaging over the frequency profile of the emission coefficient gives the mean escape probability from the point.

For all the forbidden lines and for most of the other lines, the optical depths are so small in every direction, even at the center of the line, that the mean escape probabilities from all points are essentially unity. However, for lines of larger optical depth we must examine the probability of escape quantitatively.

Consider an idealized spherical homogeneous nebula, with optical radius in the center of line τ_{0l}. So long as $\tau_{0l} \leq 10^4$, only the Doppler core of the line-absorption cross section need be considered. The photons are emitted with the same Doppler profile, and the mean probability of escape must therefore be averaged over this profile. If, at a particular normalized frequency x, the optical radius of the nebula is τ_x, the mean probability of escape averaged over all directions and volumes is

$$p(\tau_x) = \frac{3}{4\tau_x}\left[1 - \frac{1}{2\tau_x^2} + \left(\frac{1}{\tau_x} + \frac{1}{2\tau_x^2}\right)e^{-2\tau_x}\right], \tag{4.46}$$

as shown in Appendix 2. When we average over the Doppler profile, the mean escape probability for a photon emitted in the line is

$$\epsilon(\tau_{0l}) = \frac{1}{\sqrt{\pi}}\int_{-\infty}^{\infty} p(\tau_x)e^{-x^2}\,dx, \tag{4.47}$$

where τ_{0l} is the optical radius in the center of the line. This integral must be evaluated numerically, but for optical radii ($\tau_{0l} \leq 50$) that are not too large, the results can be fitted fairly accurately with $\epsilon(\tau_{0l}) = 1.72/(\tau_{0l} + 1.72)$.

If we consider a Lyman line Ln, photons emitted in this line that do not escape from the nebula are absorbed, and each absorption process represents an excitation of the n^2P level of H I. This excited level very quickly undergoes a radiative decay, and the result is either resonance scattering or resonance fluorescence excitation of another H I line. If the photon emitted with the n^2P level decays is a $1\,^2S - n^2P$ transition, the process is resonance scattering of an Ln photon; if it is emitted in the $2\,^2S - n^2P$ transition, the process is conversion of Ln into Hn plus excitation of $2\,^2S$, leading to emission of two photons in the continuum; if it is emitted in the $3\,^2S - n^2P$ transition, the process is conversion of Ln into Pn plus excitation of $3\,^2S$, leading to emission of Hα plus Lα, and so on. The probabilities of each of these processes may be found directly from the probability matrices $C_{nL,n'L'}$ and $P_{nL,n'L'}$, defined in Section 4.2. If we define $P_n(\mathrm{L}m)$ and $P_n(\mathrm{H}m)$ as the probabilities that absorption of an Ln photon results in emission of an Lm photon and of an Hm photon, respectively, then

$$P_n(\mathrm{L}m) = C_{n1,m1}P_{m1,10} \tag{4.48}$$

and

$$P_n(\mathrm{H}m) = C_{n1,m0}P_{m0,21} + C_{n1,m1}P_{m1,20} + C_{n1,m2}P_{m2,21}. \tag{4.49}$$

We can now use these probabilities to calculate the emergent Lyman-line spectrum emitted from a model nebula. It is easiest to work in terms of numbers of photons emitted. If we write R_n for the total number of Ln photons generated in the nebula per unit time by recombination and subsequent cascading, and A_n as the total number of Ln photons absorbed in the nebula per unit time, then J_n, the total number of Ln photons emitted in the nebula per unit time, is the sum of the contributions from recombination and from resonance fluorescence plus scattering:

$$J_n = R_n + \sum_{m=n}^{\infty} A_m P_m(\mathrm{L}n). \tag{4.50}$$

Since each Ln photon emitted has a probability ϵ_n of escaping, the total number of Ln photons escaping the nebula per unit time is

$$E_n = \epsilon_n J_n = \epsilon_n \left[R_n + \sum_{m=n}^{\infty} A_m P_m(\mathrm{L}n) \right]. \tag{4.51}$$

Finally, in a steady state the number of Ln photons emitted per unit time is equal to the sum of the numbers absorbed and escaping per unit time,

$$J_n = A_n + E_n = A_n + \epsilon_n J_n. \tag{4.52}$$

Thus, eliminating J_n between (4.51) and (4.52),

$$A_n = (1 - \epsilon_n) \left[R_n + \sum_{m=n}^{\infty} A_m P_m(\mathrm{L}n) \right], \tag{4.53}$$

and since the R_n and $P_m(\mathrm{L}n)$ are known from the recombination theory and the ϵ_n are known from the radiative-transfer theory, equation (4.53) can be solved for the A_n by a systematic procedure, working downward from the highest n at which ϵ_n differs appreciably from unity. Then from these values of R_n, the E_n may be calculated from (4.51), giving the emergent Lyman-line spectrum.

Next we will investigate the Balmer-line spectrum, which requires further analysis. Let us write S_n for the number of Hn photons generated in the nebula per unit time by recombination and subsequent cascading. Suppose that there is no absorption of these Balmer-line photons, so that K_n, the total number of Hn photons emitted in the nebula per unit time, is the sum of contributions from recombination and from resonance fluorescence due to Lyman-line photons,

$$K_n = S_n + \sum_{m=n}^{\infty} A_m P_m(\mathrm{H}n).$$

Then, since the S_n and $P_m(\mathrm{H}n)$ are known from the recombination theory and the A_m are known from the previous Lyman-line solution, the K_n can be calculated immediately to obtain the emergent Balmer-line spectrum. Note that R_n, S_n, J_n, K_n, and A_n are proportional to the total number of photons; the equations are linear in these quantities; and the entire calculation can therefore be normalized to any S_n, for instance, to S_4, the number of Hβ photons that would be emitted if there were no absorption effects. The results, in the form of calculated ratios of Hα/Hβ and Hβ/Hγ intensities, are shown in Figure 4.3 as a function of $\tau_{0l}(\mathrm{L}\alpha)$, the optical radius of the nebula at the center of Lα; and the transition from Case A ($\tau_{0l} \to 0$) to Case B ($\tau_{0l} \to \infty$) can be seen clearly.

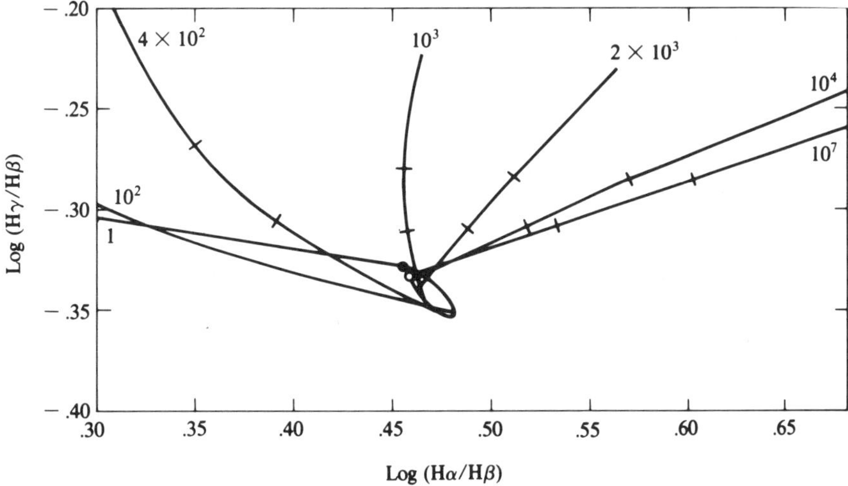

FIGURE 4.3
Radiative-transfer effects caused by finite optical depths in Lyman and Balmer lines. Ratios of total emitted fluxes Hα/Hβ are shown for homogeneous static isothermal model nebulae at $T = 10{,}000°$ K. Each line connects a series of models with the same $\tau_{0l}(\mathrm{L}\alpha)$, given at the end of the line; along it $\tau_{0l}(\mathrm{H}\alpha) = 5$ and 10 at the two points along each line indicated by bars for $\tau_{0l}(\mathrm{L}\alpha) \geq 4\times10^2$.

Although in most nebulae, the optical depths in the Balmer lines are small, there could be situations in which the density $N_{\mathrm{H}^0}(2\ ^2S)$ is sufficiently high that some self-absorption does occur in these lines. The optical depths in the Balmer lines can again be calculated from equation (4.45), and since they are proportional to $N_{\mathrm{H}^0}(2\ ^2S)$, the radiative-transfer problem is now a function of two variables, $\tau_{0l}(\mathrm{L}\alpha)$, giving the optical radius in the Lyman lines, and another, say, $\tau_{0l}(\mathrm{H}\alpha)$, giving the optical radius in the Balmer lines. Although the equations are much more complicated, since now Balmer-line photons may be scattered or converted into Lyman-line photons and vice versa, there is no new effect in principle, and the same general type of formulation developed previously for the Lyman-line absorption can still be used. We will not examine the details here, but will simply discuss physically the calculated results shown in Figure 4.3. For τ_{0l} $(\mathrm{H}\alpha) = 0$ the first effect of increasing $\tau_{0l}(\mathrm{L}\alpha)$ is that $\mathrm{L}\beta$ is converted into $\mathrm{H}\alpha$ plus the two-photon continuum. This increases the $\mathrm{H}\alpha/\mathrm{H}\beta$ ratio of the escaping photons, corresponding to a move of the representative point to the right in Figure 4.3. However, for slightly larger $\tau_{0l}(\mathrm{L}\alpha)$, $\mathrm{L}\gamma$ photons are also converted into $P\alpha$, $\mathrm{H}\alpha$, $\mathrm{H}\beta$, $\mathrm{L}\alpha$, and two-photon continuum photons, and since the main effect is to increase the strength of $\mathrm{H}\beta$, this corresponds to a move downward and to the left in Figure 4.3. For still larger $\tau_{0l}(\mathrm{L}\alpha)$, as still higher $\mathrm{L}n$ photons are converted, $\mathrm{H}\gamma$ is also strengthened, and the calculation, which takes into account all of these effects, shows that the representative point describes the small loop of Figure 4.3 as the conditions change from Case A to Case B. For large $\tau_{0l}(\mathrm{L}\alpha)$, the effect of increasing $\tau_{0l}(\mathrm{H}\alpha)$ is that, although $\mathrm{H}\alpha$ is merely scattered (because any $\mathrm{L}\beta$ photons it forms are quickly absorbed and converted back to $\mathrm{H}\alpha$), $\mathrm{H}\beta$ is absorbed and converted to $\mathrm{H}\alpha$ plus $\mathrm{P}\alpha$. This increases $\mathrm{H}\alpha/\mathrm{H}\beta$ and decreases $\mathrm{H}\beta/\mathrm{H}\gamma$ as shown quantitatively in Figure 4.3.

4.6 Radiative Transfer Effects in He I

The recombination radiation of He I singlets is very similar to that of H I, and Case B is a good approximation for the He I Lyman lines. However, the recombination radiation of the He I triplets is modified by the fact that the He^0 $2\ ^3S$ term is considerably more metastable than H^0 $2\ ^2S$, and as a result self-absorption effects are quite important (as is collisional excitation from $2\ ^3S$, to be discussed later). As the energy-level diagram of Figure 4.4 shows, $2\ ^3S$ is the lowest triplet term in He, and all recaptures to triplets eventually cascade down to it. Depopulation occurs only by photoionization, especially by H I $\mathrm{L}\alpha$, by collisional transitions to $2\ ^1S$ and $2\ ^1P$, or by the strongly forbidden $2\ ^3S - 1\ ^1S$ radiative transition, as discussed in Section 2.4. As a result $N_{2\ ^3S}$ is large, which in turn makes the optical depths in the lower $2\ ^3S - n\ ^3P$ lines significant. Figure 4.4 shows that $\lambda 10830$ $2\ ^3S - 2\ ^3P$ photons are simply scattered, but that absorption of $\lambda 3889$ $2\ ^3S - 3\ ^3P$ photons can lead to their conversion to $\lambda 4.3\ \mu\ 3\ ^3S - 3\ ^3P$, plus $2\ ^3P - 3\ ^3S\ \lambda 7065$, plus $2\ ^3S - 2\ ^3P\ \lambda 10830$. The probability of this conversion is

$$\frac{A_{3\ ^3S,3\ ^3P}}{A_{3\ ^3S,3\ ^3P} + A_{2\ ^3S,3\ ^3P}} \approx 0.10$$

per absorption. At larger $\tau_{0l}(\lambda 10830)$, still higher members of the $2\ ^3S - n^3P$ series are converted into longer wavelength photons.

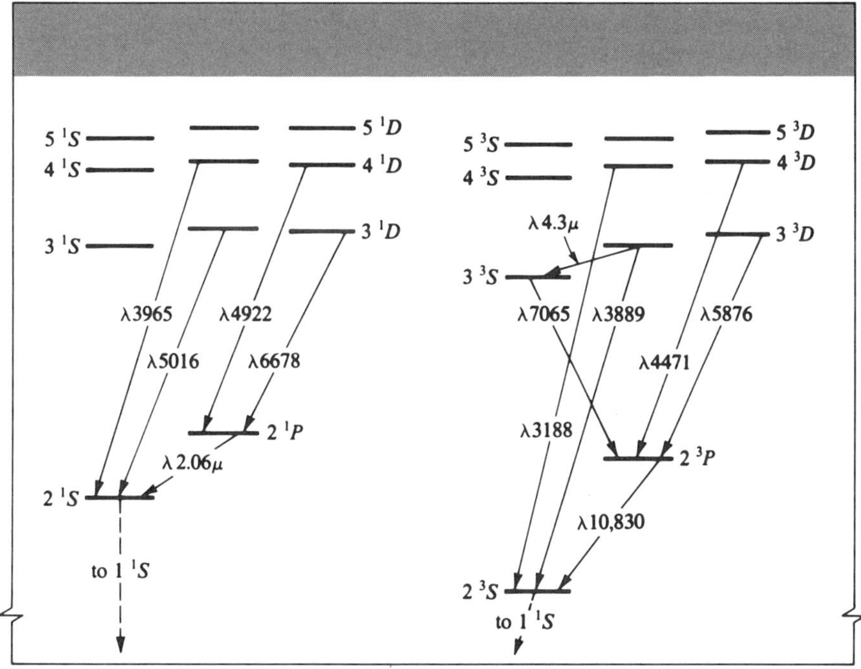

FIGURE 4.4
Partial energy-level diagram of He I, showing strongest optical lines observed in nebulae. Note that $1\ ^2S$ has been omitted, and terms with $n \geq 6$ or $L \geq 3$ have been omitted for the sake of space and clarity.

The radiative-transfer problem is very similar to that for the Lyman lines discussed in Section 4.5, and may be handled by the same kind of formalism. Calculated ratios of the intensities of $\lambda 3889$ (which is weakened by self-absorption) and of $\lambda 7065$ (which is strengthened by resonance fluorescence) relative to the intensity of $\lambda 4471$ $2\ ^3P - 4\ ^3D$ (which is only slightly affected by absorption) are shown for spherically symmetric homogeneous model nebulae in Figure 4.5.

The thermal Doppler widths of He I lines are smaller than those of H I lines, because of the larger mass of He, and therefore whatever turbulent or expansion velocity there may be in a nebula is relatively more important in broadening the He I lines. The simplest example to consider is a model spherical nebula expanding with a linear velocity of expansion,

$$V_{\rm exp}(r) = wr \qquad 0 \le r \le R; \tag{4.54}$$

for then between any two points r_1 and r_2 in the nebula, the relative radial velocity is

$$v(r_1, r_2) = ws, \tag{4.55}$$

where s is the distance between the points and w is the constant velocity gradient. Thus photons emitted at r_1 will have a line profile centered about the line frequency ν_L in the reference system in which r_1 is at rest. However, they will encounter at r_2 material absorbing with a profile centered on the frequency

$$\nu'(r_1,\ r_2) = \nu_L\left(1 + \frac{ws}{c}\right), \tag{4.56}$$

and the optical depth in a particular direction to the boundary of the nebula for a photon emitted at r_1 with frequency ν may be written

$$\tau_\nu(r_1) = \int_0^{r_2=R} N_{2\,^3S}\, k_{0l} \exp\left\{-\left[\frac{\nu - \nu'(r_1, r_2)}{\Delta\nu_D}\right]^2\right\} ds. \tag{4.57}$$

It can be seen that increasing velocity of expansion tends, for a fixed density $N_{2\,^3S}$, to decrease the optical depth to the boundary of the nebula and thus to decrease the self-absorption effects. This effect can be seen in Figure 4.5, where some calculated results are shown for various ratios of the expansion velocity $V_{\rm exp}(R) = wR$ to the thermal velocity $V_{\rm th} = [2kT/M({\rm He})]^{1/2}$, as functions of $\tau_{0l}(\lambda 3889) = N_{2\,^3S}\, k_{0l}(\lambda 3889) R$, the optical radius at the center of the line for zero expansion velocity. Note that the calculated intensity ratios for large $V_{\rm exp}/V_{\rm th}$ and large τ_0 are quite similar to those for smaller $V_{\rm exp}/V_{\rm th}$ and smaller τ_0.

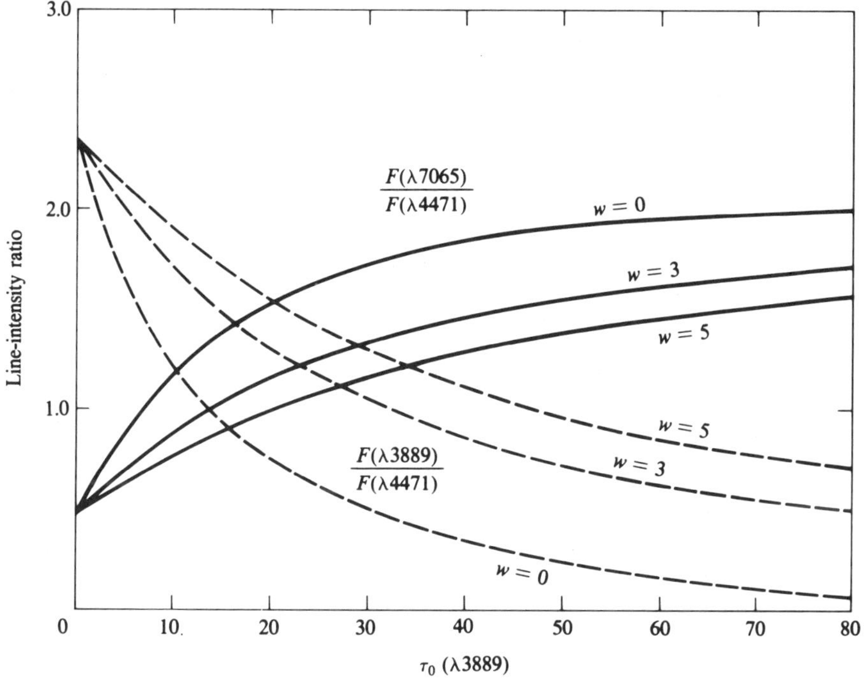

FIGURE 4.5
Radiative-transfer effects due to finite optical depths in He I $\lambda 3889$ $2\,^3S - 3\,^3P$. Ratios of emergent fluxes of $\lambda 7065$ and $\lambda 3889$ to flux of $\lambda 4471$ shown as function of optical radius $\tau_0(\lambda 3889)$ of homogeneous static ($w = 0$) and expanding ($w \neq 0$) isothermal nebulae at $T = 10,000°$ K.

4.7 The Bowen Resonance-Fluorescence Mechanism for O III

There is an accidental coincidence between the wavelength of the He II Lα line at $\lambda 303.78$ and the O III $2p^2\,^3P_2 - 3d\,^3P_2^0$ line at $\lambda 303.80$. As we have seen, in the He^{++} zone of a nebula there is some residual He$^+$, so the He II Lα photons emitted by recombination are scattered many times before they escape. As a result, there is a high density of He II Lα photons in the He^{++} zone, and since O^{++} is also present in this zone, some of the He II Lα photons are absorbed by it and excite the $3d\,^3P_2^0$ level of O III. This level then quickly decays by a radiative transition, most frequently (relative probability 0.74) by resonance scattering in the $2p^2\,^3P_2 - 3d\,^3P_2^0$ line, that is, by emitting a photon. The next most likely decay process (probability 0.24) is emission of $\lambda 303.62$ $2p^2\,^3P_1 - 3d\,^3P_2^0$, which may then escape or may be reabsorbed by another O^{++} ion, again populating $3d\,^3P_2^0$. Finally (probability 0.02), the $3d\,^3P_2^0$ level may decay by emitting one of the six longer wavelength photons $3p\ ^3L_J - 3d\ ^3P_2^0$ indicated in Figure 4.6 and listed in Table 4.13. These levels $3p\ ^3L_J$ then decay to $3s$ and ultimately back to $2p^2\ ^3P$, as shown in the figure and table (or to $2p^3$ and then back to $2p^2\ ^3P$, emitting two far-ultraviolet line photons, as shown in the figure). This is the Bowen resonance-fluorescence mechanism, the conversion of He II Lα to those lines that arise from $3d\ ^3P_2^0$ or from the levels excited by its decay. These lines are observed in many planetary nebulae, and their interpretation requires the solution of the problem of the scattering, escape, and destruction of He II Lα with the complications introduced by the O^{++} scattering and resonance fluorescence.

A competing process that can destroy He II Lα photons before they are converted to O III Bowen resonance-fluorescence photons is absorption in the photoionization of H^0 and He^0. To take this process into account quantitatively, a specific model of the ionization structure is necessary, and the available calculations refer to two spherical-shell model planetary nebulae, ionized by stars with $T_* = 63{,}000°$ K and 100,000° K, respectively. The assumed relative abundances of He and O relative to H are 0.15 and 9×10^{-4} respectively, by number. The optical depths in the center of the He II Lα line are quite large in these models, τ_{0l}(He II Lα) $\gtrsim 2 \times 10^5$, and indeed are large in all models in which the He^+ ionizing photons are completely absorbed, because the optical depth in the line center is much greater than in the continuum. Therefore, practically no He II Lα photons escape directly at the outer surface of the nebula, and this result is quite insensitive to the exact value of τ_{0l} (He II Lα).

The detailed radiative-transfer solutions show that, depending somewhat on the specific model, a little less than half of the He II Lα photons generated by recombination in the He^{++} zone are converted to Bowen resonance-fluorescence photons, and approximately one-third escape at the inner edge of the nebula, as shown numerically in Table 4.14. The photons that "escape" at the inner edge of the nebula at an angle θ to the normal simply cross the central hollow sphere of the model and enter the nebular shell again at another point at an angle θ to the normal there. However, taking account of the expansion of the nebula, these photons are redshifted by an amount $\Delta\nu = \nu_0(2\ V_{\rm exp}/c)\cos\theta$, where $V_{\rm exp}$ is the expansion velocity at the inner edge of the nebula. If, for instance, $V_{\rm exp} \approx 10$ km sec^{-1}, this can amount to a shift of several Doppler widths, since $V_{\rm th} = 6.5$ km sec^{-1} at $T = 10{,}000°$ K.

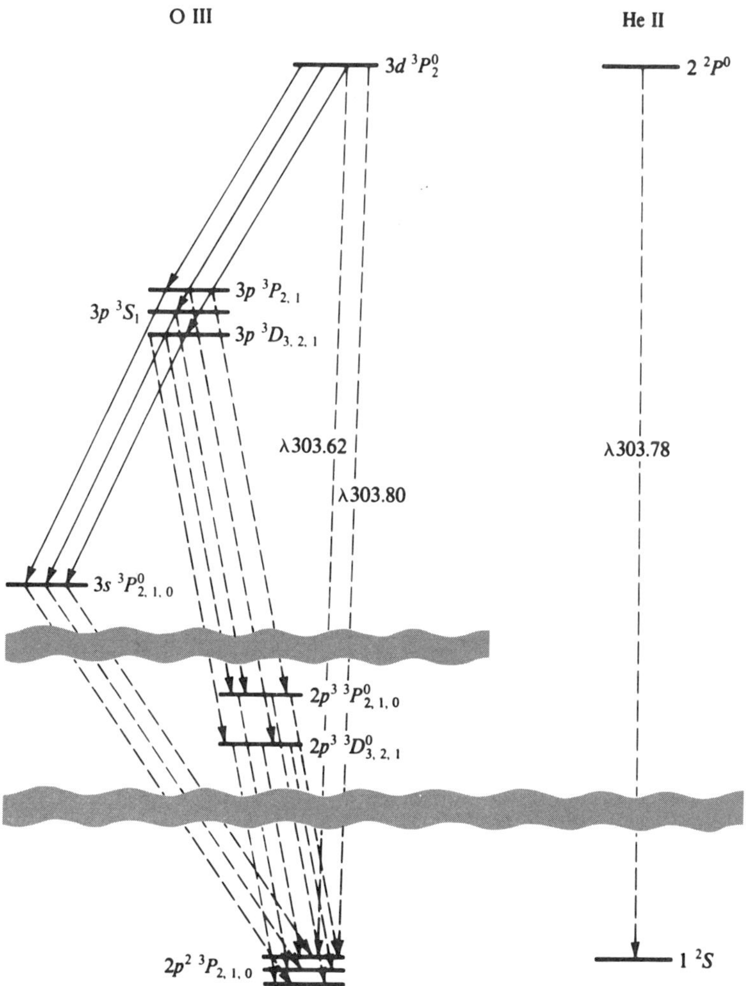

FIGURE 4.6
Schematized partial energy-level diagrams of [O III] and He II showing coincidence of He II Lα and [O III] $2p^2\ ^3P_2 - 3d\ ^3P_2^0$ $\lambda303.80$. The Bowen resonance-fluorescence lines in the optical and near-ultraviolet are indicated by solid lines, and the far ultraviolet lines that lead to excitation or decay are indicated by dashed lines.

TABLE 4.13
O III Resonance-fluorescence lines

Transition	Wavelength (Å)	Relative probability	Relative intensity
$3p\ ^3P_2 - 3d\ ^3P_2^0$	3444.10	3.74×10^{-3}	0.277
$3p\ ^3P_1 - 3d\ ^3P_2^0$	3428.67	1.25×10^{-3}	0.093
$3p\ ^3S_1 - 3d\ ^3P_2^0$	3132.86	1.23×10^{-2}	1.000
$3p\ ^3D_3 - 3d\ ^3P_2^0$	2837.17	1.16×10^{-3}	0.104
$3p\ ^3D_2 - 3d\ ^3P_2^0$	2819.57	2.08×10^{-4}	0.019
$3p\ ^3D_1 - 3d\ ^3P_2^0$	2808.77	1.38×10^{-5}	0.0013
$3s\ ^3P_2^0 - 3p\ ^3S_1$	3340.74	1.79×10^{-3}	0.136
$3s\ ^3P_1^0 - 3p\ ^3S_1$	3312.30	1.07×10^{-3}	0.082
$3s\ ^3P_0^0 - 3p\ ^3S_1$	3299.36	3.57×10^{-4}	0.028
$3s\ ^3P_2^0 - 3p\ ^3P_2$	3047.13	2.14×10^{-3}	0.179
$3s\ ^3P_1^0 - 3p\ ^3P_2$	3023.45	7.12×10^{-4}	0.060
$3s\ ^3P_2^0 - 3p\ ^3P_1$	3059.30	3.95×10^{-4}	0.033
$3s\ ^3P_1^0 - 3p\ ^3P_1$	3035.43	2.37×10^{-4}	0.020
$3s\ ^3P_0^0 - 3p\ ^3P_1$	3024.57	3.17×10^{-4}	0.027
$3s\ ^3P_0^0 - 3p\ ^3D_1$	3757.21	3.57×10^{-6}	0.0002
$3s\ ^3P_1^0 - 3p\ ^3D_1$	3774.00	2.67×10^{-6}	0.0002
$3s\ ^3P_2^0 - 3p\ ^3D_1$	3810.96	1.80×10^{-7}	0.0001
$3s\ ^3P_1^0 - 3p\ ^3D_2$	3754.67	7.22×10^{-5}	0.0049
$3s\ ^3P_2^0 - 3p\ ^3D_2$	3791.26	2.41×10^{-5}	0.0016
$3s\ ^3P_2^0 - 3p\ ^3D_3$	3759.87	5.39×10^{-4}	0.037

Those photons that enter the nebula at the inner edge may be scattered or absorbed, and a certain fraction of them again escape at the inner edge. All the photons that are absorbed within the Doppler core of the line ($|x| \leq 3$) are redistributed in frequency, while those absorbed in the wing ($|x| \geq 3$) are scattered more or less coherently. Eventually, some photons can be redshifted so much that, for them, the entire optical depth of the nebula is small and they then escape. The detailed calculation has been carried through for only one of the models of Table 4.14 (the model with $T_* = 20{,}000°$ K), and the result is that, of the 0.36 He II Lα photons that escape at the inner edge of the model, 0.53 ultimately escape by redshifting, 0.20 are ultimately converted to O III $^3P_1 - P_2^0$ photons, 0.22 are ultimately converted to Bowen resonance-fluorescence photons, and 0.05 are absorbed in H^0 or He^0 photoionization. The overall fraction of He II Lα photons converted into Bowen resonance-fluorescence photons is thus 0.43 (directly) + 0.36 × 0.22 (following at least

TABLE 4.14
Probability of escape or absorption of He II Lα

Process	Planetary nebula model		
	$T_* = 6.3 \times 10^4$ °K	$T_* = 1.0 \times 10^5$ °K	
	$T = 1.0 \times 10^4$ °K	$T = 1.0 \times 10^4$ °K	$T = 2.0 \times 10^4$ °K
Bowen conversion	0.49	0.42	0.43
He II $L\alpha$ escape	0.40	0.33	0.36
O III $^3P_1 - ^3P_2^0$ escape	0.09	0.08	0.12
H^0 or He^0 ionization	0.02	0.17	0.09

one escape into the central hole) = 0.51. These photons are distributed among the various individual lines as indicated in Table 4.13, in which the relative intensities are normalized to λ3133.

4.8 Collisional Excitation in He I

Collisional excitation of H is negligible in comparison with recombination in populating the excited levels in planetary nebulae and H II regions, because the threshold for even the lowest level, $n = 2$ at 10.2 eV, is large in comparison with the thermal energies at typical nebular temperatures. This can be confirmed quantitatively using collision strengths listed in Table 3.12. However, in He^0 the 2 ^3S level is highly metastable, and collisional excitation from it can be important, particularly in exciting 2 ^3P and thus leading to emission of He I λ10,830. To fix our ideas, let us consider a nebula sufficiently dense ($N_e \gg N_c$) that the main mechanism for depopulating 2 ^3S is collisional transitions to 2 ^1S and 2 ^1P, as explained in Section 2.4. The equilibrium population in 2 ^3S is then given by the balance between recombinations to all triplet levels, which eventually cascade down to 2 ^3S, and collisional depopulation of 2 ^3S,

$$N_e N_{\rm He^+} \alpha_B({\rm He}^0,\ n^3L) = N_e N_{2\,^3S}(q_{2\,^3S,\,2\,^1S} + q_{2\,^3S,\,2\,^1P}). \qquad (4.58)$$

The rate of collisional population of 2 ^3P is thus

$$N_e N_{2\,^3S} q_{2\,^3S,\,2\,^3P} = \frac{N_e N_{\rm He^+} q_{2\,^3S,\,2\,^3P}}{(q_{2\,^3S,\,2\,^1S} + q_{2\,^3S,\,2\,^1P})} \alpha_B({\rm He}^0, n\,^3L), \qquad (4.59)$$

so the relative importance of collisional to recombination excitation of $\lambda 10830$ is given by the ratio

$$\frac{N_e N_{2\,^3S} q_{2\,^3S,\,2\,^3P}}{N_e N_{\mathrm{He}^+} \alpha^{eff}_{\lambda 10830}} = \frac{q_{2\,^3S,\,2\,^3P}}{(q_{2\,^3S,\,2\,^1S} + q_{2\,^3S,\,2\,^1P})} \frac{\alpha_B(\mathrm{He}^0,\ n\,^3L)}{\alpha^{eff}_{\lambda 10830}} \tag{4.60}$$

Computed values for $q_{2^3S,2^3P}$ are listed in Table 4.15; they are much larger than those for $q_{2^3S,2^1S}$ and $q_{2^3S,2^1P}$ (listed in Table 2.5), because the cross section for the strong allowed $2\,^3S - 2\,^3P$ transition is much larger than the exchange cross sections to the singlet levels. At a representative temperature

TABLE 4.15
Collisional excitation rate[a] *of He I 2* ^3P

T(° K)	$q_{2\,^3S,\,2\,^3P}$
5,000	3.87×10^{-8}
6,000	6.58×10^{-8}
8,000	1.30×10^{-7}
10,000	1.99×10^{-7}
15,000	3.57×10^{-7}
20,000	4.79×10^{-7}
25,000	5.66×10^{-7}

[a] In cm^3 sec^{-1}.

$T = 10{,}000°$ K, the first factor in equation (4.60) has the numerical value 6.0; the second, 1.4; and the ratio of collisional to recombination excitation is thus about 8. In other words, collisional excitation from $2\,^3S$ completely dominates the emission of $\lambda 10830$, and the factor by which it dominates depends only weakly on T, and can easily be seen to decrease with N_e below N_c.

Though the collisional transition rates from $2\,^3S$ to $2\,^1S$ and $2\,^1P$ are smaller than to $2\,^3P$, the recombination rates of population of these singlet levels are also smaller, and the collisions are therefore also important in the population of $2\,^1S$ and $2\,^1P$. The cross sections for collisions to the higher singlets and triplets are smaller but not negligible; from the best available cross sections it appears likely that collisional population of $3\,^3P$ is significant and somewhat affects the strength of $\lambda 3889$. The cross sections from $2\,^3S$ to

$n\ ^3D$ are less accurately known; however, the available information seems to indicate that there may be a non-negligible collisionally excited component in the observed strength of $\lambda 5876$ in planetary nebulae. Some calculated collisional excitation rates for He I levels with $n = 3$ are listed in Table 4.16.

TABLE 4.16
Collisional excitation coefficients from $He^0(2\ ^3S)$ to $n = 3$*

T(° K)	$q_{2\ ^3S,\ 3\ ^3S}$	$q_{2\ ^3S,\ 3\ ^3P}$	$q_{2\ ^3S,\ 3\ ^3D}$	$q_{2\ ^3S,\ 3\ ^1D}$
6,000	3.50×10^{-10}	1.44×10^{-10}	1.05×10^{-10}	2.05×10^{-11}
8,000	1.19×10^{-9}	5.78×10^{-10}	5.27×10^{-10}	8.79×10^{-11}
10,000	2.45×10^{-9}	1.30×10^{-9}	1.41×10^{-9}	2.08×10^{-10}
15,000	6.15×10^{-9}	3.79×10^{-9}	5.43×10^{-9}	6.32×10^{-10}
20,000	9.60×10^{-9}	6.34×10^{-9}	1.08×10^{-8}	1.07×10^{-9}
25,000	1.24×10^{-8}	8.45×10^{-9}	1.61×10^{-8}	1.42×10^{-9}

[a] In $cm^3\ sec^{-1}$.

Similar collisional excitation effects occur from the metastable $He^0\ 2\ ^1S$ and $H^0\ 2\ ^2S$ levels, but they decay so much more rapidly than $He^0\ 2\ ^3S$ that their populations are much smaller and the resulting excitation rates are negligibly small.

References

A good general summary of the emission processes in gaseous nebulae is given by

Seaton, M. J. 1960. *Reports on Progress in Physics* **23,** 313, 1960.

The theory of the recombination-line spectrum of H I goes back to the early 1930's and was developed in papers by H. H. Plaskett, G. Cillie, D. H. Menzel, L. H. Aller, L. Goldberg, and others. In more recent years it has been refined and worked out more accurately by M. J. Seaton, A. Burgess, R. M. Pengelly, M. Brocklehurst, D. G. Hummer, P. J. Storey, and others. The treatment in Chapter 4 follows most closely the following definitive references:

Seaton, M. J. 1959. *M.N.R.A.S.* **119,** 90.

Pengelly, R. M. 1964. *M.N.R.A.S.* **127,** 145.

The second reference treats the low-density limit for H I and He II in detail. (Tables 4.1, 4.2, and 4.3 are derived from it.)

Pengelly, R. M., and Seaton, M. J. 1964. *M.N.R.A.S.* **127,** 165.

The effects of collisions in shifting L at fixed n are discussed in this reference.

Brocklehurst, M. 1971. *M.N.R.A.S.* **153,** 471.

Hummer, D. G., and Storey, P. J. 1987. *M.N.R.A.S.* **224,** 801.

These references include the definitive results for H I and He II at finite densities in the optical region, taking full account of the collisional transitions. (Tables 4.4 and 4.5 are based on the second of these papers.)

Giles, K. 1977. *M.N.R.A.S.* **180,** 57P.

Seaton, M. J. 1978. *M.N.R.A.S.* **185,** 5P.

These papers give the extensions (in the low-density limit) to the infrared Brackett lines of H I that are included in Table 4.2, and the ultraviolet lines of He II that are included in Table 4.3, respectively.

Robbins, R. R. 1968. *Ap. J.* **151,** 497.
Robbins. R. R. 1970. *Ap. J.* **160,** 519.
Robbins, R. R., and Robinson, E. L. 1971. *Ap. J.* **167,** 249.
Brocklehurst, M. 1972. *M.N.R.A.S.* **157,** 211.
Ferland, G. J. 1980. *M.N.R.A.S.* **191,** 243.

The first two papers by Robbins work out the theory in detail for the He I triplets; the third is concerned with the singlets (for Case *A* only). The Brocklehurst article is complete, for it describes both triplet and singlet results. (Table 4.6 is based on it.) The paper by Ferland analyzes the applicability of Case B to the He I singlets.

Brown, R. L., and Mathews, W. G. 1970. *Ap. J.* **160,** 939.

This reference collects previous references and material on the H I and He II continuum, and includes the most detailed treatment of the He I continuum. (Tables 4.7-4.9 and 4.11 are taken from this reference.)

Ferland, G. J. 1980. *P.A.S.P.* **92,** 596.

This reference gives the continuous spectrum of the H I continuum (for specific wavelengths and filters), and recombination coefficients for Cases *A* and *B* over a very wide range in temperature (500° K $\leq T \leq$ 2,000,000° K).

A very complete reference on free-free emission in the radio frequency region is:

Scheuer, P. A. G. 1960. *M.N.R.A.S.* **120,** 231.

The first published prediction that the radio-frequency recombination lines of H I would be observable was made by

Kardashev, N. S. 1959. *Astron. Zhurnal* **36,** 838 (English translation, *Soviet Astronomy A. J.* 1960. **3,** 813.

The key reference on the importance of maser action and on the exact variation of b_n with n is:

Goldberg, L. 1966. *Ap. J.* **144,** 1225.

The radiative transfer treatment in this chapter essentially follows this reference.

The equilibrium equations for the populations of the high levels are worked out in

Seaton, M. J. 1964. *M.N.R.A.S.* **127,** 177.
Sejnowski, T. J., and Hjellming, R. H. 1969. *Ap. J.* **156,** 915.
Brocklehurst, M. 1970. *M.N.R.A.S.* **148,** 417.

The last of these three references is the definitive treatment and makes full allowance for all collisional effects. (Figure 4.2 is based on it.)

A good deal of theoretical work has been done by several authors on radiative-transfer problems in nebulae. The portions of this research used in this chapter is summarized (with complete references) in

Osterbrock, D. E. 1971. *J.Q.S.R.T.* **11,** 623.

Some of the key references concerning the H I lines are:

Capriotti, E. R. 1964. *Ap. J.* **139,** 225 and **140**, 632.

Capriotti, E. R. 1966. *Ap. J.* **146,** 709.

Cox, D. P., and Mathews, W. G. 1969. *Ap. J.,* **155,** 859.

(Figure 4.3 is based on the last reference.)

Actually, the radiative transfer problem of He I lines was worked out earlier in

Pottasch, S. R. 1962. *Ap. J.* **135,** 385.

A more complete treatment is:

Robbins, R. R. 1968. *Ap. J.* **151,** 511.

(Figure 4.5 is derived from calculations in this reference.)

The Bowen resonance-fluorescence mechanism was first described by

Bowen, I. S. 1924. *Ap. J.* **67,** 1.

The radiative transition probabilities necessary for tracing all the downward radiative decays following excitation of O III and $3d\ ^3P^0_1$ are given by

Saraph, H. E., and Seaton, M. J. 1980. *M.N.R.A.S.* **193,** 617.

(Table 4.13 is derived from it.) The most complete radiative-transfer theory of this problem is due to

Weymann, R. J., and Williams, R. E. 1969. *Ap. J.* **157,** 1201.

The calculations described in Sections 4.7 are contained in the preceding reference. (Table 4.14 is derived from it.) Still further details of the Bowen resonance-fluorescence process are included in

Harrington, J. P. 1972. *Ap. J.* **176,** 127.

A very good general treatment, applicable to a wide range of physical situations, is

Kallman, T., and McCray, R. 1980. *Ap. J.* **242,** 615.

The importance of collisional excitation from $He^0\ 2\ ^3S$ was discussed by

Mathis, J. S. 1957. *Ap. J.* **125,** 318.

Pottasch, S. R. 1961. *Ap. J.* **135,** 93.

Osterbrock, D. E. 1964. *Ann. Rev. Astr. Ap.* **2,** 95.

Cox, D. P., and Daltabuit, E. 1971. *Ap. J.,* **167,** 257.

Ferland, G. J. 1986. *Ap. J.* **310,** L67.

Peimbert, M., and Torres-Peimbert, S. 1987. *Rev. Mex. Astr. Ap.* **14,** 540.

Clegg, R. E. S. 1987. *M.N.R.A.S.,* **229,** 31P.

See also the references by Robbins, and especially by Brocklehurst, above.

The He I cross sections and the collisional excitation rates calculated from them and listed in Tables 4.15 and 4.16 are from

Berrington, K. A., and Kingston, A. E. 1987, *J. Phys.B*, **20**, 6631.

5

Comparison of Theory with Observations

5.1 Introduction

In the preceding three chapters much of the available theory on gaseous nebulae has been discussed, so that we are now in a position to compare it with the available observations. The temperature in a nebula may be determined from measurements of ratios of intensities of pairs of emission lines – specifically, those emitted by a single ion from two levels with considerably different excitation energies. Although the relative strengths of H recombination lines do not vary as greatly as T varies, the ratio of the intensity of a line to the intensity of the recombination continuum varies more rapidly and can be used to measure T. Further information on the temperature may be derived from radio observations, combining long- and short-wavelength continuum observations (large and small optical depths, respectively) or long-wavelength continuum and optical line observations. The electron density in a nebula may be determined from measured intensity ratios of other pairs of lines: those emitted by a single ion from two levels with nearly the same energy but with different radiative-transition probabilities. Likewise, measurements of relative strengths of the radio recombination lines give information on both the density and the temperature in nebulae. These methods, as well as the resulting information on the physical parameters of characteristic nebulae, are discussed in the first sections of this chapter.

In addition, information on the stars that provide the ionizing photons may be derived from nebular observations. For example, if a nebula is optically thick to a certain type of ionizing radiation (for instance, in the H Lyman continuum), then the total number of photons of this type emitted by the star can be deduced from the properties of the nebula. By combining these nebular observations, which basically measure the far-ultraviolet-ionizing radiation from the involved stars, with optical measurements of the same stars, a long base-line color index that gives information on the temperature of the stars can be determined. This scheme and the information derived from it, about main-sequence O stars and about planetary-nebula central stars, are discussed in Section 5.7.

Once the temperature and density in a nebula are known, the observed strength of a line allows us to deduce the total number of ions in the nebula that are responsible for the emission of that line. Thus information is derived on the abundances of the elements in H II regions and planetary nebulae.

Each of the next eight sections of this chapter discusses a specific kind of observational analysis or diagnostic measurement of a nebula. Each method gives some specific detailed information, integrated through whatever structure there may be along the line of sight through the nebula, and also over whatever area of the nebula is covered by the analyzing device used for the observations, such as the spectrograph slit or the radio-telescope beam pattern. A more detailed comparison, in integrated form, may be made by calculating models of nebulae intended to represent their entire structure, and comparing the properties of these models with observations. A discussion of these types of models, and of the progress that has been made with them and then a brief explanation of the filling-factor concept used in them, close the chapter.

5.2 Temperature Measurements from Emission Lines

A few ions, of which [O III] and [N II] are the best examples, have energy-level structures that result in emission lines from two different upper levels with considerably different excitation energies occurring in the observable wavelength region. The energy-level diagrams of these two ions are shown in Figure 3.1, where it can be seen that, for instance, [O III] $\lambda 4363$ comes from the upper ^1S level, while $\lambda 4959$ and $\lambda 5007$ come from the intermediate ^1D level. ($^3P_0 - ^1D_2$ $\lambda 4931$, which can occur only by an electric-quadrupole transition, has a much smaller transition probability, and is so weak that it can be ignored.) It is clear that the relative rates of excitation to the ^1S and ^1D levels depend very strongly on T, so the relative strength of the lines emitted by these levels may be used to measure electron temperature.

An exact solution for the populations of the various levels, and for the relative strengths of the lines emitted by them, may be carried out along the lines of the discussion in Section 3.5. However, it is simpler and more instructive to proceed by direct physical reasoning. In the low-density limit (collisional de-excitations negligible), every excitation to the ^1D level results in emission of a photon either in $\lambda 5007$ or $\lambda 4959$, with relative probabilities given by the ratio of the two transition probabilities, which is very close to 3 to 1. Every excitation of ^1S is followed by emission of a photon in either $\lambda 4363$ or $\lambda 2321$, with the relative probabilities again given by the transition probabilities. Each emission of a $\lambda 4363$ photon further results in the population of ^1D, which again is followed by emission of either a $\lambda 4959$ photon or a

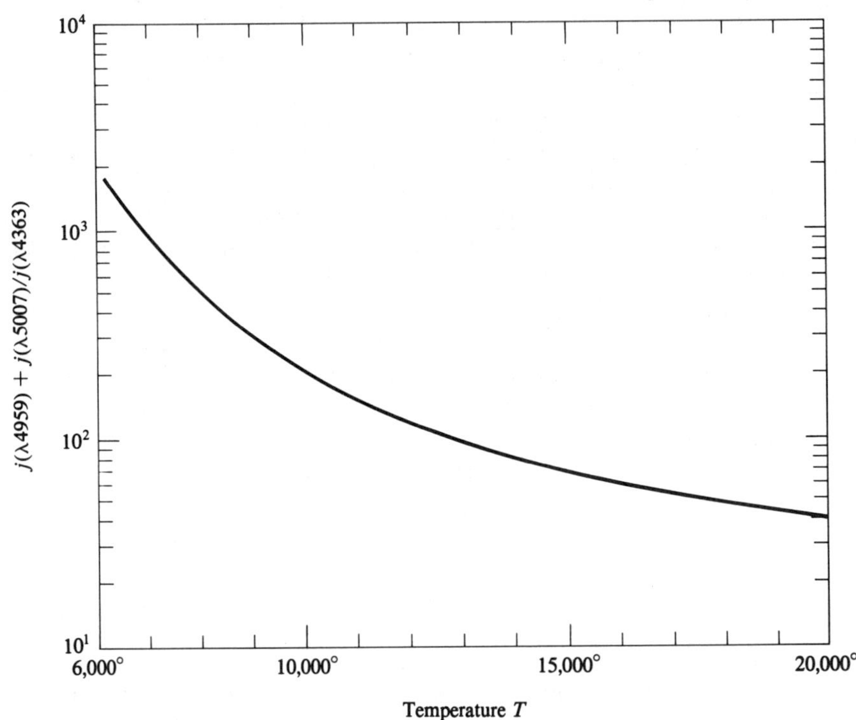

FIGURE 5.1
[O III] ($\lambda4959$ + $\lambda5007$)/$\lambda4363$ intensity ratio (in low-density limit $N_e \rightarrow 0$) as a function of temperature. Accurately calculated values shown here are indistinguishable from approximation of equation (5.4) using mean values of Ω.

$$\frac{j_{\lambda4959} + j_{\lambda5007}}{j_{\lambda4363}} = \frac{\Omega(^3P,\ ^1D)}{\Omega(^3P,\ ^1S)} \left[\frac{A_{^1S,\ ^1D} + A_{^1S,\ ^3P}}{A_{^1S,\ ^1D}}\right] \frac{\bar{\nu}(^3P,\ ^1D)}{\nu(^1D,\ ^1S)} e^{\Delta E/kT}, \tag{5.1}$$

where

$$\bar{\nu}(^3P,\ ^1D) = \frac{A_{^1D_2,\ ^3P_2}\nu(\lambda5007) + A_{^1D_2,\ ^3P_1}\nu(\lambda4959)}{A_{^1D_2,\ ^3P_2} + A_{^1D_2,\ ^3P_1}}, \tag{5.2}$$

and ΔE is the energy difference between the $^1D_2$ and $^1S_0$ levels.

Equation (5.1) is a good approximation up to $N_e \approx 10^5$ cm^{-3}. However, at higher densities collisional de-excitation begins to play a role. The lower ^1D term has a much longer radiative lifetime than the ^1S term, so it is collisionally de-excited at lower electron densities than ^1S, thus weakening $\lambda4959$ and $\lambda5007$. In addition, under these conditions collisional excitation of ^1S from the excited ^1D level begins to strengthen $\lambda4363$. The full statistical equilibrium equations (3.28) can be worked out numerically for any N_e and T, but an analytic solution correct to the first order in N_e and to the first order in $e^{-\Delta E/kT}$ is that the right-hand side of (5.1) is divided by a factor

$$f = \frac{1 + \dfrac{C(^1D,\,^3P)C(^1S,\,^1D)}{C(^1S,\,^3P)A_{^1D,\,^3P}} + \dfrac{C(^1D,\,^3P)}{A_{^1D,\,^3P}}}{1 + \dfrac{C(^1S,\,^3P) + C(^1S,\,^1D)}{A_{^1S,\,^3P} + A_{^1S,\,^1D}}} \tag{5.3}$$

where

$$C(i,j) = 8.63 \times 10^{-6} \frac{N_e}{T^{1/2}} \frac{\Omega(i,j)}{\omega_i}.$$

Inserting numerical values of the collision strengths and transition probabilities from Chapter 3, this becomes

$$\frac{j_{\lambda4959} + j_{\lambda5007}}{j_{\lambda4363}} = \frac{7.73 \exp\left[\,(3.29 \times 10^4)/T\,\right]}{1 + 4.5 \times 10^{-4}\,(N_e/T^{1/2}\,)}. \tag{5.4}$$

Here the representative values of the collision strengths from Table 3.4 have been used to calculate the numerical coefficients, but actually in O^{++} there are several resonances, and the resulting average collision strengths vary appreciably with temperature; so equation (5.4) is not exact. However, in Figure 5.1 the intensity ratio is plotted (in the low-density limit) using the correct collision strengths at each T, but to within the thickness of the line it is the same as the result of equation (5.4), so very little error results from the use of mean collision strengths.

An exactly similar treatment may be carried out for [N II], and the resulting equation analogous to (5.4) is

$$\frac{j_{\lambda6548} + j_{\lambda6583}}{j_{\lambda5755}} = \frac{6.91 \exp\left[(2.50 \times 10^4)/T\,\right]}{1 + 2.5 \times 10^{-3}\,(N_e/T^{1/2})}. \tag{5.5}$$

These two equations form the basis for optical temperature determinations in gaseous nebulae. Since the nebulae are optically thin in forbidden-line radiation, the ratio of the integrals of the emission coefficients along a ray through the nebula is observed directly as the ratio of emergent intensities, so if the nebula is assumed to be isothermal and to have sufficiently low density that the low-density limit is applicable, the temperature can be directly determined. Alternatively, the ratio of the fluxes from the whole nebula may be measured for smaller nebulae. No information need be known about the distance of the nebula, the amount of O^{++} present, and so on, because all these factors cancel out. If collisional de-excitation is not completely negligible, even a rough estimate of the electron density substituted into the correction term in the denominator provides a good value of T. The observed strengths of the lines must be corrected for interstellar extinction, but this correction is usually not too large because the temperature-sensitive lines in both [O III] and [N II] are relatively close in wavelength.

The [O III] line-intensity ratio ($\lambda4959 + \lambda5007)/\lambda4363$ is quite large and is therefore rather difficult to measure accurately. Although $\lambda\lambda4959$, 5007 are strong lines in many gaseous nebulae, $\lambda4363$ is relatively weak, and furthermore is close to Hg I $\lambda4358$, which unfortunately is becoming stronger and stronger in the spectrum of the sky. Large-intensity ratios are difficult to measure accurately, and reasonably precise temperature measurements therefore require carefully calibrated photoelectric measurements with fairly high-resolution spectral analyzers. Until now most work has been done on the [O III] lines, partly because they occur in the blue spectral region in which photomultipliers are most sensitive, and partly because [O III] is quite bright in typical high-surface-brightness planetary nebulae. The [N II] lines are stronger in the outer parts of H II regions, where the ionization is lower and the O mostly emits [O II] lines, but not too many measurements are yet available. No doubt CCD spectrographs will soon improve this situation.

Let us first examine optical determinations of the temperatures in H II regions, some selected results of which are collected in Table 5.1. Note that in this and other tables, the observed intensity ratio has been corrected for interstellar extinction in the way outlined in Chapter 7, and the temperature has been computed using numerical values from this book – in the present case Figure 5.1 and equation (5.5).

It can be seen that all the temperatures of these H II regions are in the range 7,000 to 13,000° K. Of the four slit positions in NGC 1976, the Orion Nebula, 2b and 1a are in the densest, brightest part near the Trapezium, while 5b and 5a are in less dense regions somewhat further from the center. NGC 1976 is in the same spiral arm of our Galaxy as the Sun, M 8 and M 17 are in the next spiral arm inward, and NGC 2467 and NGC 2359 are significantly further outward. The abundances of the heavy elements increase inward in our Galaxy, as we will see in Chapter 8, resulting in the differences in temperatures shown in this table.

TABLE 5.1

Temperature determinations in H II regions

Nebula	[N II] $\frac{I(\lambda 6548) + I(\lambda 6583)}{I(\lambda 5755)}$	[N II] $T(° \text{K})$	[N II] $N_e/T^{1/2}$	[O III] $\frac{I(\lambda 4959) + I(\lambda 5007)}{I(\lambda 4363)}$	[O III] $T(° \text{K})$
NGC 1976 2b	81	10,000	51	338	8,700
NGC 1976 1a	102	9,100	68	371	8,500
NGC 1976 5b	111	8,900	21	310	8,900
NGC 1976 5a	189	7,500	12	263	9,300
M 8 I	162	7,900	(10)	445	8,100
M 17 I	257	6,900	(10)	330	8,700
NGC 2467 1a	46	13,000	(1)	129	11,600
NGC 2467 1b	53	12,200	(1)	137	11,400
NGC 2359 av	—	—	(1)	90	13,200

Planetary nebulae have higher surface brightness than typical H II regions, and as a result there is a good deal more observational material available for planetaries, particularly [O III] determinations of the temperature. Most planetaries are so highly ionized that [N II] is relatively weak, but some measurements of it are also available. A selection of the best observational material is collected in Table 5.2, which shows that the temperatures in planetary nebulae are typically somewhat higher than those in H II regions. This is partly because the higher effective temperatures of the central stars in planetary nebulae (to be discussed in Section 5.8) lead to a higher input of energy per photoionization, and partly because the higher electron densities in typical planetaries result in collisional de-excitation and decreased efficiency of radiative cooling. The last three planetaries in Table 5.2 are halo objects that have relatively low heavy-element contents, as discussed in Chapter 9, and that are consequently somewhat above average in temperature. Discrepancies between T as determined from [O III] and [N II] lines in the same nebula do not necessarily indicate an error in either method; these lines are emitted in different zones of the nebula because their ionization potentials are different.

TABLE 5.2
Temperature determinations for planetary nebulae

Nebula	T[N II] (° K)	T[O III] (° K)
NGC 650	9,500	10,700
NGC 4342	10,100	11,300
NGC 6210	10,700	9,700
NGC 6543	9,000	8,100
NGC 6572	—	10,300
NGC 6720	10,600	11,100
NGC 6853	10,000	11,000
NGC 7027	—	12,400
NGC 7293	9,300	11,000
NGC 7662	10,600	12,800
IC 418	—	9,700
IC 5217	—	11,600
BB 1	10,500	12,900
Haro 4-1	—	12,000
K 648	—	13,100

From Tables 5.1 and 5.2, it is reasonable to adopt $T \approx 10{,}000^\circ$ K as an order-of-magnitude estimate for any nebula with near-solar abundances; with somewhat greater precision we may adopt representative values $T \approx 9{,}000^\circ$ K in the brighter parts of an H II region like NGC 1976, and $T \approx 11{,}000^\circ$ K in a typical bright planetary nebula.

Another method that may be used in principle to determine the temperature in a nebula is to compare the relative strength of a collisionally excited line, such as C III] λ1909, with a recombination line of the next lower state of ionization, such as C II λ4267, since both depend on the product of densities $N_{C^{++}} N_e$, which therefore cancels out from their ratio. The difficulties are that the effective recombination rate for emission of λ4267 has not been accurately calculated, and that it is difficult to measure accurately and compare lines in the optical and satellite ultraviolet spectral regions. Both of these difficulties will probably be surmounted in future years.

5.3 Temperature Determinations from Optical Continuum Measurements

Although it might be thought that the temperature in a nebula could be measured from the relative strengths of the H lines, in fact their relative strengths are almost independent of temperature, as Table 4.4 shows. The physical reason for this behavior is that all the recombination cross sections to the various levels of H have approximately the same velocity dependence, so the relative numbers of atoms formed by captures to each level are nearly independent of T, and since the cascade matrices depend only on transition probabilities, the relative strengths of the lines emitted are also nearly independent of T. These calculated relative line strengths are in good agreement with observational measurements, as will be discussed in detail in Section 7.2.

However, the temperature in a nebula can be determined by measuring the relative strength of the recombination continuum relative to a recombination line. Physically, the reason this ratio does depend on the temperature is that the emission in the continuum (per unit frequency interval) depends on the width of the free-electron velocity-distribution function, that is, on T.

The theory is straightforward, for we may simply use Table 4.4 to calculate the H-line emission, and Tables 4.7 to 4.9 and 4.12 to calculate the continuum emission, and thus find their ratio as a function of T. Table 5.3 lists the calculated ratios for two choices of the continuum, first at Hβ λ4861, which includes the H I recombination and two-photon continua as well as the He I recombination continuum. (A nebula with $N_{\mathrm{He}^+} = 0.10\ N_{\mathrm{H}^+}$ and $N_{\mathrm{He}^{++}} = 0$ has been assumed, but any other abundances or ionization conditions deduced from line observations of the nebula could be used.) The second choice is the Balmer discontinuity, $j_\nu(\lambda 3646-) - j_\nu(\lambda 3646+)$, which eliminates everything except the H I recombination continuum resulting from recaptures into $n = 2$. (The He II recombination, of course, would also contribute if $N_{\mathrm{He}^{++}} \neq 0$.) Note that the $\lambda 4861$ continuum has been calculated in the limit $N_p \rightarrow 0$ (no collisional de-excitation of H I $2\,^2S$ and hence maximum relative strength of the H I two-photon continuum) and also for $N_p = 10^4$ cm^{-3}, $N_{\mathrm{He}^+} = 10^3$ cm^{-3}, taking account of collisional de-excitation, while the Balmer discontinuity results are independent of density and N_{He^+}.

The continuum at $\lambda 4861$ is made up chiefly of the H I Paschen and higher-series continua, whose sum increases slowly with T, and the two-photon continuum, whose strength decreases slowly with T; the sum hence is roughly independent of T, and the ratio of this continuum to Hβ therefore increases with T. On the other hand, the strength of the Balmer continuum at the series limit decreases approximately as $T^{-3/2}$, and its ratio to Hβ therefore decreases slowly with T, as Table 5.3 shows.

The observations of the continuum are difficult because it is weak and can be seriously affected by weak lines. High-resolution spectrophotometric measurements with high-sensitivity detectors are necessary. To date the most accurate published data seem to be measurements of the Balmer continuum, which is considerably stronger than the continuum near Hβ. A difficulty in measuring the Balmer continuum, of course, is that the higher Balmer lines are crowded just below the limit, so the intensity must be measured at longer wavelengths and extrapolated to $\lambda 3646+$. Furthermore, continuous radiation emitted by the stars in the nebulae and scattered by interstellar dust may have a sizable Balmer discontinuity, which is difficult to disentangle from the true nebular-recombination Balmer discontinuity. Some of the best published results for H II regions and planetary nebulae are collected in Table 5.4, which shows that the temperatures measured by this method are generally somewhat smaller than the temperatures for the same objects measured from forbidden-line ratios. The most accurately observed nebula of all is NGC 7027, in which the observed continuum over the range $\lambda\lambda 3300$-11,000, including both the Balmer and Paschen discontinuities, as well as the Balmer-line and Paschen-line ratios, matches very well the calculated spectrum for $T = 17{,}000°$ K, somewhat higher than the temperatures indicated by forbidden-line ratios. These discrepancies will be discussed again in the next section following the discussion of temperature measurements from the radio-continuum observations.

TABLE 5.3
Ratio[a] of continuum to line emission

	T			
	5,000° K	10,000° K	15,000° K	20,000° K
lim $N_p \to 0, N_{He^+} = 0.10\ N_p$				
$\frac{j_\nu(\lambda 4861)}{j_{\mathrm{H}\beta}}$	3.45×10^{-15}	5.82×10^{-15}	7.31×10^{-15}	9.21×10^{-15}
$N_p = 10^4$ cm^{-3}, $N_{\mathrm{He}^+} = 10^3$ cm^{-3}, $N_e = 1.1 \times 10^4$ cm^{-3}				
$\frac{j_\nu(\lambda 4861)}{j_{\mathrm{H}\beta}}$	2.44×10^{-15}	4.81×10^{-15}	6.35×10^{-15}	8.20×10^{-15}
N_p, N_{He^+}, N_e arbitrary				
$\frac{j_\nu(Bac)}{j_{\mathrm{H}\beta}}$	3.10×10^{-14}	1.89×10^{-14}	1.41×10^{-14}	1.25×10^{-14}

[a] All ratios are in units of Hz^{-1}.

TABLE 5.4
Observations of Balmer discontinuity in nebulae

Nebula[a]	$\dfrac{I_\nu(\lambda 3646-) - I_\nu(\lambda 3646+)}{I(H\beta)}$ (Hz^{-1})	T(° K)
NGC 1976 I	2.32×10^{-14}	7,300
NGC 1976 II	2.47×10^{-14}	6,800
NGC 1976 III	2.56×10^{-14}	6,500
NGC 6572 A	2.06×10^{-14}	8,700
NGC 6572 B	2.45×10^{-14}	6,900
NGC 7009 B	2.27×10^{-14}	7,700
IC 418 A	2.26×10^{-14}	7,700

[a] In NGC 1976, I, II, and III are different slit positions. In the planetary nebulae, A represents an entrance diaphragm 30″ in diameter; B represents an entrance slit 8″ × 80″ oriented east-west.

5.4 Temperature Determinations from Radio-Continuum Measurements

Another completely independent temperature determination can be made from radio-continuum observations. The idea is quite straightforward, namely, that at sufficiently low frequencies any nebula becomes optically thick, and therefore at these frequencies (assuming an isothermal nebula) the emergent intensity is the same as that from a blackbody, the Planck function $B_\nu(T)$; or, equivalently, the measured brightness temperature is the temperature within the nebula

$$T_{b\nu} = T(1 - e^{-\tau_\nu}) \to T \text{ as } \tau_\nu \to \infty \tag{5.6}$$

as in equation (4.37). Note that if there is background nonthermal synchrotron radiation (beyond the nebula) with brightness temperature $T_{bg\nu}$ and foreground radiation (between the nebula and the observer) with brightness temperature $T_{fg\nu}$, this equation becomes

$$T_{b\nu} = T_{fg\nu} + T(1 - e^{-\tau_\nu}) + T_{bg\nu}e^{-\tau_\nu} \to T_{fg\nu} + T \text{ as } \tau_\nu \to \infty. \tag{5.7}$$

The difficulty in applying this method is that at frequencies low enough so that the nebulae are optically thick ($\nu \approx 3 \times 10^8$ Hz or $\lambda \approx 10^2$ cm for many dense nebulae), even the largest radio telescopes have beam sizes that are comparable to or larger than the angular diameters of typical H II regions.

Therefore, the nebula does not completely fill the beam, and a correction must be made for the projection of the nebula onto the antenna pattern.

The antenna pattern of a simple parabolic or spherical dish is circularly symmetric about the axis, where the sensitivity is at a maximum. The sensitivity decreases outward in all directions, and in any plane through the axis it has a form much like a Gaussian function with angular width of order λ/d, where d is the diameter of the telescope. The product of the antenna pattern with the brightness-temperature distribution of the nebula then gives the mean brightness temperature, which is measured by the radio-frequency observations. The antenna pattern thus tends to broaden the nebula and to wipe out much of its fine structure. To determine the temperature of a nebula that is small compared with the width of the antenna pattern, we must know its angular size accurately. But of course, no nebula really has sharp outer edges, inside which it has infinite optical depth and outside which it has zero optical depth. In a real nebula, the optical depth decreases more or less continuously but with many fluctuations, from a maximum value somewhere near the center of the nebula to zero just outside the edge of the nebula, and what is really needed is the complete distribution of optical depth over the face of the nebula.

This can be obtained from high radio-frequency measurements of the nebula, for in the high-frequency region, the nebula is optically thin, and the measured brightness temperature gives the product $T\tau_1$,

$$T_{b1} = T(1 - e^{-\tau_1}) \rightarrow T\tau_1 \text{ as } \tau_1 \rightarrow 0, \tag{5.8}$$

as in equation (4.37). (In the remainder of this section, the subscript 1 is used to indicate a high frequency, and subscript 2 is used to indicate a low frequency.) At high frequencies, the largest radio telescopes have much better angular resolution than at low frequencies because of the smaller values of λ/d, so if the nebula is assumed to be isothermal, the high-frequency measurements can be used to prepare a map of the nebula, giving the product $T\tau_1$ at each point. Thus for any assumed T, the optical depth τ_1 is determined at each point from the high-frequency measurements. The ratio of optical depths, τ_1/τ_2, is known from equation (4.31), so τ_2 can be calculated at each point, and then the expected brightness temperature T_{b2} can be calculated at each point:

$$T_{b2} = T(1 - e^{-\tau_2}). \tag{5.9}$$

Integrating the product of this quantity with the antenna pattern gives the expected mean-brightness temperature at the low frequency. If the assumed T is not correct, this expected result will not agree with the observed mean brightness temperature, and another assumed temperature must be tried until agreement is reached. This, then, is a procedure for correcting the radio-frequency continuum measurements for the effects of finite beam size at low frequencies.

Some of the most accurate available radio-frequency measurements of temperature in H II regions are collected in Table 5.5. At 408 MHz, most of the nebulae listed have central optical depths $\tau_2 \approx 1$ to 10, but at 85 MHz, the optical depths are much larger. The beam size of the antenna is larger at 85 MHz ($\sim 50'$ with the original Mills Cross), and in addition the background (nonthermal) radiation is larger, so many of the nebulae are measured in absorption at this lower frequency. The uncertainties are probably of order $\pm$ 1,000° K.

TABLE 5.5
Electron temperatures in H II *regions from radio frequency continuum observations*

Nebula	T(° K)	
	408 MHz	85 MHz
NGC 1976	8,550°	7,000°
RCW 38	7,500°	4,000°
RCW 49	7,750°	6,000°
NGC 6334	7,000°	10,000°
NGC 6357	6,900°	10,000°
M 17	7,850°	8,000°
M 16	–	5,000°
NGC 6604	–	4,000°

The two frequencies give fairly consistent temperature determinations, and comparison of Table 5.5 with Table 5.1 shows that the two quite independent ways of measuring the mean temperature in a nebula agree fairly well. There are, however, some slight remaining differences; the temperature as determined from the radio-frequency observations is usually lower than the temperature determined from forbidden-line ratios. The probable explanation of this discrepancy is that the temperature is not constant throughout the nebula, as has been tacitly assumed, but instead differs from point to point because of variations in the local heating and cooling rates. Under this interpretation, a more complicated comparison between observation and theory is necessary. An ideal method would be to know the entire temperature structure of the nebula, to calculate from it the expected forbidden-line ratios and radio-frequency continuum brightness temperatures, and then to compare them with observation; this model approach will be discussed in Section 5.10.

However, the general type of effects that are expected can easily be understood. The forbidden-line ratios determine the temperature in the region in which these lines themselves are emitted. That is, the [O III] ratio measures a mean temperature in the O^{++} zone and the [N II] ratio measures the mean temperature weighted in a different way in the N^+ zone. The emission coefficient for the forbidden lines increases strongly with increasing temperature, and therefore the mean they measure is strongly weighted toward high-temperature regions. On the other hand, the free-free emission coefficient decreases with increasing temperature, and therefore the mean it measures is weighted toward low-temperature regions. We thus expect a discrepancy in the sense that the forbidden lines indicate a higher temperature than do the radio-frequency measurements, as is in fact confirmed by observation. It is even possible to get some information about the range of variation of the temperature along a line through the nebula from comparison of these various temperatures, but since the result depends on the ionization distribution also, we will not consider this method in detail.

Exactly the same method can be used to measure the temperatures in planetary nebulae, but because they are very much smaller than the antenna beam size at the frequencies at which they are optically thick, the correction for this effect is quite important. Nearly all the planetary nebulae are too small for mapping at even the shortest radio-frequency wavelengths with single-dish radio telescopes, but it is possible to use the surface brightness in a hydrogen recombination line such as Hβ, since it is also proportional to the integral

$$I(H\beta) \propto \int N_p N_e ds = E_p,$$

to get the relative values of τ_2 at each point in the nebula. Even the optical measurements have finite angular resolution because of the broadening effects of seeing. Then, for any assumed optical depth of the nebula at one point and at one frequency, the optical depths at all other points and at all frequencies can be calculated. For any assumed temperature, the expected flux at each frequency can thus be calculated and compared with the radio measurements, which must be available for at least two (and preferably more) frequencies, one in the optically thin region and one in the optically thick region. The two parameters T and the central optical depth must be varied to get the best fit between calculations and measurements. The uncertainties exist largely because accurate optical isophotes, from which the distribution of brightness temperature over the nebula could be calculated, are not available.

With the Very Large Array (VLA) radio interferometer it is possible to achieve angular resolution as small as 0.″05 at high frequency (λ = 1.3 cm), ideal for these measurements. Mean temperatures determined in this way for planetary nebulae include T = 8,300° K for NGC 6543, T = 8,500° K for IC 418, and T = 14,000° K for NGC 7027. The high-resolution radio images, obtained at these high frequencies, provide excellent information on the spatial structure of planetary nebulae, as discussed further in Chapter 9.

5.5 Electron Densities from Emission Lines

The average electron density in a nebula may be measured by observing the effects of collisional de-excitation. This can be done by comparing the intensities of two lines of the same ion emitted by different levels with nearly the same excitation energy so that the relative excitation rates at the two levels depend only on the ratio of collision strengths. If the two levels have different radiative transition probabilities or different collisional de-excitation rates, the relative populations of the two levels will depend on the density, and the ratio of intensities of the lines they emit will likewise depend on the density. The best examples of lines that may be used to measure the electron density are [O II] $\lambda3729/\lambda3726$, and [S II] $\lambda6716/\lambda6731$, with energy-level diagrams shown in Figure 5.2.

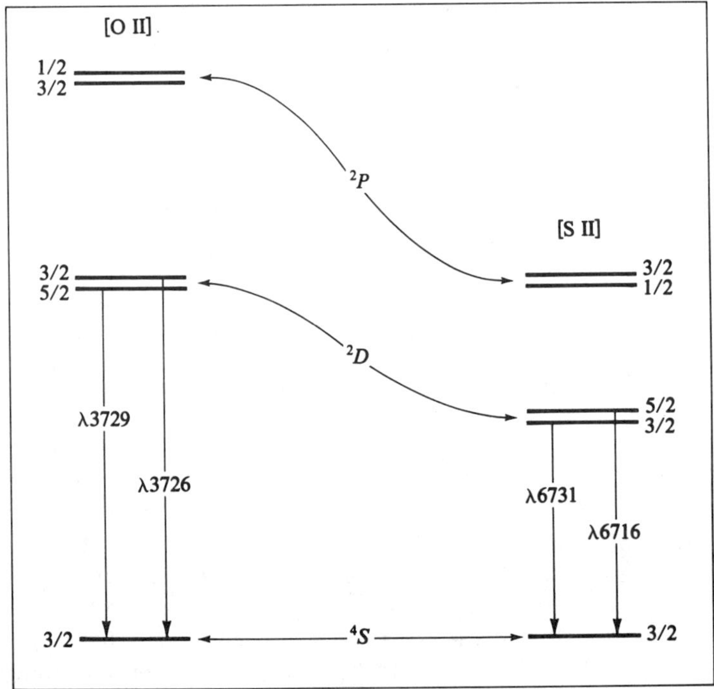

FIGURE 5.2
Energy-level diagrams of the $2p^3$ ground configuration of [O II] and $3p^3$ ground configuration of [S II].

The relative populations of the various levels and the resulting relative line-emission coefficients may be found by setting up the equilibrium equations for the populations of each level as described in Section 3.5. However, direct physical reasoning easily shows what effects are involved.

Consider the example of [O II] in the low-density limit $N_e \to 0$, in which every collisional excitation is followed by emission of a photon. Since the relative excitation rates of the $^2D_{5/2}$ and $^2D_{3/2}$ levels are proportional to their statistical weights (see equation 3.22), the ratio of strengths of the two lines is simply $j_{\lambda 3729}/j_{\lambda 3726} = 1.5$. On the other hand, in the high-density limit, $N_e \to \infty$, collisional excitations and de-excitations dominate and set up a Boltzmann population ratio. Thus the relative populations of the two levels $^2D_{5/2}$ and $^2D_{3/2}$ are in the ratio of their statistical weights, and therefore the relative strengths of the two lines are in the ratio

$$\frac{j_{\lambda 3729}}{j_{\lambda 3726}} = \frac{N_{^2D_{5/2}}}{N_{^2D_{3/2}}} \frac{A_{\lambda 3729}}{A_{\lambda 3726}} = \frac{3}{2} \frac{3.6 \times 10^{-5}}{1.8 \times 10^{-4}} = 0.30.$$

The transition between the high- and low-density limits occurs in the neighborhood of the critical densities (see equation 3.31), which are $N_c \approx 3 \times 10^3$ cm^{-3} for $^2D_{5/2}$ and $N_c \approx 1.6 \times 10^4$ cm^{-3} for $^2D_{3/2}$. The full solution of the equilibrium equations, which also takes into account all transitions, including excitation to the ^2P levels with subsequent cascading downward, gives the detailed variation of intensity ratio with the electron density that is plotted in Figure 5.3. Note from the collisional transition rates that the main dependence of this ratio is on $N_e/T^{1/2}$. There is also a very slight temperature dependence (as a consequence of the cascading from ^2P) that cannot be seen on this graph.

An exactly similar treatment holds for [S II]; the calculated ratio $j_{\lambda 6716}/j_{\lambda 6731}$ is also shown in Figure 5.3. Other pairs of lines from ions with the same type of structure, which may also be used for measuring electron densities, are [N I], [Cl III], [Ar IV], and [K V] in the optical region, as well as [Ne IV] $\lambda\lambda$2422, 2424 in the satellite ultraviolet.

From the observational point of view, it is unfortunate that the [O II] $\lambda\lambda$3726, 3729 are so close in wavelength; a spectrograph, spectrometer, or interferometer with good wavelength resolution must be used to separate the lines. However, a fair amount of data is available on both H II regions and planetary nebulae.

Some results in H II regions, all derived from [O II] intensity ratios, are listed in Table 5.6. Since only the brightest H II regions can be observed spectrophotometrically, the data of Table 5.6 are biased toward relatively high electron densities. Nevertheless, it can be seen that typical densities in several H II regions are of order $N_e \approx 10^2$ cm^{-3}. (NGC 1976 M is a position in the outer part of the Orion Nebula.) Several H II regions have dense condensations in them, though – for instance, the central part of the Orion Nebula, near the Trapezium (NGC 1976 A), with $\lambda 3729/\lambda 3726 = 0.50$, corresponding to $N_e \approx 3.0 \times 10^3$ cm^{-3}. In fact, observations of the [O II] ratio at many points in NGC 1976, of which only A and M are listed in Table 5.6, show that the mean electron density is highest near the center of the

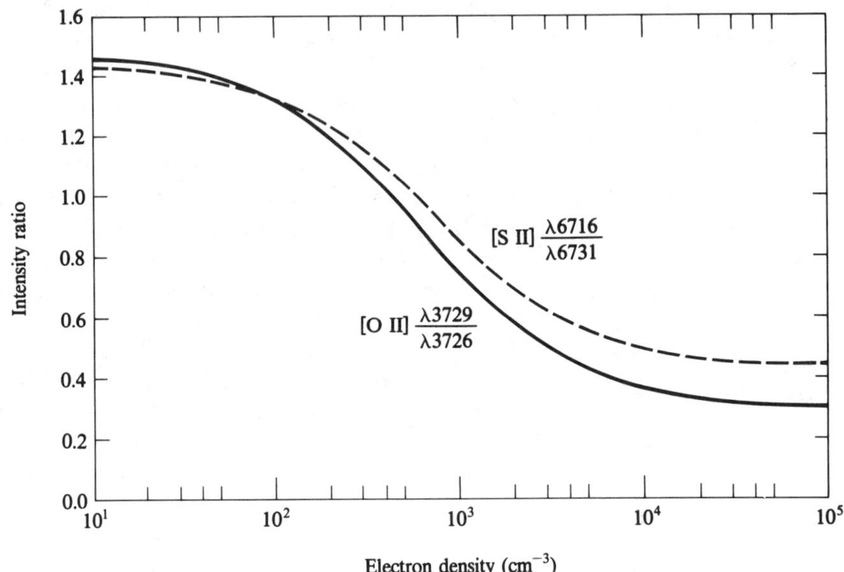

FIGURE 5.3
Calculated variation of [O II] (*solid line*) and [S II] (*dashed line*) intensity ratios as function of N_e at $T = 10,000°$ K. At other temperatures the plotted curves are very nearly correct if the horizontal scale is taken to be $N_e(10^4/T)^{1/2}$.

TABLE 5.6
Electron densities in H II *regions*

Object	$\frac{I(\lambda 3729)}{I(\lambda 3726)}$	$N_e(cm^{-3})$
NGC 1976 A	0.50	3.0×10^3
NGC 1976 M	1.26	1.4×10^2
M 8 Hourglass	0.65	1.5×10^3
M 8 outer	1.26	1.5×10^2
NGC 281	1.37	7×10
NGC 7000	1.38	6×10

nebula and decreases relatively smoothly outward in all directions. The three-dimensional structure of the nebula thus presumably must have a density maximum, and the intensity ratio observed at the center results from emission all along the line of sight, so the actual central density must be higher than 4.5 $\times\ 10^3\ \text{cm}^{-3}$. A model can be constructed that approximately reproduces all the measured [O II] ratios in NGC 1976; this model has $N_e \approx 1.7 \times 10^4\ \text{cm}^{-3}$ at the center and decreases to $N_e \approx 10^2\ \text{cm}^{-3}$ in the outer parts. Furthermore, measurements of the [S II] ratio at many points in the inner bright core of NGC 1976 (about 8′ diameter) show good agreement between the electron densities determined from the [S II] lines and the [O II] lines.

Similarly, in M 8 the [O II] measurements show that the density falls off outward from the Hourglass, a small dense condensation in which $N_e \approx 2 \times 10^3\ \text{cm}^{-3}$. Comparable published [S II] measurements do not exist for M 8, or indeed for any H II region except NGC 1976, which is unfortunate, especially since $\lambda\lambda6716$, 6731 are relatively bright and easy to observe in many of these objects.

Some information on electron densities in planetary nebulae derived from [O II] and [S II] is listed in Table 5.7, in which the densities derived from these two ions are mostly within a factor of two of each other. In most planetaries the degree of ionization is high, and most of the [O II] and [S II] lines that arise in fairly low stages of ionization are emitted either in the outermost parts of the nebula or in the densest parts, where recombination depresses the ionization the most. Thus the densities derived from these ions may not be representative of the entire nebula. The higher stages of ionization, [Ar IV], [K V], and so on, are more representative, but their lines are weaker and more difficult to measure. An example is NGC 7662, for which the [Ar IV] lines give $N_e = 1.0 \times 10^4\ \text{cm}^{-3}$, while the [Ne IV] pair at $\lambda\lambda2422$, 2424 give $9.6 \times 10^3\ \text{cm}^{-3}$, both at an assumed $T = 10^4\ ^\circ$ K.

The electron densities derived from these line ratios may be used in equations (5.4) and (5.5) to correct the observations of the temperature-sensitive lines of [O III] and [N II] for the slight collisional de-excitation effect; and actually these corrections have already been taken into account in Tables 5.1 and 5.2. Though the electron density derived from [O II] line measurements may not exactly apply in the [O III] emitting region, the density effect is small enough that an approximate correction should be satisfactory.

In the densest planetaries known, collisional de-excitation of [O III] $^1D_2$ is strong enough that the $(\lambda4959 + \lambda5007)/\lambda4363$ ratio is significantly affected. The best example is IC 4997, with $\lambda3729/\lambda3726 = 0.34$, corresponding to N_e poorly determined in the high-density limit but certainly greater than

10^5 cm^{-3}. The measured ($\lambda 4959 + \lambda 5007)/\lambda 4363 \approx 22$, which would correspond to $T \approx 4 \times 10^4$ ° K if there were no collisional de-excitation. This temperature is far too large to be understood from the known heating and cooling mechanisms, and the ratio is undoubtedly strongly affected by collisional de-excitation. If it is assumed that $T \approx 12{,}000°$ K, the [O III] ratio gives $N_e \approx 10^6$ cm^{-3}; higher assumed temperatures correspond to somewhat lower electron densities, and vice versa.

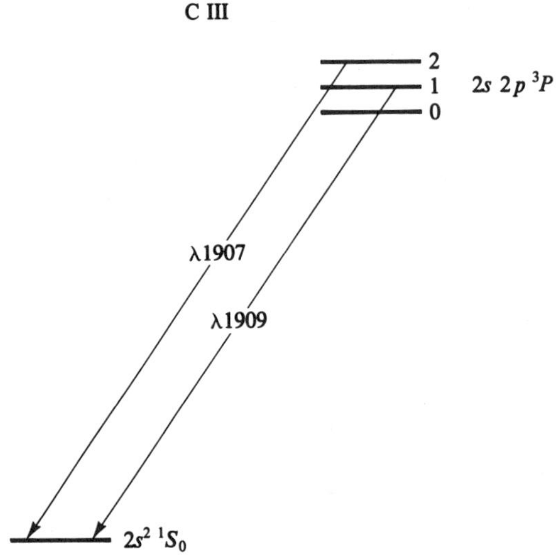

FIGURE 5.4
Energy-level diagram of lowest terms of C III $2s^2$ and $2s2p$ configurations, and resulting C III] and [C III] emission lines. The splitting within the $2s2p\ ^3P$ term is exaggerated in this diagram.

The relatively high mean electron density in IC 4997 is confirmed by measurements in the satellite ultraviolet spectral region. The ion C III, whose energy-level diagram is shown in Figure 5.4, has two observed emission lines, [C III] $3s^2\ ^1S_0 - 3s\ 3p\ ^3P_2$ $\lambda 1907$, a highly forbidden magnetic quadrupole transition, and C III] $3s^2\ ^1S_0 - 3s3p\ ^3P_1$ $\lambda 1909$, an intercombination or "semi-forbidden" electric-dipole transition. The two lines therefore have a ratio of intensities fixed in the low-density limit by collision strengths alone. It is approximately $I(^1S_0 - {}^3P_2)/I(^1S_0 - {}^3P_1) \approx 5/3$ by equation (3.22), but more nearly exactly $= 1.53$, because the downward radiative transition $^1S_0 - {}^3P_0$ is completely forbidden; hence collisional excitation of $^3P_0$ is always followed by a further collisional process, either de-excitation to $^1S_0$, or excitation to $^3P_1$ or $^3P_2$. In the high-density limit, on the other hand, the ratio is fixed by the ratio of statistical weights and transition probabilities,

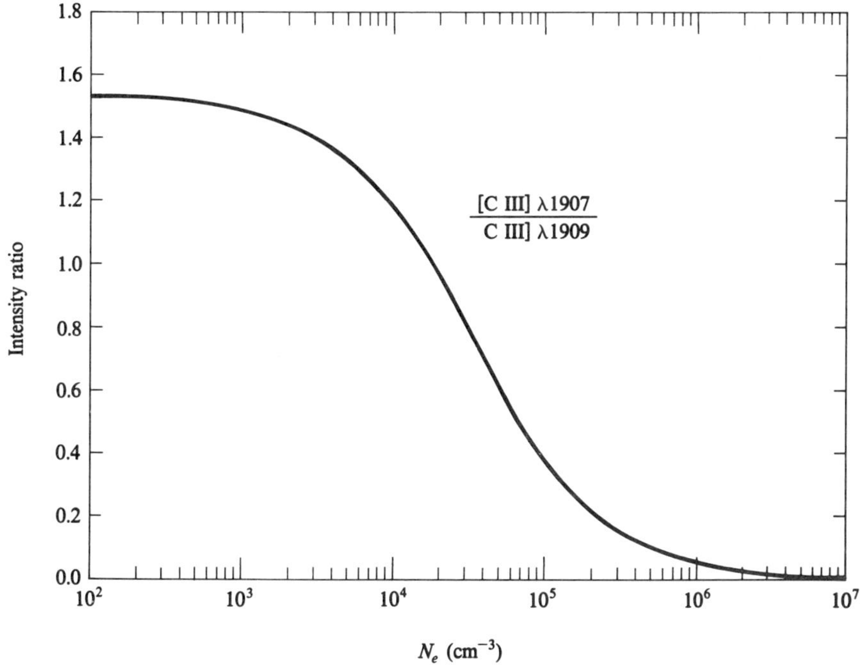

FIGURE 5.5
Calculated variation of [C III] $\lambda1907$/C III] $\lambda1909$ intensity ratio as function of electron density N_e at $T = 10,000^\circ$ K.

and $\approx 9 \times 10^{-5}$. The detailed form of its variation with electron density is shown in Figure 5.5.

To date the $\lambda1907/\lambda1909$ ratio has been measured for only a few bright planetary nebulae. They mostly have ratios in the range 0.8 to 1.4, corresponding to $N_e = 3 \times 10^4$ to 3×10^3 cm^{-3}. The large, low-density nebulae like NGC 650/1, 3587, 6720, 6853, and 7293 are too faint to have yet been measured in these lines. But the small, high-surface-brightness planetary IC 4997 has $\lambda1907/\lambda1909 \approx 0.03$, showing that its mean electron density $N_e \approx 10^6$ cm^{-3}, agreeing quite well with the [O II] results.

5.6 Electron Temperatures and Densities from Emission Lines

The development in recent years of sensitive infrared detectors, and of airborne telescopes that can be flown to altitudes above most of the infrared absorption in the Earth's atmosphere, have made it possible to measure "fine-structure" lines such as [O III] $^3P_0 - {}^3P_1$ $\lambda88$ μ and $^3P_1 - {}^3P_2$ $\lambda52$ μ (see

TABLE 5.7
Electron densities in planetary nebulae

Nebula	[O II] $\frac{\lambda 3729}{\lambda 3726}$	[O II] N_e [a] (cm^{-3})	[S II] $\frac{\lambda 6716}{\lambda 6731}$	[S II] N_e [a] (cm^{-3})
NGC 40	0.78	1.1×10^3	0.69	2.1×10^3
NGC 650/1	1.23	2.1×10^2	1.08	4.0×10^2
NGC 2392	0.78	1.1×10^3	0.88	9.1×10^2
NGC 2440	0.64	1.9×10^3	0.62	3.2×10^3
NGC 3242	0.62	2.2×10^3	0.64	2.8×10^3
NGC 3587	1.30	1.4×10^2	1.25	1.8×10^2
NGC 6210	0.47	5.8×10^3	0.66	2.5×10^3
NGC 6543	0.44	7.9×10^3	0.54	5.9×10^3
NGC 6572	0.38	2.1×10^4	0.51	8.9×10^3
NGC 6720	1.04	4.7×10^2	1.14	3.2×10^2
NGC 6803	0.57	2.8×10^3	–	–
NGC 6853	1.16	2.9×10^2	–	–
NGC 7009	0.50	4.6×10^3	0.61	3.3×10^3
NGC 7027	0.48	5.2×10^3	0.59	4.0×10^3
NGC 7293	1.32	1.3×10^2	1.28	1.6×10^2
NGC 7662	0.56	3.0×10^3	0.64	2.8×10^3
IC 418	0.37	3.2×10^5	0.49	9.5×10^3
IC 2149	0.56	3.0×10^3	0.57	4.6×10^3
IC 4593	0.63	2.0×10^3	–	–
IC 4997	0.34	1.0×10^6	0.45	1.0×10^5

[a] N_e given for assumed $T = 10^4$ °K; for any other T divide listed value by $(T/10^4)^{1/2}$.

Figure 3.1 and Table 3.8). These far-infrared lines have much smaller excitation potentials than the optical lines such as $^3P_2 - {}^1D_2$ $\lambda 5007$. Thus a ratio like $j_{\lambda 5007}/j_{\lambda 52\mu}$ depends strongly on temperature but, since the $^3P_2$ level has a much-lower critical electron density than $^1D_2$ does, the ratio depends on density also. On the other hand, the ratio $j_{\lambda 52\mu}/j_{\lambda 88\mu}$ hardly depends on temperature at all (since both excitation potentials are so low in comparison with typical nebular temperatures), but does depend strongly on density (since the two upper levels have different critical densities). Hence by measuring two [O III] ratios, we can determine the average values of the two parameters

T, N_e. Figure 5.6 shows calculated curves of the values of the two [O III] intensity ratios for various values of temperature and electron density. Observed values of the line ratios are entered on the diagram for several planetary nebulae, from which the average T and N_e can be immediately read off. They agree reasonably well with values determined independently from optical lines alone. Including the infrared lines makes determinations of temperature and density possible for many more ions than the optical lines. The chief difficulty, given an airborne telescope and spectrograph, is to be certain that exactly the same area is measured in both spectral regions. For this reason the measurements available to date are chiefly for entire planetary nebulae.

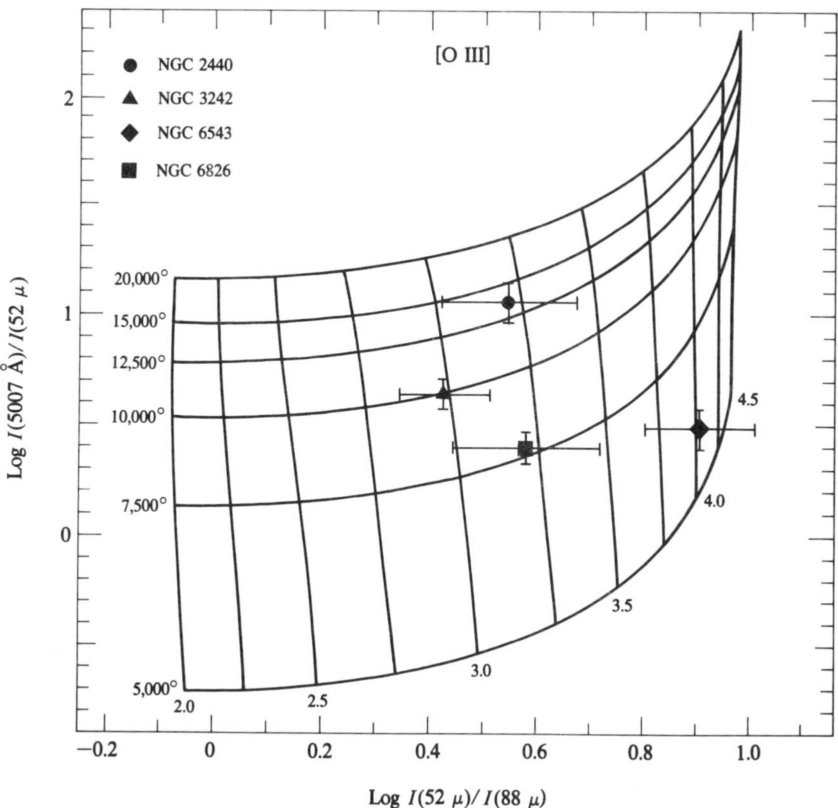

FIGURE 5.6
Calculated variation of [O III] forbidden-line relative-intensity ratios as functions of $T(5000^\circ$ to $20,000^\circ$ K) and N_e. Observed planetary-nebula ratios plotted with indication of probable errors.

5.7 Electron Temperatures and Densities from Radio Recombination Lines

Information can be obtained on the temperature and density in gaseous nebulae from measurements of the radio recombination lines. Practically all the observational results refer to H II regions, which have considerably larger fluxes than planetary nebulae and hence can be much more readily observed with radio telescopes. The populations of the high levels of H depend on T and N_e, as explained in Section 4.4, and the strengths of the lines emitted by these levels relative to the continuum and to one another therefore depend on N_e, T, and the optical depth, which is conventionally expressed in terms of the emission measure E defined in equation (4.32). Comparison of measured and calculated relative strengths thus can be used to calculate mean values of N_e, T, and E.

To calculate the expected strengths, we must solve the equation of radiative transfer, since the maser effect is often important, as was shown in Section 4.4. Furthermore, the continuum radiation is not weak in comparison with the line radiation and therefore must be included in the equation of transfer. The observations are generally reported in terms of brightness temperature. We will use T_C for the measured temperature in the continuum near the line and $T_L + T_C$ for the measured brightness temperature at the peak of the line (see Figure 5.7), so that T_L is the excess brightness temperature due to the line.

We will consider an idealized homogeneous isothermal nebula; the optical depth in the continuum, which we will write τ_C, is given by equation (4.32). The optical depth in the center of the line is

$$\tau_{cL} = \tau_L + \tau_C, \tag{5.10}$$

where τ_L is the contribution from the line alone,

$$d\tau_L = \kappa_L \, ds,$$

and

$$\kappa_L = N_n k_{0L}. \tag{5.11}$$

Here we consider an $n, \Delta n$ line between an upper level $m = n + \Delta n$ and a lower level n; the central line-absorption cross section, corrected for stimulated emission as in equation (4.43), is

$$\begin{aligned} \kappa_{0L} &= \frac{\omega_m}{\omega_n} \frac{\lambda^2}{8\pi^{3/2}\Delta\nu_D} A_{m,n} \left(1 - \frac{b_m}{b_n} e^{-h\nu/kT}\right) \\ &= \frac{\omega_m}{\omega_n} \frac{\lambda^2 (\ln 2)^{1/2}}{4\pi^{3/2}\Delta\nu_L} A_{m,n} \left(1 - \frac{b_m}{b_n} e^{-h\nu/kT}\right). \end{aligned} \tag{5.12}$$

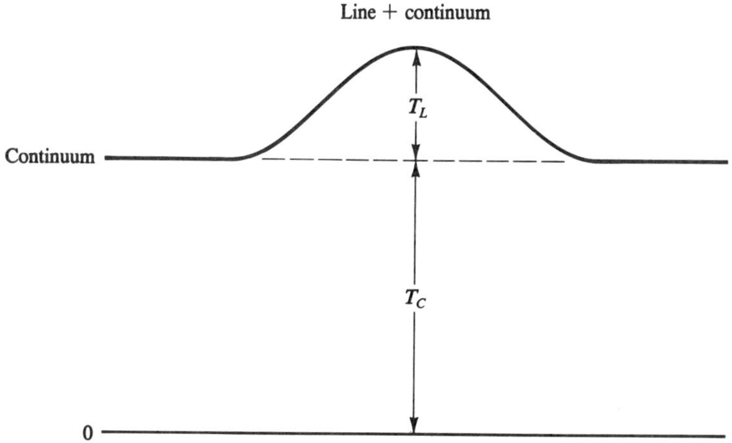

FIGURE 5.7
A radio-frequency line superimposed on the radio-frequency continuum, showing the brightness temperatures at the center of the line and in the nearby continuum; T_L and T_C, respectively.

In this equation a Doppler profile has been assumed, with $\Delta\nu_D$, the half width at e^{-1} of maximum intensity, and $\Delta\nu_L$ the full width at half-maximum intensity, the conventional quantity used in radio astronomy. Combining (5.12) with

$$N_m = b_n n^2 \left(\frac{h^2}{2\pi mkT}\right)^{3/2} e^{X_n/kT} N_p N_e, \qquad (5.13)$$

and using exp $(X_n/kT) \approx 1$ to a good approximation for all observed radio-frequency recombination lines, expressing $A_{m,n}$ in terms of the corresponding f-value f_{nm}, and expanding the stimulated-emission correction as in equation (4.43) gives, for the special case of local thermodynamic equilibrium ($b_m = b_n = 1$, which we will denote by an asterisk throughout this section),

$$\begin{aligned}\tau_L^* &= 1.53 \times 10^{-9} \frac{n^2 f_{nm} \nu}{\Delta\nu_L T^{2.5}} E_p \\ &= 1.01 \times 10^7 \frac{\Delta n f_{nm}}{n \Delta\nu_L T^{2.5}} E_p. \qquad (5.14)\end{aligned}$$

The proton-emission measure

$$E_p = \int N_p N_e ds \qquad (5.15)$$

is expressed in cm^{-6} pc in both forms of equation (5.14), and

$$\nu = \frac{\nu_0}{n^2} - \frac{\nu_0}{m^2} = \frac{2\nu_0 \Delta n}{n^3}.$$

In the true nebular case,

$$\begin{aligned} \tau_L &= \tau_L^* b_n \frac{(1 - (b_m/b_n)e^{-h\nu/kT})}{(1 - e^{-h\nu/kT})} \\ &= \tau_L^* b_m \left(1 - \frac{kT}{h\nu}\frac{d \ln b_n}{dn}\Delta n\right), \end{aligned} \qquad (5.16)$$

by the power-series expansion, and the continuum optical depth is the same as in thermodynamic equilibrium, because the free electrons have a Maxwellian distribution.

Now we will use these expressions and the formal solution of the equation of transfer to calculate the ratio of brightness temperatures $r = T_L/T_C$ in the special case of thermodynamic equilibrium,

$$\begin{aligned} r^* &= \frac{T_L + T_C}{T_C} - 1 = \frac{T(1 - e^{-\tau_{CL}})}{T(1 - e^{-\tau_C})} - 1 \\ &= \frac{1 - e^{-(\tau_L^* + \tau_C)}}{1 - e^{-\tau_C}} - 1. \end{aligned} \qquad (5.17)$$

If $\tau_L^* \ll 1$ (this is a good approximation in all lines observed to date), and in addition, $\tau_C \ll 1$ (this is generally but not always a good approximation),

$$r^* = \frac{\tau_L^*}{\tau_C}.$$

Under the assumption of local thermodynamic equilibrium, the observed ratio of brightness temperatures in line and continuum thus gives (in the limit of small optical depth) the ratio of optical depths, which, in turn, from equations (4.32) and (5.14), measures T. Note that the continuum-emission measure E involves all positive ions, but the proton-emission measure E_p involves only H^+ ions; so their ratio depends weakly on the helium abundance, which, however, is reasonably well-known. This scheme was used in the early days of radio recombination-line observations to determine the temperatures in H II regions, but it is not correct, because in a nebula the deviations from thermodynamic equilibrium are significant, as is shown by the fact that measurements of different lines in the same nebula, when reduced in this way, give different temperatures.

To calculate the brightness-temperature ratio $r = T_L/T_C$ in the true nebular case, we note that the brightness temperature in the continuum is still given by

$$T_C = T(1 - e^{-\tau_C}).$$

However, both the line-emission and line-absorption coefficients differ from their thermodynamic equilibrium values. The line-emission coefficient depends on the population in the upper level; so

$$j_L = j_L^* b_m,$$

while the line-absorption coefficient, as shown in equation (5.16), is

$$\kappa_L = \kappa_L^* b_m \beta,$$

where

$$\beta = 1 - \frac{kT}{h\nu}\frac{d \ln b_n}{dn}\Delta n. \tag{5.18}$$

The equation of transfer in intensity units is

$$\frac{dI_\nu}{d\tau_{cL}} = -I_\nu + \frac{j_L + j_C}{\kappa_L + \kappa_C} = -I_\nu + S_\nu, \tag{5.19}$$

where

$$\begin{aligned} S_\nu &= \frac{j_L^* b_m + j_C}{\kappa_L^* b_m \beta + \kappa_C} \\ &= \frac{\kappa_L^* b_m + \kappa_C}{\kappa_L^* b_m \beta + \kappa_C} B_\nu(T) \end{aligned} \tag{5.20}$$

from Kirchoff's law, so that the brightness temperature at the center of the line is

$$T_L + T_C = \left[\frac{\kappa_L^* b_m + \kappa_C}{\kappa_L^* b_m \beta + \kappa_C}\right] T[1 - e^{-(b_m \beta \tau_L^* + \tau_C)}]. \tag{5.21}$$

Hence, finally,

$$r = \frac{T_L}{T_C} = \left[\frac{\kappa_L^* b_m + \kappa_C}{\kappa_L^* b_m \beta + \kappa_C}\right] \left[\frac{1 - e^{-(b_m \beta \tau_L^* + \tau_C)}}{1 - e^{-\tau_C}}\right] - 1, \tag{5.22}$$

which depends only on one optical depth, say, τ_C, on the ratio of optical depths, $\tau_L^*/\tau_C = \kappa_L^*/\kappa_C$, given by equations (4.42) and (5.14), and on the b_n factors, which, in turn, depend on N_e and T.

Thus, when the deviations from thermodynamic equilibrium are taken into account, r depends not only on T, but also on N_e and τ_C (or equivalently, E). Therefore observations of several different lines in the same nebula are necessary to determine T, N_e, and E from measurements of radio-frequency recombination lines. The procedure is to make the best possible match between all measured lines in a given nebula and the theoretical calculations for a given T, N_e and E, using the $b_n(T, N_e)$ calculations described in Chapter 4. There are observational problems connected with the fact that the radio recombination lines, coming as they do from levels with large n and thus large atomic radii, suffer significant impact broadening even at the low densities of nebulae. This makes the wings of the line difficult to define observationally except with very good signal-to-noise ratio data. Otherwise significant contributions from the wings may easily be overlooked. Another problem is that measurements are made at different frequencies and with different radio telescopes, so the antenna beam patterns are not identical for all lines. Model calculations show that over a wide range of nebular conditions, measurements of lines with $\Delta n = 1$ at frequencies near 10 GHz (such as 109α at 5.009 GHz) are only slightly affected by maser effects and by deviations from thermodynamic equilibrium, and therefore are especially suitable for determining nebular temperatures. Data on some of the best observed nebulae are collected in Table 5.8. Note that the last seven nebulae in this table are all large distant H II regions, observed as bright sources in the radio region, but so strongly affected by interstellar extinction that they are completely unobservable in the optical region. W 43 and the four G nebulae are all significantly closer to the center of the Galaxy than the other objects, and NGC 1976 is the most distant from it. The measured temperatures show a clear increase with increasing distance from the center, which is consistent with the decrease in heavy-element abundance outward from the center that we shall discuss in Chapter 8.

Average electron densities can also be found from the radio recombination-line measurements. The best procedure is to compare lines of two different frequencies, such as 85α and 109α, or 66α and 85α. Naturally it is important to match the antenna beamwidths as closely as possible. Very high n lines cannot be used, because impact (Stark) broadening becomes important, and makes the wings difficult to define and measure accurately. The mean electron density derived in this way for the Orion Nebula, NGC 1976, $N_e = 2.4 \times 10^3$ cm^{-3}, is comparable with an emission-weighted average of the [O II] determinations. Likewise, radio recombination-line measurements for a few of the highest surface brightness planetary nebulae give mean electron densities ranging from $N_e = 8.5 \times 10^3$ cm^{-3} in NGC 6543 to 1.6×10^5 cm^{-3} in NGC 7027.

TABLE 5.8
Temperatures of nebulae from radio recombination-line measurements

Nebula	T(° K)
NGC 1976	9,400
M 16	6,900
M 17	7,300
W 43	6,700
W 49	8,200
W 51	7,500
G 18.9−0.5	5,600
G 24.5+0.5	6,200
G 23.4−0.2	5,900
G 24.5−0.2	5,700

5.8 Ionizing Radiation from Stars

Observations of gaseous nebulae may be used to find the numbers of ionizing photons emitted by a star, and thus to determine a long base-line color index for it between the Lyman ultraviolet region and an ordinary optical region, from which the effective temperature of the star can be derived. The idea of the method is quite straightforward. If the nebula around the star is optically thick in the Lyman continuum, it will absorb all the ionizing photons emitted by the star. Thus the total number of ionizations in the nebula per unit time is just equal to the total number of ionizing photons emitted per unit time, and since the nebula is in equilibrium, these ionizations are just balanced by the total number of recaptures per unit time; so

$$\int_{\nu_0}^{\infty} \frac{L_\nu}{h\nu} d\nu = Q(\mathrm{H}^0) = \int_0^{r_1} N_p N_e \alpha_B(\mathrm{H}^0, T)\, dV,$$

where L_ν is the luminosity of the star per unit frequency interval. Note that by using the recombination coefficient α_B, we have included the ionization processes due to diffuse ionizing photons emitted in recaptures within the nebula; see equation (2.19). The luminosity of the entire nebula in a particular emission line, say Hβ, also depends on recombinations throughout its volume:

$$L(H\beta) = \int_0^{r_1} 4\pi j_{\mathrm{H}\beta}\, dV$$

$$= h\nu_{\mathrm{H}\beta} \int_0^{r_1} N_p N_e \alpha_{\mathrm{H}\beta}^{eff}(\mathrm{H}^0, T) dV.$$

Thus dividing

$$\frac{\frac{L(H\beta)}{h\nu_{H\beta}}}{\int_{\nu_0}^{\infty} \frac{L_\nu}{h\nu} d\nu} = \frac{\int_0^{r_1} N_p N_e \alpha_{H\beta}^{eff}(H^0, T) dV}{\int_0^{r_1} N_p N_e \alpha_B(H^0, T) dV}$$

$$\approx \frac{\alpha_{H\beta}^{eff}(H^0, T)}{\alpha_B(H^0, T)} \tag{5.23}$$

gives the result that the number of photons emitted by the nebula in a specific recombination line such as Hβ is directly proportional to the number of photons emitted by the star with $\nu \geq \nu_0$. Note that the proportionality between the number of ionizing photons absorbed and the number of line photons emitted does not depend on any assumption about constant density, and that replacing the ratio of integrals by the ratio of recombination coefficients is a good approximation because $\alpha_{\mathrm{H}\beta}^{eff}/\alpha_B$ depends only weakly on T. Note further that any other emission line could have been used instead of Hβ. Alternatively, the radio-frequency continuum emission at any frequency at which the nebula is optically thin could have been used; however, then the ratio of nebular photons emitted to ionizing photons would involve the ratio of the number of protons to the total number of positive ions, which depends weakly on He abundance. The number of ionizing photons may be compared with the luminosity of the star at a particular frequency ν_f in the observable region,

$$\frac{L_{\nu_f}}{\int_{\nu_0}^{\infty} \frac{L_\nu}{h\nu} d\nu} = \frac{L_{\nu_f}}{\frac{L(H\beta)}{h\nu_{H\beta}}} \frac{\frac{L(H\beta)}{h\nu_{H\beta}}}{\int_{\nu_0}^{\infty} \frac{L_\nu}{h\nu} d\nu}$$

$$= h\nu_{H\beta} \frac{\alpha_{H\beta}^{eff}(H^0, T)}{\alpha_B(H^0, T)} \frac{\pi F_{\nu_f}}{\pi F_{H\beta}}, \tag{5.24}$$

where the ratio of luminosities has been expressed in terms of the ratio of the observed fluxes at the earth from the star at ν_f and from the nebula at Hβ. This ratio is independent not only of the distance, but also of the interstellar extinction if the nebula and the star are observed at the same effective wavelength by choosing $\nu_f = \nu_{\mathrm{H}\beta}$.

It is often more convenient to make the stellar measurements with a fairly wide filter of the type ordinarily used for photometry (for instance, the V filter of the UBV system), and we can then write a similar equation in terms of

$$L_V = \int_0^\infty s_\nu(V) L_\nu \, d\nu$$

and

$$\pi F_V = \int_0^\infty s_\nu(V) \, \pi F_\nu d\nu,$$

where $s_\nu(V)$ is the sensitivity function of the telescope-filter-photocell combination, known from independent measurements. For measurements of stars in bright nebulae, it is advantageous to use a narrower-band filter that isolates a region in the continuum between the brightest nebular emission lines, to minimize the correction for the "sky" background. In principle, any observable frequency ν_f can be used, and likewise any observable recombination line, for instance Hα, might be measured instead of Hβ. The method of using the nebular observations to measure the stellar ultraviolet radiation was first proposed by Zanstra, who assumed that the flux from a star could be approximately represented by the Planck function $B_\nu(T_*)$, so that

$$\frac{L_{\nu_f}}{\int_{\nu_0}^\infty \frac{L_\nu}{h\nu} d\nu} = \frac{B_{\nu_f}(T_*)}{\int_{\nu_0}^\infty \frac{B_\nu(T_*)}{h\nu} d\nu}$$

and the measurements thus determine T_*, the so-called Zanstra temperature of a star that ionizes a nebula. However, modern theoretical work on stellar atmospheres shows that there are important deviations between the emergent fluxes from stars and Planck functions, particularly in the regions where there are large changes in opacity with frequency, such as at the Lyman limit itself and at the various limits due to other ions at shorter wavelengths; so it is not a very good approximation to set $F_\nu = B_\nu(T_*)$. As illustrations, Figures 5.8 and 5.9 show calculated models for stars with $T_* = 40,000^\circ$ K, log $g = 4$, approximately an O6 main-sequence star, and $T_* = 100{,}000^\circ$ K, log $g = 6$, a fairly typical planetary-nebula star. Thus the ratios

$$\frac{L_{\nu_f}}{\int_{\nu_0}^\infty \frac{L_\nu}{h\nu} d\nu} = y(T_*) = \frac{\pi F_\nu(T_*, g)}{\int_{\nu_0}^\infty \frac{\pi F_\nu(T_*, g)}{h\nu} d\nu} \tag{5.25}$$

should be determined from the best available sequences of model stellar atmospheres, and it can be seen that there is a one-parameter relationship $y = y(T_*)$ for a fixed value of g or along a fixed line in the T_*, log g plane.

We will first use these relationships to examine the effective temperatures of Population I O stars in H II regions, and then generalize these equations and use them to describe the higher-temperature planetary-nebula central stars.

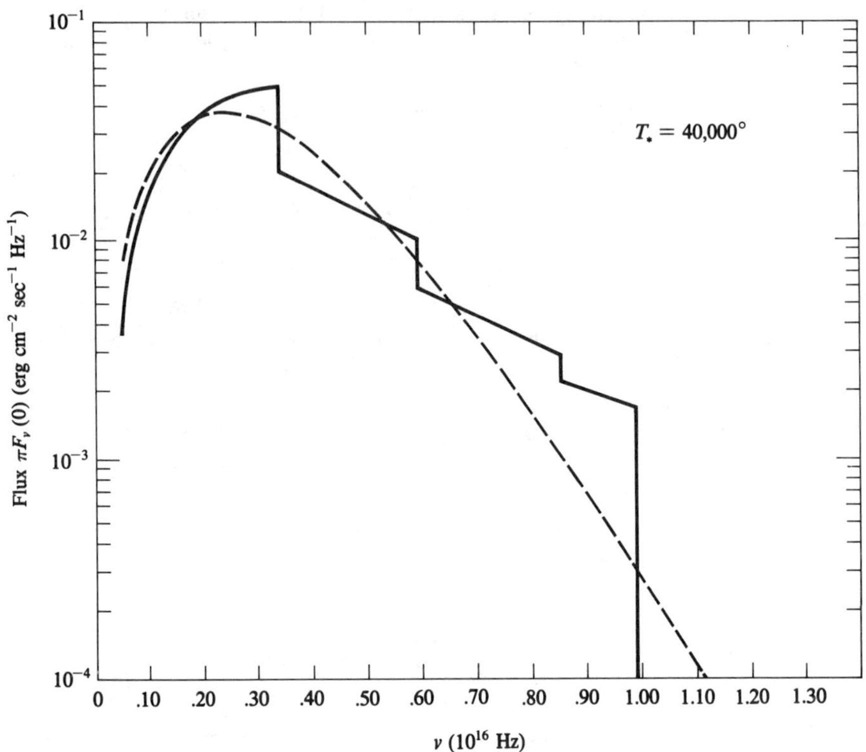

FIGURE 5.8
Calculated flux from a model O6 star with $T_* = 40{,}000°$ K, log $g = 4$ (*solid line*), compared with blackbody flux for same temperature (*dashed line*).

Many H II regions are observed, but a fairly large fraction of them contain several O stars that contribute to the ionization and thus complicate the determination of the effective temperature of individual stars. The best nebulae for measurement are clearly those with only a single involved hot star. Furthermore, the basic assumption of the method is that the nebula completely absorbs the stellar ionizing radiation and is a true Strömgren sphere (radiation-bounded rather than density-bounded). It is difficult to be certain that this assumption is fulfilled in any specific nebula, though well-defined ionization fronts at the outer edge of a nebula suggest that it is and thus indicate that it is a good candidate to be measured. However, as we will see in Chapter 7, absorption of ionizing photons by dust can still cause serious errors in the results.

Relatively few measurements of the total fluxes of nebulae in Hα or Hβ are available, but some photoelectric measurements have been made with narrow filter-plate combinations, particularly near Hα. These measurements have the defect that they include the [N II] $\lambda\lambda$6548, 6583 lines within the filter band pass, but we can correct for the contribution of these lines, at least statistically. There are considerably more nebulae for which radio-frequency continuum measurements are available, although they have the defect that interstellar extinction enters the ratio of optical stellar flux to radio-frequency nebular flux in full force. From these measurements of about 25 nebulae, the best available model stellar atmospheres were used to derive the temperature scales for the main-sequence stars shown in Table 5.9.

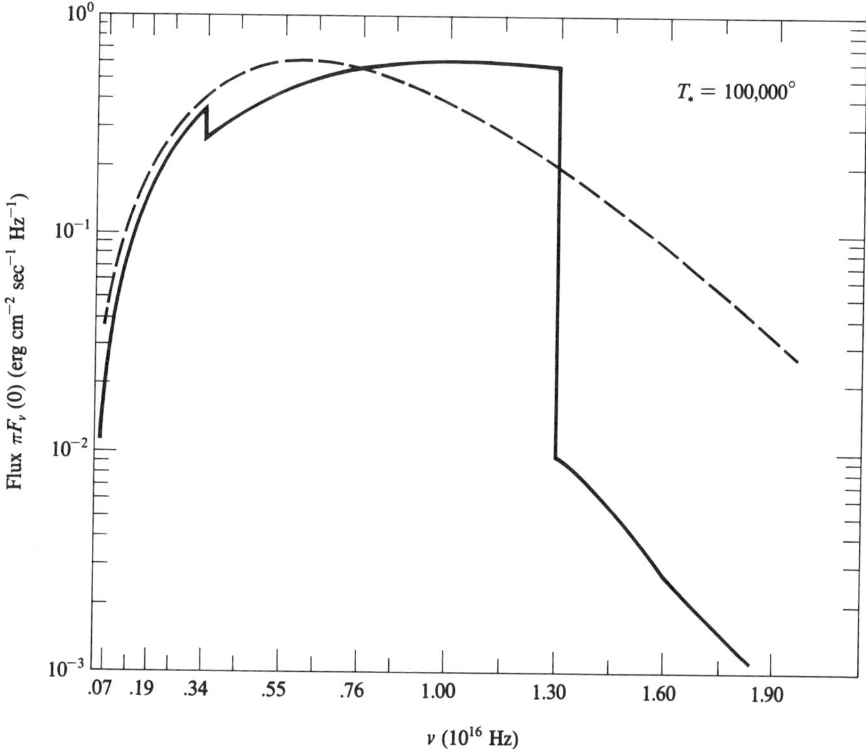

FIGURE 5.9
Calculated flux from a model planetary-nebulas central star with $T_* = 100{,}000°$ K, log $g = 6$ (*solid line*), compared with blackbody flux for same temperature (*dashed line*).

TABLE 5.9
Temperatures of hot main-sequence stars derived from Zanstra method

Spectral type MK	$(B-V)_0$	T_*(° K)
O5	−0.32	48,000
O6	−0.32	40,000
O7	−0.32	35,000
O8	−0.31	33,500
O9	−0.30	32,000
O9.5	−0.30	31,000
B0	−0.30	30,000
B0.5	−0.28	26,200

In many planetary nebulae the number of ionizing photons emitted by the star beyond the He^+ limit can also be measured from λ4686, the strongest He II recombination line. This line is too weak to be observed in any H II region, indicating that the flux of He^+-ionizing photons is small in all main-sequence O stars, and confirming the calculated models in this respect. Many of the planetary-nebula central stars, however, are considerably hotter, and emit many photons with $h\nu \geq 4h\nu_0 = 54.4$ eV. Thus, from the He II observations and from equation (2.29), we have the relation analogous to that in equation (5.24),

$$\frac{L_{\nu_f}}{\int_{4\nu_0}^{\infty} \frac{L_\nu}{h\nu} d\nu} = h\nu_{\lambda 4686} \frac{\alpha_{\lambda 4686}^{eff}(\mathrm{He}^+, T)\pi F_{\nu_f}}{\alpha_B(\mathrm{He}^+, T)\pi F_{\lambda 4686}}. \tag{5.26}$$

Hence from the measured H I and He II line fluxes of the nebula, together with the measured stellar flux at some observable frequency, two independent determinations of T_* can be made by the Zanstra method. In some nebulae these two values for T_* agree, but in other nebulae they disagree considerably. For instance, in NGC 7662, the H I measurement indicates $T_* = 70{,}000°$ K, but the He II measurement indicates that $T_* = 113{,}000°$ K, which would require more than 100 times more He^+-ionizing photons than the lower temperature. The discrepancy may be understood as resulting from the fact that the nebula is not optically thick to the H-ionizing radiation as equation (5.24) assumes. If the nebula is density-bounded rather than ionization-bounded, then we must replace equation (5.24) with

$$\frac{L_{\nu_f}}{\int_{\nu_0}^{\infty} \frac{L_\nu}{h\nu} d\nu} = \eta_H h\nu_{\mathrm{H}\beta} \frac{\alpha_{\mathrm{H}\beta}^{eff}(\mathrm{H}^0, T)\pi F_{\nu_f}}{\alpha_B(\mathrm{H}^0, T)\pi F_{\mathrm{H}\beta}}, \tag{5.27}$$

where $\eta_{\rm H}$ represents the fraction of the H-ionizing photons that are absorbed in the nebula. We could also imagine that all the $\rm He^+$-ionizing photons are not absorbed within the nebula, that is, that even the $\rm He^{++}$ zone is density-bounded rather than ionization-bounded. However, this does not seem to be the case in most observed planetaries, because nearly all observed planetaries have He I lines in their observed spectra, indicating the existence of an outer $\rm He^+$ zone, which, as the discussion of Chapter 2 shows, is certainly optically thick to $\rm He^+$-ionizing radiation. In a similar way, if [O I] lines are observed in a nebula, they indicate the presence of $\rm O^0$, and therefore also of $\rm H^0$, which has the same ionization potential as $\rm O^0$, and thus indicate that the nebula is optically thick to H-ionizing radiation and that $\eta_{\rm H} = 1$ (assuming spherical symmetry).

One further item of information can be obtained from measurements of the flux in a He I recombination line, such as $\lambda 4471$ or $\lambda 5876$, namely, the number of photons emitted that can ionize $\rm He^0$. This condition is:

$$\frac{L_{\nu_f}}{\int_{\nu_2}^{\infty} \frac{L_\nu}{h\nu} d\nu} = \eta_{\rm He} h\nu_{\lambda 5876} \frac{\alpha^{eff}_{\lambda 5876}({\rm He}^0, T)\pi F_{\nu_f}}{\alpha_B({\rm He}^0, T)\pi F_{\lambda 5876}}. \qquad (5.28)$$

If the nebula is known to be optically thick to the He-ionizing radiation, either because the $\rm He^+$ zone is observed to be smaller than the $\rm H^+$ zone, or because the apparent abundance ratio $N_{\rm He^+}/N_p \leq 0.1$ (presumably indicating that the $\rm He^+$ zone is smaller than the $\rm H^+$ zone, even though this was not directly observed), then $\eta_{\rm He} = 1$.

Although the integrals giving the numbers of photons that can ionize $\rm He^+$, $\rm He^0$, and $\rm H^0$ in equations (5.26), (5.28), and (5.27), respectively, overlap, the equations are nevertheless essentially correct, because (as indicated in Chapter 2), nearly every recombination of a $\rm He^{++}$ ion leads to emission of a photon that can ionize $\rm He^0$ or $\rm H^0$, and nearly every recombination of a $\rm He^+$ ion leads to emission of a photon that can ionize $\rm H^0$.

The observational data on the fluxes in $\rm H\beta$, $\lambda 4686$, and $\lambda 4471$ are fairly complete and fairly accurate for planetary nebulae. The measurements of the stellar continuum fluxes are less accurate, because the stars are faint and must be observed on the bright background of the nebula. The best measurements are those that avoid the wavelengths of the nebular emission lines.

Those planetary nebulae considered to be optically thick to $\rm H^0$-ionizing radiation (because of the presence of [O I] lines in their spectra) have Zanstra temperatures derived both from H I and from He I or He II lines that are in good agreement, a confirmation of the theory. For the other planetaries, T_* is derived from the He II measurements and equation (5.26), and the H I measurements are then used in equation (5.27) to calculate $\eta_{\rm H}$. Some observational results are given in Table 5.10. The left-hand side of the table is based on the approximation $F_\nu(T_*, g) = B_\nu(T_*)$ for the fluxes of the central stars, while the right-hand side is based on values of πF_ν obtained from a series of model atmospheres calculated for planetary-nebula stars. Note that the calculated T_* are not very sensitive to the assumption, but the model atmospheres in general indicate somewhat higher temperatures.

TABLE 5.10
Temperatures of central stars of planetary nebulae derived from Zanstra method

Nebula	T_*(° K)	
	Black-body approximation	Model atmosphere
NGC 2392	65,000	80,000
NGC 3242	93,000	109,000
NGC 3587	105,000	123,000
NGC 6210	67,000	–
NGC 6543	66,000	82,000
NGC 6572	61,000	77,000
NGC 6826	69,000	85,000
NGC 6853	132,000	148,000
NGC 7009	81,000	98,000
NGC 7662	113,000	130,000
IC 351	91,000	108,000
IC 418	42,000	–

Another, related method of determining the temperature of the central star is to measure directly all the cooling radiation. If the fluxes in all the emission lines plus the continuum, from the ultraviolet to the infrared, are added together, this gives directly the energy radiated by the nebula. Comparing this quantity with the flux in Hβ gives the energy radiated per recombination (through the ratio $\alpha_B/\alpha_{\mathrm{H}\beta}^{eff}$), that is, the energy input to the nebula per photoionization. This is exactly the quantity the Zanstra method gives, and determines the effective temperature of the star, either in the blackbody approximation, or from a series of model atmospheres. This method of temperature determination is called Stoy's method and, as observations into the ultraviolet and infrared spectral regions become more straightforward, it will become increasingly useful. Note that if the continuum flux is not measured, it can be estimated as the energy lost (in the Balmer, Paschen, etc. continua) by the recombining electrons as outlined in Section 3.3.

5.9 Abundances of the Elements in Nebulae

It is clear that abundances of the observed ions in nebulae can be derived from measurements of the relative strengths of their emission lines. All the individual nebular lines are optically thin; so no curve-of-growth effects of the kind that complicate stellar atmosphere abundance determinations occur. Many light elements are observable in the optical spectra of nebulae, including H, He, N, O, and Ne, although unfortunately C is not. However it can be observed in the satellite ultraviolet spectral region. On the other hand, the strengths of collisionally excited lines depend strongly on temperature, which complicates the determination of relative abundances. Furthermore, all stages of ionization of an element are generally not observable in the optical spectral region; for instance, though [O II] and [O III] have strong lines in diffuse nebulae, O IV and O V do not. However, there is an [O IV] line in the far infrared, and O IV] and O V lines in the satellite ultraviolet. Opening up these new wavelength regions has greatly aided abundance determinations.

In general, as we have seen in Chapter 4, the observed intensity I_l of an emission line is given by the integral

$$I_l = \int j_l ds = \int N_i N_e \epsilon_l(T) ds \tag{5.29}$$

taken along the line of sight through the nebula, where N_i and N_e are the density of the ion responsible for the emission and the electron density, respectively.

For the recombination lines, the emission coefficients have been discussed in Chapter 4, and we have, for instance,

$$I_{\mathrm{H}\beta} = \frac{1}{4\pi} \int N_p N_e h\nu_{\mathrm{H}\beta} \alpha^{eff}_{\mathrm{H}\beta}(\mathrm{H}^0, T) ds,$$

$$I_{\lambda 5876} = \frac{1}{4\pi} \int N_{He^+} N_e h\nu_{\lambda 5876} \alpha^{eff}_{\lambda 5876}(\mathrm{He}^0, T) ds,$$

$$I_{\lambda 4686} = \frac{1}{4\pi} \int N_{\mathrm{He}^{++}} N_e h\nu_{\lambda 4686} \alpha^{eff}_{\lambda 4686}(\mathrm{He}^+, T) ds.$$

For all the recombination lines $\epsilon_l(T) \propto T^{-m}$ can be fitted over a limited range of temperature, with $m \approx 1$. For instance, for Hβ, $m = 0.90$, while for He I λ5876, $m = 1.13$. Thus the recombination-emission coefficients are not particularly temperature-sensitive and the abundances derived from them do not depend strongly on the assumed T.

Less abundant ions, such as C II, O IV, and O V, have weak permitted emission lines as observed in planetary nebulae, and these lines have often been interpreted as resulting from recombination, and have been used to derive abundances of the parent ions. However, many of these lines are actually excited by resonance-fluorescence, and their emission coefficients therefore depend not only on temperature and density but on the local radiation field as well, so they cannot be used to derive abundances in any straightforward way. Some, however, such as C II 3 ^2D – 4 ^2F $\lambda4267$, cannot be excited by resonance-fluorescence, and are suitable for abundance determinations. Only fragmentary calculations of their effective recombination coefficients are now available, however.

It is also possible to measure relative abundances of He^+ in H II regions from relative strengths of the radio recombination lines of H I and He I. At the very high n of interest in the radio region, both H and He are nearly identical one-electron systems except for their masses, so that the relative strengths of their lines (separated by the isotope effect) are directly proportional to their relative abundances, as long as the lines are optically thin, and the nebula is a complete H^+, He^+ region, with no H^+, He^0 zone.

For abundance determinations of elements other than H and He, only collisionally excited lines are available, and for these lines, in contrast to the recombination lines, the emission coefficient depends more sensitively on the temperature,

$$
\begin{aligned}
I_\nu &= \frac{1}{4\pi} \int N_i N_e h\nu q_{1,2}(T)\; b\; ds \\
&= \frac{1}{4\pi} \int N_i N_e h\nu \frac{8.63 \times 10^{-6}}{T^{1/2}} \frac{\Omega(1,2)}{\omega_1}\, e^{-\chi/kT}\; b\; ds,
\end{aligned}
$$

in the low-density limit, where b is the fraction of excitations to level 2 that are followed by emission of a photon in the line observed.

The temperature must be determined from observational data of the kind discussed in Section 5.2 through 5.7. From the measured relative strengths of the lines and the known emission coefficients, the abundances can be determined from a model of the structure of the nebula. The simplest model treats the nebula as homogeneous with constant T and N_e, and thus might be called a one-layer model. From each observed relative line strength, the abundance of the ion that emits it can be determined. In some cases two successive stages of ionization of the same element are observed, such as O^+ and O^{++}, and their relative abundances can be used to construct an empirical ionization curve giving $N(A^{+m+1})/N(A^{+m})$ as a function of ionization potential. Thus, finally, the relative abundance of every element with at least one observed line can be determined. Discrepancies (for instance, in N_e and T) determined from different line ratios indicate that this model is too simplified to give highly accurate results, though the abundances determined from it are generally thought to be correct to within a factor of order two or three.

A somewhat more sophisticated scheme takes into account the spatial variations of temperature and uses the observations themselves to get as much information as possible on these variations. The emission coefficient is expanded in a power series

$$\epsilon_l(T) = \epsilon_l(T_0) + (T - T_0)\left(\frac{d\epsilon_l}{dT}\right)_0 + \frac{1}{2}(T - T_0)^2\left(\frac{d^2\epsilon_l}{dT^2}\right)_0, \tag{5.30}$$

correct to the second order. It is clear that for recombination lines with

$$\epsilon_l(T) = CT^{-m}$$

or for collisionally excited lines with

$$\epsilon_l(T) = \frac{De^{-\chi/kT}}{T^{1/2}},$$

(in the low-density limit) the necessary derivatives can be worked out analytically. Then integrating along the line of sight,

$$\int N_i N_e \epsilon_l(T)\, ds = \epsilon_l(T_0) \int N_i N_e \;\; ds + \frac{1}{2}\left(\frac{d^2\epsilon_l}{dT^2}\right)_0 \int N_i N_e (T - T_0)^2\, ds, \tag{5.31}$$

where T_0 is chosen so that

$$T_0 = \frac{\int N_i N_e T\, ds}{\int N_i N_e\, ds}. \tag{5.32}$$

If all ions had the same space distribution $N_i(s)$, then from two line ratios, such as [O III] ($\lambda4959 + \lambda5007)/\lambda4363$ and [N II] ($\lambda6548 + \lambda6583)/\lambda5755$, both T_0 and

$$t^2 = \frac{\int N_i N_e (T - T_0)^2\, ds}{T_0^2 \int N_i N_e\, ds}$$

could be determined instead of the one constant T_0 from one line ratio, as in the single-layer model. Then T_0 and t^2 could be used to find the abundances of all the ions with measured lines. The difficulty with this method is that all ions do not have the same distribution. For instance, O^{++} is more strongly concentrated to the source of ionizing radiation than N^+, so other more or less arbitrary assumptions must be made.

The most sophisticated method of all to determine the abundances from the observations is to calculate a complete model of the nebula, in an attempt to reproduce all its observed properties; this approach will be discussed in the next section.

Let us turn now to the observational results. The He/H abundance ratio has been measured in many nebulae. Perhaps the most exhaustively measured nebula is the Orion Nebula, for which the most recent results for N_{He^+}/N_p range from 0.060 to 0.090 in various slit positions. A measurement in its nearby companion nebula NGC 1982 gives $N_{He^+}/N_p = 0.009$ and definitely shows that this slit position is in an H^+, He^0 zone, where He is neutral. The exciting star of NGC 1982 is a B1 V star, so the fact that the nebula is a He^0 zone is understood from Figure 2.5. This observation shows that some correction of the abundance of He for the unobserved He^0 is probably necessary at all the observed slit positions in the Orion Nebula. Empirically, the correction can be based on the observed strength of [S II] $\lambda\lambda 6716$, 6731, because their emitting ion S^+ has an ionization potential of 23.4 eV, approximately the same as the ionization potential of He^0, 24.6 eV, so that to a first approximation

$$\frac{N_{He^0}}{N_{He^+}} = \frac{N_{S^+}}{N_{S^{++}}}$$

yields the abundance of He^0. A more sophisticated procedure is to interpolate between the ionization of S^+ and of O^+ (ionization potential 35.1 eV) in such a way that the corrections at all slit positions yield as nearly as possible the same final He/H ratio. The final result for NGC 1976 is $N_{He}/N_H = 0.10$; two other H II regions observed optically, M 8 and M 17, have essentially this same relative He abundance.

Radio measurements of He^+/H^+ abundance ratios are available for many diffuse nebulae. These determinations are in fairly good agreement with the optical measurement for nebulae common to both sets of observations, as shown in Table 5.11. The average abundance ratio from seven H II regions observed in the radio-frequency region is $N_{He^+}/N_p = 0.08$. However, there is no known way in which the correction for He^0 can be obtained from radio measurements alone, and the fact that at least two nebulae, NGC 2024 and NGC 1982, are observed to have $N_{He^+}/N_p \approx 0$ shows that this correction certainly is needed. Radio measurements of H II regions very near the Galactic center give quite low N_{He^+}/N_p ratios, but probably they indicate that the ionizing stars are predominantly rather cool, producing H^+, He^0 zones, rather than low helium abundance.

TABLE 5.11
Comparison of optical and radio helium abundance determinations

Nebula	He^+/H^+ optical	He^+/H^+ radio
NGC 1976	0.075	0.080
NGC 6618	0.097	0.086

Both He II and He I recombination lines are observed in many planetary nebulae, showing the presence of both He^{++} and He^+, though some planetaries, like H II regions, have only He I lines. Nearly all planetaries have central stars that are so hot that they have no outer H^+, He^0 zones, though a few exceptions have been discussed in previous sections of this text. Thus no correction is necessary for unobserved He^0 in most planetary nebulae.

If only the most accurate photoelectric measurements of planetary nebulae are used, the derived helium abundances have only a small scatter, as shown in Table 5.12. The average for nine well-observed planetaries in this table is $N_{He}/N_H = 0.11$, while the range is from 0.09 from 0.13. Only in IC 418 was a correction for the He^0 necessary. Some other planetaries have still higher He abundances, as will be described in Chapter 9. Since the accuracy of the measurements, as judged from the relative intensities of He I $\lambda\lambda$4471, 5876, is about 0.01, the differences between the nebulae are real. Some special planetaries with well-determined helium abundances are K 648 in the globular cluster M 15 with $N_{He}/N_H = 0.10$ and Ha 4-1 = PK 49 +88° 1 near

TABLE 5.12
Helium abundances in planetary nebulae

Nebula	N_{He^+}/N_p	$N_{He^{++}}/N_p$	N_{He}/N_H
NGC 1535	0.08	0.02	0.10
NGC 6572	0.11	0.00	0.11
NGC 6720	0.08	0.03	0.11
NGC 6803	0.13	0.00	0.13
NGC 6884	0.10	0.02	0.12
NGC 7009	0.10	0.01	0.11
NGC 7027	0.08	0.04	0.12
NGC 7662	0.06	0.04	0.10
IC 418	0.07	0.00	0.09

the Galactic pole and approximately 10 kpc distant, with $N_{He}/N_H = 0.11$. Thus these extreme Population II objects have helium abundances that are indistinguishable from the more common observed bright planetary nebulae near the Sun.

Among the H II regions, the most complete abundance determinations of the heavy elements are available for NGC 1976. The results given in Table 5.13 show that it has fairly normal abundances of N, O, Ne, and S. Similar heavy-element abundance determinations are available for M 8 and M 17, with results nearly the same as for NGC 1976. Abundance determinations have been made for rather more planetary nebulae, and the average results for these planetaries, taken from a recent summary paper, are also shown in Table 5.13. There are real variations from these mean values, as will be discussed further in Chapter 9. The main difficulty with the abundance determinations for the heavy elements is that large and rather uncertain corrections are required for unseen ions, that is, ions without observable lines. Ultraviolet and infrared measurements have improved this situation in recent years. Still, the calculations of model planetary nebulae that are described in the next section probably represent the most nearly accurate method for finding abundances of the heavy elements.

TABLE 5.13
Abundances of elements in gaseous nebulae

Element	Logarithm of relative abundance	
	Average planetary nebula	NGC 1976
H	12.00	12.00
He	11.04	11.00
C	8.85	8.52
N	8.11	7.76
O	8.62	8.75
F	4.6:	–
Ne	8.02	7.90
Na	6.05	–
S	6.99	7.41
Cl	5.19	5.15
Ar	6.40	6.7
K	4.85	–
Ca	4.92	–

5.10 Calculations of the Structure of Model Nebulae

The basic idea of a calculation of a model H II region or a model planetary nebula is quite straightforward. It is:

(a) to make reasonable assumptions about the physical parameters of the ionizing star, the density distribution, and the relative abundances of the elements in the nebula (its size, geometrical structure, and so on);

(b) to calculate, from these assumptions, the resulting complete physical structure, that is, the ionization, temperature, and emission coefficients as functions of position; and thus

(c) to calculate the expected emergent radiation from the nebula at each point in each emission line.

Comparing this predicted model with the observed properties of a nebula provides a check on whether the initial assumptions are consistent with the observations; if they are not, then the assumptions must be varied until a match with the observational data is obtained. In principle, if all the emission lines were accurately measured at every point in the nebula, and if the central star's radiation were measured at each observable frequency, it might be possible to specify accurately all the properties of the star and of the nebula in this way. Of course, in practice the observations are not sufficiently complete and accurate, and do not have sufficiently high angular resolution to enable us to carry out this ambitious program. Nevertheless, quite important information is derived from the model-nebula calculation.

Let us write down in simplified form the equations used in calculating the structure of a model nebula. For computational reasons, practically all work to date has assumed spherical model nebulae, and we will write the equation in these terms. The basic equations are described in Chapters 2 and 3, so we will simply quote them here. The equation of transfer is

$$\frac{dI_\nu}{ds} = -I_\nu \frac{d\tau_\nu}{ds} + j_\nu, \tag{5.33}$$

where the increment in optical depth at any frequency is given by a sum

$$\frac{d\tau_\nu}{ds} = \sum N_j a_{\nu_j} \tag{5.34}$$

over all atoms and ions with ionization potentials $h\nu_j < h\nu$. In practice, because of their great abundance, only H^0, He^0, and He^+ are important except possibly at the very highest frequencies. However, with present-day computation facilities it is simple to include contributions from all ions in the optical depth and thus be absolutely safe. Likewise, the emission coefficient j_ν is a sum of terms of which again only those due to recombinations of H^+, He^+, and He^{++} are important.

The ionization equation that applies between any two successive stages of ionization of any ion is

$$N(X^{+i}) \int_{\nu_i}^{\infty} \frac{4\pi J_\nu}{h\nu} a_\nu(X^{+i})\, d\nu = N(X^{+i+1}) N_e \alpha_G(X^{+i}, T) \tag{5.35}$$

as in equation (2.30), and the total number of ions in all stages of ionization is

$$\sum_{i=0}^{\max} N(X^{+i}) = N(X).$$

The energy-equilibrium equation is

$$G = L_R + L_{FF} + L_C \tag{5.36}$$

as in equation (3.32), where the gain term and each of the loss terms is a sum over the contributions of all ions; again, in practice, usually only H and He are important in L_R and L_{FF}. Collisionally excited line radiation from the less-abundant heavy elements dominates the cooling, however, and many terms must be included in L_C.

For any assumed input-radiation source at the origin, taken to be a star with either blackbody spectrum or a spectrum calculated from a model stellar atmosphere, these equations can be integrated. If the on-the-spot approximation described in Chapter 2 is used, they can be integrated outward. If instead the detailed expressions for the emission coefficients are used and the diffuse radiation field is explicitly calculated, it is necessary to use an iterative procedure. The on-the-spot approximation can be used as a first approximation from which the ionization at each point in the nebula and the resulting emission coefficients can be calculated. Then the diffuse radiation field can be calculated working outward from the origin, and using the then more-nearly accurate total radiation field, the ionization and T can be recalculated at each point. This process can be repeated as many times as needed until it converges to the desired accuracy.

As an example, let us examine two recent models of what is probably the most studied planetary nebula, NGC 7662, a bright, northern object with a highly symmetric double-ring structure. These models are based on ground-based optical measurements, and ultraviolet measurements from the IUE satellite. They are spherically symmetric models, with radial density profiles chosen to match the observed Hβ surface-brightness contours, averaged over angle, as accurately as possible. The density distribution derived in this way has its maximum at a radius of about 6.″5 from the central star on the sky, corresponding to 0.045 pc at the assumed distance of 1.5 kpc. For one model (I) the central star is taken to be a blackbody, with temperature $T_* = 107{,}000°$ K, from the Zanstra method together with the continuum flux of the star measured at $\lambda 1400$. The second model (II) uses a model stellar atmosphere with $T_* = 120{,}000^\circ$ K, $\log g = 6$ based on the same data. The effects of dust (to be discussed in Chapter 7) within the nebula are taken into account in model II but not in model I.

Models with homogeneous density, or with smoothly varying density, will not fit the observed data at all well. The densities indicated by relative line-intensity measurements (such as [O II] $\lambda\lambda 3726$, 3729, [S II] $\lambda\lambda 6716$, 6731), and by the ionization, are much higher than the mean densities derived from the Hβ surface-brightness measurements; so we must assume that there are density fluctuations, or small dense condensations, within the nebula. Direct photographs of nebulae, including NGC 7662, show the largest of these condensations clearly. In fact, the [O I] emission particularly can be observed to occur in dense condensations that must be "neutral" (H^0, O^0) zones, surrounded by partly ionized edges in which both O^0 and free electrons are present. These density fluctuations are idealized in the model as condensations with densities N_e filling a fraction ϵ of the volume, the rest of which is taken to be empty. The density profile then gives the variation of N_e (in the condensations) with nebular radius. For NGC 7662, model I assumes $\epsilon = 0.34$ throughout the nebula; in model II $\epsilon = 1$ in the inner part of the model ($1'' < r < 10''$), but $\epsilon = 0.2$ in the outer part ($10'' < r < 15''$), to mimic approximately the observational result that the density fluctuations are more pronounced in the outer part of the nebula. These specific values of ϵ were determined by adjusting them to optimize the agreement of the calculated line strengths with the observational data. The strengths of the lines from singly ionized atoms, and of the density-sensitive lines, are especially important for determining these filling factors.

For both models the maximum electron density is about 4.5×10^3 cm^{-3}. The calculated electron temperature varies from a high of 23,000° K (model I) or 25,000° K (model II) close to the central star to a low of about 11,000° K (both models) at $10''$ radius (0.07 pc) just outside the He^{++} zone, and then rises slightly to about 12,000° K (model I) or 13,000° K (model II) at the outer edge of the nebula, at radius $15''$ (0.11 pc).

All the relative abundances of the elements were also adjusted to achieve the best overall agreement between the observed and calculated line strengths. In Table 5.14 the observed line strengths are compared with those calculated; in general the models agree well with the observed data. This is the most accurate, but most complicated, method of using the observed line strengths to determine abundances in planetary nebulae. The results for NGC 7662 are listed in Tables 5.15 and 5.16. Model II, which takes into account the effects of dust within the nebula, is probably more nearly physically correct.

Models of the type discussed here have also been applied to H II regions. They are apparently the most realistic available representations of these highly chaotic objects, but there are few observations of the optical-line radiation with which they may be compared. The fundamental problem, however, is

TABLE 5.14
Observed and calculated line strengths for NGC 7662 models

Ion	λ	Relative line strength		
		Observed	Model I	Model II
H I	4861	100.00	100.00	100.00
He I	4471	2.6:	2.60	2.75
He I	5876	7.1	7.17	7.32
He II	4686	44	44.9	42.6
C II	1335	<3	9.5	–
C II]	2326	10.9 ± 2.3	17.4	18.8
C II	4267	0.76	0.16	0.18
C III]	1908	335 ± 7	352	365
C III	2297	8.5 ± 2.0	3.4	9.0
C IV	1549	641 ± 11	741	665
[N II]	5755	0.2	0.15	0.079
[N II]	6583	4.1	5.75	2.99
N III]	1751	10 ± 3	22.8	9.5
N IV]	1487	31 ± 6	32.2	36.4
N V	1240	22 ± 11	11.1	12.1
[O II]	3727	13.5	11.25	9.45
[O II]	7325	1.4:	0.91	0.76
O III]	1663	19 ± 3	22.8	24.1
[O III]	2321	(3.7)	3.6	3.7
[O III]	4363	16.2	14.5	15.4
[O III]	5007	1160	1140	1180
O IV]	1402	27 ± 3	27.8	30.1
[Ne III]	3869	71	88.0	81.2
[Ne IV]	2423	93 ± 7	101	109
[Ne IV]	4720	2.0	0.95	1.02
[Ne V]	3426	11.2	8.47	30.8
Mg I]	4571	0.18–1.8	0.042	0.069
Mg II]	2800	≤0.8	0.9	0.9
Si III]	1883	8 ± 5	2.2	2.7
Si IV	1397	3 ± 1	3.5	3.9
[S II]	4073	0.6:	0.41	0.31
[S II]	6725	1.0	2.25	1.69
[S III]	3722	13.5	11.25	9.45
[S III]	9069	9.2	23.4	12.5
[S IV]	10.52μ	112	106	93

that the assumption of spherical symmetry, and in particularly of homogeneity, or else of a density distribution that depends only on distance from the center, possibly including some simplified form of density fluctuations, is far too simple to describe real nebulae. Though some planetaries have a fairly symmetric form and smooth structure, no H II regions do. The photographs in this book show the very complicated structure of actual nebulae, and the example of the planetary nebula NGC 6853 in Figure 5.10 demonstrates how strongly the local structure can affect line-emission coefficients. We must develop models that represent this structure in some realistic way before we can consider the nebulae to be adequately represented.

TABLE 5.15
Observed and calculated density-sensitive line ratios for NGC 7662 models

Ion	Ratio	Observed	Model I	Model II
C III]	$\lambda1907/\lambda1909$	1.5 ± 0.2	1.35	1.37
[Ne IV]	$\lambda2424/\lambda2422$	1.1 ± 0.1	1.08	1.20
[O II]	$\lambda3729/\lambda3726$	0.56	0.56	0.57
[S II]	$\lambda6716/\lambda6731$	0.66	0.66	0.66

TABLE 5.16
Relative Abundances of Elements from NGC 7662 Models

Element	Model I	Model II	Sun
H	1.00	1.00	1.00
He	0.094	0.094	–
C	3.3×10^{-4}	6.2×10^{-4}	4.7×10^{-4}
N	8.0×10^{-5}	6.0×10^{-5}	9.8×10^{-5}
O	3.4×10^{-4}	3.6×10^{-4}	8.3×10^{-4}
Ne	7.0×10^{-5}	7.0×10^{-5}	1.5×10^{-4}
Mg	7.0×10^{-7}	8.0×10^{-7}	4.2×10^{-5}
Si	1.8×10^{-6}	6.0×10^{-6}	4.3×10^{-5}
S	1.5×10^{-5}	1.5×10^{-5}	1.7×10^{-5}

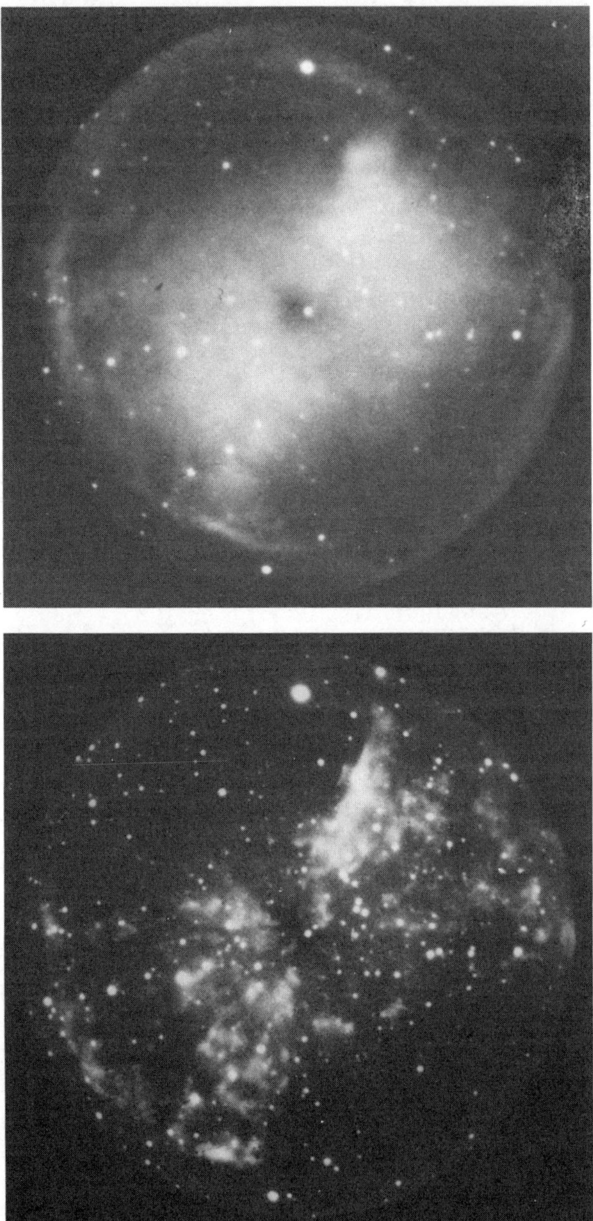

FIGURE 5.10
Monochromatic photographs of the planetary nebula NGC 6853 in the light of Hβ λ4861 (*top*), and in the light of [O I] λ6300 (*bottom*). Note that, in contrast to Hβ, the [O I] emission is strongly concentrated to many bright spots, which are high-density neutral condensations (surrounded by partly ionized edges) in which both O^0 and free electrons are present (*Steward Observatory photograph*).

5.11 Filling Factor

The density fluctuations in NGC 7662 described in the previous section are not unique to that planetary nebula. Direct photographs of nearly all planetaries and H II regions show chaotic structure of the same general type. Some nebulae appear to have large hollow central regions. As we will see in Chapter 6, these can be understood as resulting from a highly ionized, low-density, high-velocity "wind," flowing out from the central star or stars. These density condensations, low-density hollows, etc., are an important feature of the structure of gaseous nebulae.

They can be detected quantitatively if the densities derived from [O II] line ratios in a large, well-resolved nebula of known distance, such as NGC 1976, are used to predict the expected high-frequency radio continuum brightness temperature, combining equations (4.32) and (4.37). They give, in the limit of small optical depth, which is a good approximation for high-frequency observations,

$$T_{b\nu} = 8.24 \times 10^{-2} T^{-0.35} \nu^{-2.1} E, \tag{5.37}$$

in the units mentioned in Section 4.4, with

$$E = \int N_+ N_e \, ds,$$

the emission measure, in cm^{-6} pc. Note that the predicted brightness temperature depends only very weakly on the nebular temperature.

The measured values of $T_{b\nu}$ are invariably smaller than those predicted in this way, typically by a factor of order ten. This difference can only be understood in terms of density fluctuations. The line-ratio density measurements are heavily weighted toward the regions of strongest emission, that is, of highest density. These measured densities thus deviate greatly from the average density along a typical path or ray through the nebula. Fluctuations in density must be taken into account in describing the structure of the nebula.

The simplest, though extreme, way to do so is to idealize the nebula as containing gas in small clumps or condensations, with electron density N_e within the condensations, but with zero electron density between them. The "filling factor" ϵ is then the fraction of the total volume occupied by the condensations. The space between the condensations, in this simple picture, is vacuum, which makes no contribution to the emission, mass, opacity, etc., of the nebula. The filling factor may be assumed to be constant throughout a nebula, as in model I for NGC 7662, described in the Section 5.10, or to vary with position, as in model II. For NGC 1976, if we assume a constant filling factor the comparison of density and radio-continuum measurements gives $\epsilon \approx 0.03$, and values ranging from 0.01 to 0.5 or so have been determined for other H II regions and planetary nebulae.

Note that under the filling-factor description of nebulae the intensity of an emission line is given by

$$I_l = \int j_l \ ds = \int \epsilon N_i N_e \epsilon_l(T) \ ds, \tag{5.38}$$

replacing equation (5.29); the luminosity in the same line, integrated over the volume of the nebula, is

$$L_l = \int \epsilon N_i N_e \epsilon_l(T) \ dV; \tag{5.39}$$

the number of recombinations is

$$Q(\mathrm{H}^0) = \frac{4\pi}{3} r_1^3 \ \epsilon N_p N_e \alpha_B(\mathrm{H}^0), \tag{5.40}$$

replacing equation (2.19) or

$$Q(\mathrm{H}^0) = \int_0^{r_1} \epsilon N_p N_e \alpha_B(H^0) \ dV, \tag{5.41}$$

replacing its analogue in Section (5.8), and the total mass of H in the nebula is

$$M_H = m_H \int_0^{r_1} \epsilon N_p \ dV. \tag{5.42}$$

Likewise the radial optical depth becomes

$$\tau_\nu(r) = \int_0^r \epsilon N_{\mathrm{H}^0}(r') a_\nu \ dr', \tag{5.43}$$

replacing equation (2.12). Similar generalizations can be made in other equations, always on the basis that N stands for the density in the condensations, which are assumed to fill a fraction ϵ of the total volume, with vacuum (or hot, low-density invisible gas) between them.

References

An old but very good overall reference on the comparison of theory and observation of the optical radiation of nebulae, written in the context of planetary nebulae but applicable in many ways to H II regions also, is

Seaton, M. J. 1960. *Rep. Progress in Phys.* 1960. **23,** 313.

The method of measuring electron temperatures from optical emission-line intensity ratios seems to have been first suggested by

Menzel, D. H., Aller, L. H., and Hebb, M. H. 1941. *Ap. J.* **93,** 230.

This method has subsequently been used by many authors.

The form of the [O III] ratio plotted in Figure 5.1 is based on collision strengths Ω from

Baluja, K. L., Burke, P. G., and Kingston, A. E. 1980. *J. Phys. B.* **13,** 829.

The observational data used in Table 5.1 are from

Peimbert, M., and Costero, R. 1969. *Bol. Obs. Tonantzintla Tacubaya* **5,** 3.

Peimbert, M., Torres-Peimbert, S. 1977. *M.N.R.A.S.* **179,** 217.

Peimbert, M., Torres-Peimbert, S., and Rayo, J. F. 1978. *Ap. J.* **220,** 516.

The data in Table 5.2 are from

Kaler, J. B. 1986. *Ap. J.* **308,** 322.

The C III] λ1909/C II λ4267 ratio method is described by

Kaler, J. B. 1986. *Ap. J.* **308,** 337.

The observational data for the optical continuum temperature determinations of Table 5.4 are from

Peimbert, M. 1971. *Bol. Obs. Tonantzintla Tacubaya* **36,** 29.

The data for NGC 7027 described in the text of Section 5.3 are from

Miller, J. S., and Mathews, W. G. 1972. *Ap. J.* **172,** 593.

The radio-continuum method of measuring the temperature of a nebula by observing it in the optically thick region has been discussed and used by many authors. The numerical data on which Table 5.5 is based are from

Shaver, P. A. 1970. *Ap. Letters* **5,** 167.

The most complete information on radio-frequency measurements of planetary nebulae, including the most accurate temperatures derived by this method, is included in

Bignell, R. C. 1983. *Planetary Nebulae* (IAU Symposium No. 103). Dordrecht: Reidel, p. 69.

Pottasch, S. R. 1984. *Planetary Nebulae.* Dordrecht: Reidel, p. 89.

The idea of using the [O II] intensity ratio to measure electron densities in nebulae seems to have been first suggested by

Aller, L. H., Ufford, C. W., and Van Vleck, J. H. 1949. *Ap. J.* **109,** 42.

It was worked out quantitatively by

Seaton, M. J. 1954. *Ann. d'Ap.* **17,** 74.

A complete discussion including theoretical calculations and also observational data on several of the H II regions used in Table 5.6, is found in

Seaton, M. J., and Osterbrock, D. E. 1957. *Ap. J.* **125,** 66.

For many years a discrepancy existed between the calculated transition probabilities for the np^3 ions like [O II] and [S II] on which nebular electron-density determinations depend, and the values implied by the astronomical data. This discrepancy was finally resolved by the realization that a fully relativistic quantum-mechanical treatment, including Dirac relativistic wave functions and the relativistic corrections to the magnetic-dipole operator, is necessary to calculate accurately the transition probabilities for these ions. Relevant references are

Eissner, W., and Zeippen C. J. 1981. *J. Phys. B.* **14,** 2125.

Zeippen, C. J. 1982. *M.N.R.A.S.* **198,** 111.

Mendoza, C., and Zeippen, C. J. 1982. *M.N.R.A.S.* **199,** 1025.

Zeippen, C. J. 1987. *Astr. Ap.*, **173,** 410.

The observational data in Tables 5.6 and 5.7 and also the data discussed in Section 5.5 are from

Osterbrock, D., and Flather, E. 1959. *Ap. J.* **129,** 26. (NGC 1976).

Meaburn, J. 1969. *Astron. Space Science* **3,** 600 (M 8).

Danks, A. C. 1970. *Astr. Ap.* **9,** 175. [S II].

Danks, A. C., and Meaburn, J. 1971. *Astron. Space Science* **11,** 398. (NGC 1976).

Kaler, J. B. 1976. *Ap. J. Suppl.* **31,** 571. (A compilation of planetary-nebula measurements.)

Barker, T. 1978. *Ap. J.* **219,** 914. (Planetary nebulae.)

Feibelman, W. A., Boggess, A., Hobbs, R. W., and McCracken, C. W. 1980. *Ap. J.* **241,** 725. C III].

Feibelman, W. A., Boggess, A., McCracken, C. W., and Hobbs, R. W. 1981. *Ap. J.* **246,** 807. C III].

O'Dell, C. R., and Castaneda, H. O., 1984. *Ap. J.* **283,** 158. (Planetary nebulae).

A very good reference to the simultaneous determination of temperature and density by comparison of optical and infrared emission line strengths, is

Dinerstein, H. L., Lester, D. F., and Werner, M. W. 1985. *Ap. J.* **291,** 561.

Figure 5.6 is taken from this reference.

Comparisons of radio-recombination-line measurements with calculated strengths and the determination of T, N_e, and E have been investigated by many authors, beginning with

Goldberg, L. 1966. *Ap. J.* **144,** 1225.

Mezger, P. G., and Hoglund, B. 1967. *Ap. J.* **147,** 490.

Dyson, J. E. 1967. *Ap. J.* **150,** L45.

Note, however, that the definition of β used here differs slightly from that used in most of these papers. The more recent results are taken from

Brown, R. L., Lockman, F. J., and Knapp, G. R. 1978. *Ann. Rev. Astr. Ap.* **16,** 445.

Shaver, P. A. 1980. *Astr. Ap.* **91,** 279.

Shaver, P. A. McGee, R. X., Newton, L. M., Danks, A. C., and Pottasch, S. R. 1983. *M.N.R.A.S.* **204,** 53.

Odegard, N. 1985. *Ap. J. Suppl.* **57,** 571.

The temperatures of Table 5.8 and the densities discussed in the text of Section 5.7 are from the last two of these references, respectively.

The method of measuring the ultraviolet radiation of stars from the recombination radiation of the nebulae they ionize was first suggested by

Zanstra, H. 1931. *Pub. Dominion Astrophys. Obs.* **4,** 209.

The treatment in this chapter is based primarily on the following papers, which include the best available optical and radio-frequency measurements:

Harman, R. J., and Seaton, M. J. 1966. *M.N.R.A.S.* **132,** 15.

Some of the blackbody temperature determinations of Table 5.10 are taken from this reference, if not superseded by later results. The model stellar-atmosphere determinations of Table 5.10 are taken from

Capriotti, E. R., and Kovach, W. S. 1968. *Ap. J.* **151,** 991.

The temperature determinations of Table 5.9 are taken from

Morton, D. C. 1969. *Ap. J.* **158,** 629.

Good tables of the left-hand sides of equations (5.23) and (5.25) for many published model stellar atmospheres are given by

Bohlin, R. C., Harrington, J. P., and Stecher, T. P. 1982. *Ap. J.* **252,** 635.

The best magnitudes of the central stars of planetary nebulae, measured with narrow-band filters at $\lambda 4428$ and $\lambda 5500$, between the strong nebular lines, and corrected for contamination by the nebular continuum can be found in

Shaw, R. A., and Kaler, J. B. 1985. *Ap. J.* **295,** 537.

Blackbody temperature determinations from it are used in Table 5.10 in preference to earlier results.

Stoy, R. H. 1933. *M.N.R.A.S.* **93,** 588.

Kaler, J. B. 1976. *Ap. J.* **210,** 843.

Kaler, J. B. 1978. *Ap. J.* **220,** 887.

These three references describe and apply the Stoy method for temperature determination.

Recent surveys of abundance determinations in nebula include

Aller, L. H., and Czyzak, S. 1983. *Ap. J. Suppl.,* **51,** 211.

Peimbert, M., and Torres-Peimbert, S. 1977. *Rev. Mexicana Astron. Ap.* **2,** 181.

Hawley, S. A., and Miller, J. S. 1977. *Ap. J.* **212,** 94.
Hawley, S. A., and Miller, J. S. 1978. *Ap. J.* **220,** 609.

The first of these deals with NGC 1976, the best studied H II region, and all the others are on planetary nebulae.

Among the many discussions of complete models of planetary nebulae, two of the best are

Harrington, J. P., Seaton, M. J., Adams, P. S., and Lutz, J. H. 1982. *M.N.R.A.S.* **199,** 517.
Clegg, R. E. S., Harrington, J. P., Barlow, M. J., and Walsh, J. R. 1987. *Ap. J.* **314,** 551.

The results of Tables 5.14, 5.15, and 5.16 are taken from the first of these.

Model H II regions are discussed in

Stasinska, G. 1980. *Astr. Ap.* **84,** 320.
Stasinska, G. 1980. *Astr. Ap. Suppl.* **48,** 208.
Rubin, R. H. 1983. *Ap. J.* **274,** 671.
Mathis, J. S. 1985. *Ap. J.* **291,** 247.
Rubin, R. H. 1985. *Ap. J. Suppl.* **57,** 349.
Evans, I. N., and Dopita, M. A. 1985. *Ap. J. Suppl.* **58,** 125.
Simpson, J. P., Rubin, R. H., Erickson, E. F., and Haas, M. R. 1986. *Ap. J.* **311,** 895.

The Mathis paper discusses in detail corrections of the abundance ratios for unobserved stages of ionization, such as He^0, from S^+/S^{++} and O^+/O^{++} ratios, as does the earlier paper

Mathis, J. S. 1982. *Ap. J.* **261,** 195.

Dense neutral condensations in planetary nebulae are discussed by

Capriotti, E. R., Cromwell, R. H., and Williams, R. E. 1971. *Ap. Letters* **7,** 241.

(Figure 5.10 is taken from this paper.)

The filling-factor concept was stated and applied to the analysis of observational data by

Strömgren, B. 1948. *Ap. J.* **108,** 242.
Osterbrock, D. E., and Flather, E. 1959. *Ap. J.* **129,** 26.

6

Internal Dynamics of Gaseous Nebulae

6.1 Introduction

The first five chapters of this book have described gaseous nebulae entirely from a static point of view. However, this description is not complete because nebulae certainly have internal motions, whose effects on their structures cannot be ignored. It is easy to see that an ionized nebula cannot be in static equilibrium, for if it is density-bounded, it will expand into the surrounding vacuum; if it is ionization-bounded, the hot ionized gas (with $T \approx 10{,}000^\circ$ K) will initially have a higher pressure than the surrounding, cooler neutral gas ($T \approx 100^\circ$ K) and will therefore tend to expand until its density is low enough that the pressures of the two gases are in equilibrium. In addition, when the hot star in a nebula first forms and the source of the ionizing radiation is thus "turned on," the ionized volume initially grows in size at a rate fixed by the rate of emission of ionizing photons, and an ionization front separating the ionized and neutral regions propagates into the neutral gas.

Observations agree in showing that the internal velocities of nebulae are not everywhere zero. Measured radial velocities show that planetary nebulae are expanding more or less radially; mean expansion velocities are of order 25 km sec^{-1}, and the velocity gradient is positive outward. Many H II regions are observed to have complex internal velocity distributions that can best be described as turbulent.

This chapter will therefore concentrate on the internal dynamics of nebulae. First it considers the hydrodynamic equations of motion that are applicable to nebulae, particularly in the spherically symmetric form in which these equations have actually been applied to date. This discussion leads to a study of ionization fronts and of shock fronts that are generated by ionization fronts and by the expansion of the nebulae. Then the available theoretical results for planetary nebulae and H II regions are analyzed. Finally, a brief synopsis of the available observational material is given, and it will be seen that more theoretical work is necessary before the observations can be fully understood, but that progress has been made in understanding some of the complications present in nature.

6.2 Hydrodynamic Equations of Motion

The standard equation of motion for a compressible fluid, such as the gas in a nebula, may be written

$$\rho \frac{D\mathbf{v}}{Dt} \equiv \rho \left(\frac{\partial \mathbf{v}}{\partial t} + \mathbf{v} \cdot \nabla \mathbf{v} \right) = -\nabla p - \rho \nabla \phi, \tag{6.1}$$

where D stands for the time derivative following an element of the gas, and ∂ stands for the partial time derivative at a fixed point in space. On the right-hand side of the equation, the forces included are the pressure gradient and the force resulting from the gravitational potential of the involved stars and of the nebula itself. However, the dimensions of any observed structure in a nebula are so large that the gravitational forces are negligibly small, and the second term can be omitted. Note, however, that in equation (6.1) electromagnetic forces have also been omitted; there may be nebulae in which there are magnetic fields large enough that this omission is incorrect. However, there is no strong evidence for the existence of such fields in most H II regions and since they would further complicate the problem, we will make this simplification here. The gas pressure is

$$p = \frac{\rho k T}{\mu m_{\rm H}}, \tag{6.2}$$

and in most situations the radiation pressure can be neglected, because the density of radiation is so low. The hydrodynamic equation of continuity,

$$\frac{D\rho}{Dt} = -\rho \nabla \cdot \mathbf{v}, \tag{6.3}$$

also relates the density and velocity fields.

The energy equation is a generalization of the thermal balance equation (3.32),

$$\frac{DU}{Dt} \equiv \frac{D}{Dt} \left(\frac{3}{2} \sum_j N_j k T \right) = G - L + \frac{p}{\rho} \frac{D\rho}{Dt} - U \nabla \cdot \mathbf{v}, \tag{6.4}$$

where U is the internal kinetic energy per unit volume, G and L are the energy gain and loss rates per volume per unit time discussed in Chapter 3, the next term on the right-hand side of the equation gives the heating rate resulting from compression, and the last term gives the dilation effect, analogous to the term on the right-hand side of equation (6.3). Note that ionization energy is not included on either side of equation (6.4), but the kinetic energy of all particles is, so the sum includes all atoms and ions as well as electrons, and is dominated (in the ionized gas) by N_p, N_e, N_{He^+}, and $N_{He^{++}}$. It is a reasonably good approximation to assume, as we have in equation (6.4), that all the ionized species are in temperature equilibrium with one another, because the Coulomb-scattering cross sections are so large, and the relaxation times are correspondingly short.

It is somewhat more convenient to rewrite equation (6.4) in a form that includes the internal kinetic energy per unit mass, $E = U/\rho$, for it then becomes

$$\frac{DE}{Dt} = \frac{D}{Dt}\left(\frac{U}{\rho}\right) = \frac{1}{\rho}(G-L) - p\frac{D}{Dt}\left(\frac{1}{\rho}\right). \tag{6.5}$$

Finally, the ionization equation is a generalization of equation (2.30),

$$\begin{aligned}\frac{DN(X^{+i})}{Dt} = &- N(X^{+i})\int_{\nu_i}^{\infty}\frac{4\pi J_\nu}{h\nu}a_\nu(X^{+i})\,d\nu \\ &+ N(X^{+i+1})N_e\alpha_A(X^{+i},T) \\ &- N(X^{+i})N_e\alpha_A(X^{+i-1},T) \\ &+ N(X^{+i-1})\int_{\nu_{i-1}}^{\infty}\frac{4\pi J_\nu}{h\nu}a_\nu(X^{+i-1})\,d\nu \\ &- N(X^{+i})\nabla\cdot\mathbf{v}.\end{aligned} \tag{6.6}$$

The time-dependent equations are thus nonlinear integrodifferential equations, and are complicated enough that, to date, only vastly simplified problems have been solved numerically. However, except very near the edge of the nebula, the time scale for photoionization and recombination is shorter than the dynamical time scale, so it is correct to assume a static nebular model everywhere except there.

In addition to the continuous variations in ρ, $\mathbf{v}$, and so on, implied by equations (6.1) - (6.6), there may also be near-discontinuities, or shock and ionization fronts, in nebulae. Let us first consider a shock front, across which ρ, $\mathbf{v}$, and p change discontinuously, but the ionization does not change. Actually, of course, a real shock front is not an infinitely sharp discontinuity, but in many situations the mean-free path for atomic collisions (which gives the relaxation length) is so short in comparison with the dimensions of the flow that ρ, $\mathbf{v}$, and p are nearly discontinuous. For this analysis it is most convenient to use a reference system moving with the shock front, for if the motion is steady, this reference system moves with constant velocity. If we assume a plane, steady shock, and denote the physical parameters ahead of and behind the shock by subscripts 0 and 1, respectively, then the momentum and mass-conservation conditions across the front, corresponding to equations (6.1) and (6.3), respectively, are, in this special reference sytem,

$$p_0 + \rho_0 v_0^2 = p_1 + \rho_1 v_1^2, \tag{6.7}$$

$$\rho_0 v_0 = \rho_1 v_1, \tag{6.8}$$

where the velocity components are in the direction of motion perpendicular to the front.

Furthermore, the energy-conservation condition found by integrating equation (6.5) across the front, using the equation of state (6.2), is that the gas is compressed adiabatically:

$$p = K\rho^\gamma. \tag{6.9}$$

For a monatomic gas, as in an H II region, $\gamma = 5/3$, while for a diatomic gas $\gamma = 7/5$, etc. This relation is generally used by substituting it into the equation of motion (6.1), taking the dot product with $\mathbf{v}$, and integrating through the front, giving

$$\frac{1}{2} v_0^2 + \frac{\gamma}{\gamma - 1} \frac{p_0}{\rho_0} = \frac{1}{2} v_1^2 + \frac{\gamma}{\gamma - 1} \frac{p_1}{\rho_1}, \tag{6.10}$$

or, for $\gamma = 5/2$,

$$\frac{1}{2} v_0^2 + \frac{5}{2} \frac{p_0}{\rho_0} = \frac{1}{2} v_1^2 + \frac{5}{2} \frac{p_1}{\rho_1}. \tag{6.11}$$

Note that the first term on either side of equation (6.11) represents the flow-kinetic energy per unit mass, and the second term may be broken up into two contributions, $(3/2)p/\rho = (3/2)kT/\mu m_{\mathrm{H}}$, the thermal kinetic energy per unit mass, and p/ρ, the compressional contribution to the energy per unit mass. The more general form, (6.10), includes, in addition, the energy contributions of the internal degrees of freedom of the gas molecules.

Equations (6.7), (6.8), and (6.10) or (6.11) are the familiar Rankine-Hugoniot conditions on the discontinuities at a shock front. However, the physical situation in a gaseous nebula is quite different from that in a laboratory shock tube, and as a result the applicable equations often take a different form. To see this, we estimate the order of magnitudes of the various terms in equation (6.4). From the discussion of Chapter 3, and particularly Figures 3.2 and 3.3, we know that the heating and cooling rates G and L are of order $10^{-24} N_e N_p$ erg cm^{-3} sec^{-1}, and if we consider a "typical" nebula with density $N_e \approx N_p \approx 10^3$ cm^{-3}, intermediate between bright planetaries and bright H II regions, $G \approx L \approx 10^{-18}$ erg cm^{-3} sec^{-1}. At the equilibrium temperature $T \approx 10{,}000°$ K, $U \approx 10^{-9}$ erg cm^{-3}; so typical time scales for heating and cooling by radiative processes are $U/G \approx 10^9$ sec ≈ 30 yr. On the other hand, typical velocities in nebulae are of order of a few times the velocity of sound, at most 30 km sec^{-1}, which corresponds to 10^{-12} pc sec^{-1}. Since the sizes of nebulae are typically in the range 0.1 pc (planetary nebulae) to 10 pc (H II regions), the time scales for appreciable expansion or motion are considerably longer than 10^9 sec, and the heating and cooling rates due to compression and dilation in equation (6.4) are therefore considerably smaller than the heating and cooling rates due to radiation. Thus to a first approximation, the temperature in the nebula is fixed by radiative processes, independently of the hydrodynamic conditions, and a shock front in a nebula may be considered isothermal. What happens, of course, is that across the actual shock front equation (6.10) applies, and the temperature is higher behind the front than ahead of it. But in the hot region immediately behind the front, the radiation rate is large and the gas is very rapidly cooled, so that relatively close behind the shock the gas is again at the equilibrium temperature, the same temperature as in the gas just ahead of the shock. (Here the simplification has been made that the equilibrium temperature does not depend on the density, as it must physically, because of the change in the ionization of the heavy elements which are responsible for the cooling. This is a small effect which we neglect here.) The jump conditions (6.7) and (6.8), instead of (6.10),

$$\frac{p_0}{\rho_0} = \frac{p_1}{\rho_1} = \frac{kT}{\mu m_{\rm H}}, \tag{6.12}$$

corresponding to $\gamma \rightarrow 1$ in (6.10), can therefore be applied between the points just ahead of the shock and the points close behind it. The thickness of this "isothermal shock front" is fixed by the radiation rate and is of order (for the conditions assumed previously) 10^{-3} pc.

Next let us consider an ionization front, across which not only ρ, v, and p, but also the degree of ionization change discontinuously. This is a good approximation at the edge of an ionization-bounded region, because, as we have seen, the ionization decreases very sharply in a distance of the order of the mean free path of an ionizing photon, about 10^{-3} pc for the density $N_{\rm H} = 10^3$ cm^{-3} assumed previously. Across this front, the momentum and mass conservation conditions (6.7) and (6.8) still apply. However, the energy-conservation condition is different from that which applies at a shock front, because energy is added to the gas crossing the ionization front. Furthermore, the rate of flow of gas through the ionization front is fixed by the flux of ionizing photons arriving at the front, since each ionizing photon produces one electron-ion pair. Thus equation (6.8) becomes

$$\rho_0 v_0 = \rho_1 v_1 = m_i \phi_i, \tag{6.13}$$

where ϕ_i is the flux of ionizing photons,

$$\phi_i = \int_{\nu_0}^{\infty} \frac{\pi F_\nu}{h\nu} \, d\nu, \tag{6.14}$$

and m_i is the mean mass of the ionized gas per newly created electron-ion pair. Let us write the excess kinetic energy per unit mass transferred to the gas in the ionization process as $q^2/2$, defined by the equation

$$\phi_i \left(\frac{1}{2} m_i q^2 \right) = \int_{\nu_0}^{\infty} \frac{\pi F_\nu}{h\nu} (h\nu - h\nu_0) \, d\nu. \tag{6.15}$$

The conservation of energy across the ionization front may then be expressed in the form

$$\frac{1}{2} v_0^2 + \frac{5}{2} \frac{p_0}{\rho_0} + \frac{1}{2} q^2 = \frac{1}{2} v_1^2 + \frac{5}{2} \frac{p_1}{\rho_1} \tag{6.16}$$

instead of (6.11), in which the extra term on the left-hand side represents the kinetic energy per unit mass released in the photoionization process.

Once again, however, we note that ordinarily in gaseous nebulae the radiative cooling is quite rapid, and as a result a short distance behind the front the temperature reaches the equilibrium values set by the balance between radiative heating and cooling. This does not, however, lead to an isothermal ionization front, because the ionization conditions and hence the heating and cooling rates are quite different on the two sides of the front. Therefore, instead of equations (6.12) or (6.16) we have the conditions

$$\frac{p_0}{\rho_0} = \frac{kT_0}{\mu_0 m_{\rm H}}$$

and

$$\frac{p_1}{\rho_1} = \frac{kT_1}{\mu_1 m_{\rm H}}, \tag{6.17}$$

where T_0 and T_1 are constants fixed by the heating and cooling rates in the $\rm H^0$ region ahead of the shock and in the $\rm H^+$ region behind it, respectively, and μ_0 and μ_1 are the corresponding mean molecular weights. Very rough order-of-magnitude estimates are $T_0 \approx 100°$ K, $T_1 \approx 10{,}000°$ K, $\mu_0 \approx 1$, $\mu_1 \approx 1/2$.

6.3 Ionization Fronts and Expanding H^+ Regions

This section will first consider the shock fronts that can occur in the ionized H^+ regions and in the neutral H^0 regions outside them. The ionization fronts that separate the two regions will then be discussed and classified, and finally this classification will be used to describe the evolution and expansion of an idealized H^+ region formed when a hot star is formed in an initially neutral H^0 region.

The jump conditions (6.7), (6.8), and (6.10) relate ρ_0, v_0, and p_0, the physical conditions ahead of the front, with ρ_1, v_1, and p_1, the corresponding conditions behind the front. These equations may be solved to give any three of these quantities in terms of any other three; for our purposes it is most convenient to consider ρ_0 and p_0 given, and to express the ratios ρ_1/ρ_0 and p_1/p_0 in terms of the Mach number M of the shock front. This can be expressed in terms of c_0, the sound speed in the undisturbed region ahead of the shock,

$$c_0 = \sqrt{\frac{\gamma p_0}{\rho_0}} = \sqrt{\frac{\gamma k T_0}{\mu_0 m_H}}. \tag{6.18}$$

For example, for an isothermal ($\gamma = 1$) shock in an H^0 region, with $T = 100°$ K, $c_0 \approx 0.9$ km sec^{-1}, while for an adiabatic ($\gamma = 5/3$) shock in an H^+ region with $T = 10{,}000°$ K, $c_0 = 17$ km sec^{-1}. Then the Mach number is defined as

$$M = \frac{|v_0|}{c_0}, \tag{6.19}$$

the ratio of the speed of the shock, with respect to the gas ahead of the front, to the sound speed in this gas. The Mach number ranges between the limits $M \to 1$ for a weak shock, which in this limit is just an infinitesimal disturbance propagating with the velocity of sound, to $M \to \infty$ for a strong shock propagating extremely supersonically.

It is straightforward to show that, in terms of this parameter, the ratio of pressures behind and ahead of the shock is

$$\frac{p_1}{p_0} = \frac{2\gamma}{\gamma + 1} M^2 - \frac{\gamma - 1}{\gamma + 1}, \tag{6.20}$$

and the ratio of densities is

$$\frac{\rho_1}{\rho_0} = \frac{(\gamma + 1)M^2}{(\gamma - 1)M^2 + 2}. \tag{6.21}$$

and approaches infinity for an isothermal shock. Thus very great compressions occur behind strong isothermal shocks.

Across an ionization front, the jump conditions described by (6.7), (6.8), and (6.12) or (6.17) relate ρ_0, v_0, and p_0, while (6.14) gives the flux through the front. Let us consider the simplified form of the conditions described by (6.17) and correspondingly express the results in terms of the isothermal sound speeds $\gamma = 1$ in equation (6.18). Then, solving for the ratio of densities,

$$\frac{\rho_1}{\rho_0} = \frac{c_0^2 + v_0^2 \pm [(c_0^2 + v_0^2)^2 - 4c_1^2 v_0^2]^{1/2}}{2c_1^2}. \tag{6.22}$$

Physically ρ_1/ρ_0 must be real, and therefore there are two allowed ranges of speed of the ionization front,

$$v_0 \geq c_1 + \sqrt{c_1^2 - c_0^2} \equiv v_R \approx 2c_1, \tag{6.23}$$

or

$$v_0 \leq c_1 - \sqrt{c_1^2 - c_0^2} \equiv v_D \approx \frac{c_0^2}{2c_1}, \tag{6.24}$$

where the approximations apply for $c_1 \gg c_0$, which, as we have seen previously, is the case in H II regions. The higher critical velocity v_R is the velocity of an "R-critical" front; here R stands for "rare" or "low-density" gas since for a fixed ϕ_i as $\rho_0 \rightarrow 0$, $v_0 \rightarrow \infty$ and must ultimately become greater than v_R. Likewise, the lower critical velocity v_D is the velocity of a D-critical front, with D standing for "dense" or "high-density" gas. Ionization fronts with $v_0 > v_R$ are called R-type fronts, and since $c_1 > c_0$, $v_R > c_0$, and these fronts move supersonically into the undisturbed gas ahead of them, while D-type fronts with $v_0 < v_D < c_0$ move subsonically with respect to the gas ahead of them.

Let us consider the evolution of the H^+ region that would form if a hot star were instantaneously "turned on" in an infinite homogeneous H^0 cloud. Initially, very close to the star, ϕ_i is large and a (spherical) R-type ionization front moves into the neutral gas. Let us simplify greatly by omitting c_0 and expanding (6.22) for $v_0 \ll c_1$. The results are

$$\frac{\rho_1}{\rho_0} = \left\{ \begin{array}{c} \dfrac{v_0^2}{c_1^2}\left(1 - \dfrac{c_1^2}{v_0^2}\right) \gg 1 \\ \\ 1 + \dfrac{c_1^2}{v_0^2} \approx 1 \end{array} \right\} \tag{6.25}$$

for the positive and negative signs, respectively, correct to the second order in

c_1/v_0; and these two cases are called strong and weak R-type fronts, respectively. The corresponding velocities from (6.13) are

$$v_1 = \left\{ \begin{array}{l} \dfrac{c_1^2}{v_0} \ll c_1 \\ v_0\left(1 - \dfrac{c_1^2}{v_0^2}\right) \approx v_0 \gg c_1 \end{array} \right\}, \tag{6.26}$$

respectively. Thus in a strong R-type front, the velocity of the ionized gas behind the front is subsonic with respect to the front, and the density ratio is large; on the other hand, in a weak R-type front, the velocity of the ionized gas behind the front is supersonic, and the density ratio is close to unity. A strong R-type front cannot exist in nature because disturbances in the ionized gas behind it continually catch up with it and weaken it; the initial growth of the H^+ region occurs as a weak R-type front, runs out into the neutral gas, leaving the ionized gas behind it only slightly compressed and moving outward with subsonic velocity (in a reference system fixed in space)

$$v_0 - v_1 = c_1\left(\frac{c_1}{v_0}\right) \ll c_1. \tag{6.27}$$

Though these analytic results only hold to the first order in c_1/v_0 and the zeroth order in c_0/c_1, the general description is valid so long as the ionization front remains weak R-type. Physically, an R-type front is one in which there are so many ionizing photons that the front moves very rapidly into the neutral gas ahead of it. Thus the pressure behind the front rises greatly, because of the heating and the ionization, but the density does not, because there is little time in which the gas may move.

However, as the front runs out into the neutral gas, the ionizing flux ϕ_i decreases both because of geometrical dilution and because of recombinations and subsequent absorption of ionizing photons interior to the front. Thus, from equation (6.13), v_0 decreases and ultimately reaches v_R, and from this time onward the simple R-type front can no longer exist. At this point, $\rho_1/\rho_0 = 2$ and $v_1 = c_1$ (again to the zeroth order in c_0/c_1), that is, the ionized gas behind the front is moving just sonically with respect to the front. At this moment a shock front breaks off from the ionization front, and the now D-critical ionization front follows it into the precompressed neutral gas. As time progresses, the shock front gradually weakens (because of the geometrical divergence) and the ionization front continues as a strong D-type front, with a large density jump. Physically, the shock is compressed by the ram pressure of the ionized material streaming away from the ionization front toward the star. The shock front increases the density ahead of the ionization front, and therefore makes it move more slowly.

This behavior is shown graphically in Figure 6.1, the result of the numerical integration (with a simplified cooling law) of the system of partial differential equations described previously. The graphs show the velocity and density as functions of radial coordinates at nine consecutive instants of time ranging from 2.2×10^4 yr to 2.0×10^6 yr after turn-on of a model O7 star in a homogeneous H^0 region with initial density 6.5 cm^{-3}. For the first two time steps the ionization front is weak R-type, but between 7.8×10^4 yr and 9.0×10^4 yr it becomes R-critical and a shock front breaks off. It can be seen as the discontinuity just slightly ahead of the ionization front at 9.0×10^4 yr. With advancing time the shock slowly advances with respect to the ionization front, compressing the neutral gas, while the ionized gas within the H^+ zone expands and the density decreases, so that by 2.0×10^6 yr it is, on the average, only about 0.2 of the density in the undisturbed H^0 zone. If the integration were carried further in time, the shock would weaken still further and the density in the H^+ zone would decrease to the final pressure-equilibrium value

$$\frac{\rho_1}{\rho_0} = \frac{T_0}{2T_1} \approx \frac{1}{2 \times 10^2}, \tag{6.28}$$

but in real nebulae the ionizing star burns out long before this stage is reached.

One interesting feature of an ionization front is the high temperature near its leading edge. It results from the fact that when the free electrons are first produced by photoionization, they appear in the gas at a temperature T_i, set by the mean energy above the threshold of the ionizing photons, as described in Section 3.2. For O stars these mean energies are quite high; so in the ionization fronts which they produce, T_i is typically 2 to 4×10^4 ° K, as Table 3.1 shows. Cooling by radiation, chiefly emission-line photons, quickly reduces the electron temperature T to the equilibrium value. Since appreciable numbers only of neutral atoms and singly ionized ions are present in the front, this process greatly strengthens their emission lines, especially [O II] λ3727, [N II] $\lambda\lambda$6548, 6583, and [O I] $\lambda\lambda$6300, 6364. The thickness of the layer in which the electron temperature rises to its peak and then falls to the equilibrium temperature is the "cooling length," the distance the ionization front travels during the cooling time scale mentioned above.

For a typical calculated model of an O-star ionization front in an H II region with density $N_p = 10$ cm^{-3}, and D-type ionization-front speed 2.5 km sec^{-1}, representative of much of its lifetime, the calculated peak temperature reached is 16,000° K. The thickness of the front is about 0.03 pc, and the calculated emission-line intensity ratios for the radiation emitted from the front alone are [O II] $\lambda3729/H\beta \approx 20$, [N II] $\lambda6583/H\alpha \approx 2$ and [O I] $\lambda6300/H\alpha \approx 0.5$. All these ratios are much higher than their values in the main body of the nebula, which is in ionization and thermal equilibrium. Though the specific numerical values depend upon assumed abundances, model stellar atmospheres, etc., the general trend is certainly valid and explains the great strength of [O II] $\lambda3727$ and [N II] $\lambda\lambda6548, 6583$ in ionization fronts noted by early observers.

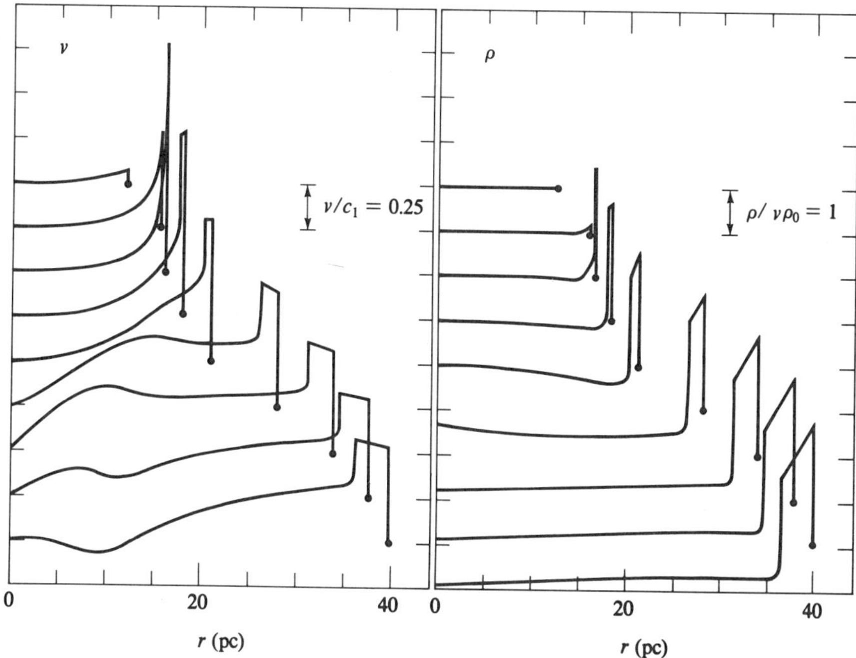

FIGURE 6.1
Simplified model of expanding H II region with initial $N_H = 6.4$ cm^{-3} and $v = 0$, around an O7 star that is turned on at $t = 0$. Left-hand side shows v/c_1 and right-hand side shows ρ/ρ_0, both as functions of r in pc. Successive time steps shown are 2.2×10^4 yr, 7.8×10^4 yr, 9.0×10^4 yr, 1.8×10^5 yr, 3.6×10^5 yr, 9×10^5 yr, 1.4×10^6 yr, 1.8×10^6 yr, and 2×10^6 yr. Each time is displaced downward by $v/c_1 = 0.25$ and by $\rho/\rho_0 = 1$ from the previous time.

6.4 Comparisons with Observational Measurements

The comparatively straightforward theory of spherically symmetric, expanding H II regions given in Section 6.3 is exceedingly difficult to check observationally, because there are few nearby nebulae to which it directly applies. As the photographs in this book show, actual nebulae have a very complicated, nonuniform density distribution; they exhibit brightness fluctuations down to the smallest observable scale. Since these brightness fluctuations are already integrated along the line of sight by the observational technique itself, the actual small-scale density variations must be even more extreme. This is readily confirmed by the measurements of electron density from [O II] and [S II] line ratios, as explained in Section 5.5. Thus the basic picture of a homogeneous "infinite" cloud ionized by a single star within it does not apply except as a very rough approximation. Furthermore, it is quite unlikely that the initial velocity is everywhere zero, as is assumed in the available calculations. Finally, the observational evidence seems to show that massive star formation occurs in the dense cores of H_2 molecular clouds, not H^0 clouds.

We may nevertheless hope that the calculated results will also be true in some very large-scale average over space. The large, faint, low-density H II region around the O6 star λ Ori has an approximately circular form, with a brightness distribution quite similar to those implied by the density distribution for some of the later stages shown in Figure 6.1. However, since the expected velocities are relatively small, of order 10 km sec^{-1} (the isothermal velocity of sound), quite high-dispersion spectral measurements are required, which are difficult to obtain because of the low surface brightness of typical nebulae. Therefore, to date only a few of the brightest objects have been studied.

By far the most complete observational study is available for NGC 1976, the Orion Nebula. In the central brightest regions, multislit spectrograms have been obtained covering an area of $4' \times 4'$ centered on θ^1 Ori, the Trapezium, the exciting multiple star, at a dispersion of 4.5 Å mm^{-1}, corresponding to an instrumental profile with full width at half maximum of approximately 9 km sec^{-1}, and a probable error of measurement of the peak of a line of approximately 1 km sec^{-1}. An example of the [O III] λ5007 images in one of these spectrograms is shown in Figure 6.2. On these spectrograms [O II] λ3726, Hγ and [O III] λ5007 were measured for velocity at a rectangular grid of points separated by $1''.3$ in each direction, and in addition the line profiles were measured at selected points. Further from the center of NGC 1976, spectrograms are available at several regions at greater distances (up to about $10'$) from θ^1 Ori at the lower dispersion of approximately 9.2 A mm^{-1}. In addition, Fabry-Perot line profiles of [O III] λ5007 with an angular resolution of $6''.5$ or $13''$ and an instrumental profile with a width of about 4 km sec^{-1} are available for the central brightest region.

The measured velocities show no very strong evidence of expansion of the nebula similar to that calculated for an initially homogeneous cloud in the previous section. As can be seen in Figure 6.3, there is no clearly apparent pattern of maximum velocity of approach and recession along the line through the center of the nebula, dropping to zero radial velocities at the edges of the nebula. The mean radial velocity of the [O III] projected in the center of the nebula is about 10 km sec^{-1} more negative than the radial velocity of the exciting stars: the ionized gas is approaching the Earth with this velocity relative to the stars. This is a clue from the optical spectral region that NGC 1976 is in fact not a spherically symmetric structure. Radio-frequency measurements of molecular emission lines, made within the past decade or two, have shown this very clearly, and that it is an ionized indentation

FIGURE 6.2

Multislit image of [O III] $\lambda 5007$ in an area of NGC 1976, the Orion Nebula, with center 7″.4 E, 27″.0 of θ^1 Ori C. Dimensions are 35″ E-W (along the slits) and 41″ N-S (along dispersion). Slits are separated by 1″.3 (*North is at right, East is at top*). Note how the broadening, doubling, and intensity change from point to point. (*Palomar Observatory photograph.*)

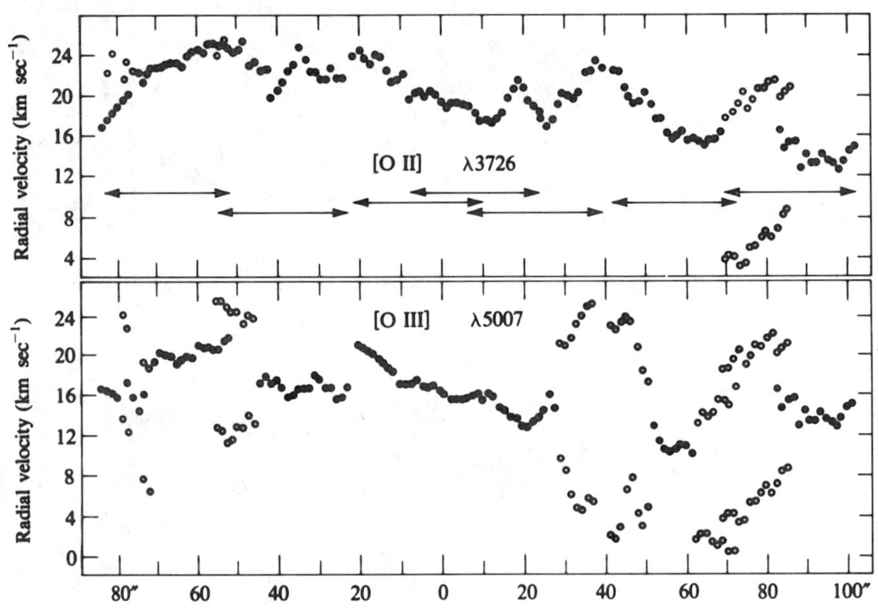

FIGURE 6.3
Radial velocities measured in [O II] and [O III] lines in NGC 1976 along an E-W line 25″ S of θ^1 Ori C. Solid circles indicate points at which line is measured single; open circles indicate points at which line is measured double. The length measured on each separate spectral plate is indicated by an arrow.

at the front face of a large molecular cloud. The O stars are outside the cloud, ionizing a bubble or blister into the face of it; ionized material is continually streaming out from it as discussed in the next section.

The main impression given by the measured velocities in NGC 1976 is one of a turbulent velocity field. First of all, the line profiles at many points in the nebula are approximately Gaussian, but have line widths greater than the thermal width, corresponding quantitatively to a radial velocity dispersion (root-mean-square deviation from the mean) of from 4 km sec $^{-1}$ to 12 km sec^{-1} after the thermal Doppler broadening for $T = 10{,}000°$ K has been removed. This dispersion represents the effect of motions along the line of sight integrated through the whole nebula. At many other points in the nebula, the emission lines are split or double; that is, the profiles are not Gaussian but show double peaks, with separations ranging from 10 km sec^{-1} to 20 km sec^{-1}. Thus, in the regions of line doubling, there are velocity differences along the line of sight through the nebula as large as twice the velocity of sound. The regions in which line doubling occurs are continuous with regions with single lines, and the weighted average velocity of the two components is continuous with the velocity measured from the "single" line just outside these regions. Further, in these regions the resolved [O II] and [O III] components often have different relative intensities, showing that there are differences between the ionization, temperature, or density in the two emitting volumes with different mean velocities. Finally, there are regions where the lines, though not resolved into two components, have strongly non-Gaussian profiles, evidently representing lines of sight with less extreme velocity variations than those that give rise to line splitting.

The measured radial velocities of the peaks of the lines vary with position in the nebula, again with no regular pattern such as would be expected from an expanding sphere, but in an irregular way, with the root-mean-square radial-velocity difference between two positions increasing approximately as the 0.33 power of the distance between them. For other well observed H II regions similar patterns have been observed, but more often with a somewhat smaller power, such as 0.1 or 0.2. The observed situation is evidently much closer to what is ordinarily called turbulence than to expansion. There is no obvious correlation between the velocity variations and the apparent surface brightness variations, though the latter show that there are tremendous density variations within the nebula.

It seems most likely that the energy source driving the observed turbulence is the primary photoionization process. Dense non-ionized cold "clouds" probably are contained within the ionized region, and as these regions become ionized, hot gas expands away from them and interacts with gas expanding from other similar clouds. The turbulence thus results from photoionization of an initially very inhomogeneous neutral-gas complex. The theoretical description of this situation is complicated, and only small parts of it have yet been worked out, in particular, the ionization of dense spherical neutral "globules" by a spherically symmetric ionizing radiation field. Undoubtedly, future progress must come from calculations of more realistic models with asymmetric structures, and must include a statistical treatment of the combined effect of many such structures.

Fabry-Perot interferometers, which have the advantage of providing high wavelength resolution on low-surface-brightness objects, have also been used for measurements of line profiles in H II regions. Several of them such as M 8, M 16, and M 17, have been measured, and the results obtained were rather similar to those obtained with NGC 1976, but the only nebula for which an expansion pattern is clearly seen is NGC 2244, the large Rosette Nebula in Monoceros. A dominantly turbulent velocity field, presumably caused by ionization and expansion of structures within an initially nonhomogeneous H I region, is found in all H II regions so far observed, and undoubtedly requires serious theoretical study. Pressures due to winds from O stars are undoubtedly also important in some H II regions.

6.5 Non-spherically Symmetric Models

As was stated above, one reason for the failure of the spherically symmetric, expanding models of H II regions has been revealed by radio-frequency measurements of molecular emission lines. They show that many optically observed H II regions, not just the Orion Nebula, are located in or near the edges of dense molecular clouds, into which the ionization fronts are propagating. Evidently the exciting stars that we see have formed near the edges of these regions. No doubt there are other H II regions about stars within the giant clouds, and still others associated with stars near the edges of the clouds or on the far side from us, that are invisible because of the presence of large amounts of interstellar dust, and hence large optical depths, within the clouds.

NGC 1976 is the prime example of an H II region at the edge of a dense cloud. A highly schematized sketch of its structure, derived from optical, infrared, and radio data, including molecular-line observations, is shown in Figure 6.4. The ionized region is a "blister" of indentation at the edge of a dense neutral molecular cloud. The region outside the cloud may be thought of as a much less dense "intercloud medium."

If an ionizing star turns on inside an idealized homogeneous, bounded cloud, at a distance $d < r_1$ (the Strömgren radius in the dense medium) from the edge of the cloud the ionized region grows more or less along the lines of the previous discussions until its radius $r = d$. Then the ionized region breaks out into the less dense medium where the weak R-type ionization front runs out more rapidly. A rarefaction wave runs back into the denser ionized region. Behind it the flow velocity is outward, toward the less dense medium. Figure 6.5 is a sketch of a late stage of this evolution, based on numerical calculations. Even if the two media are assumed to be homogeneous initially,

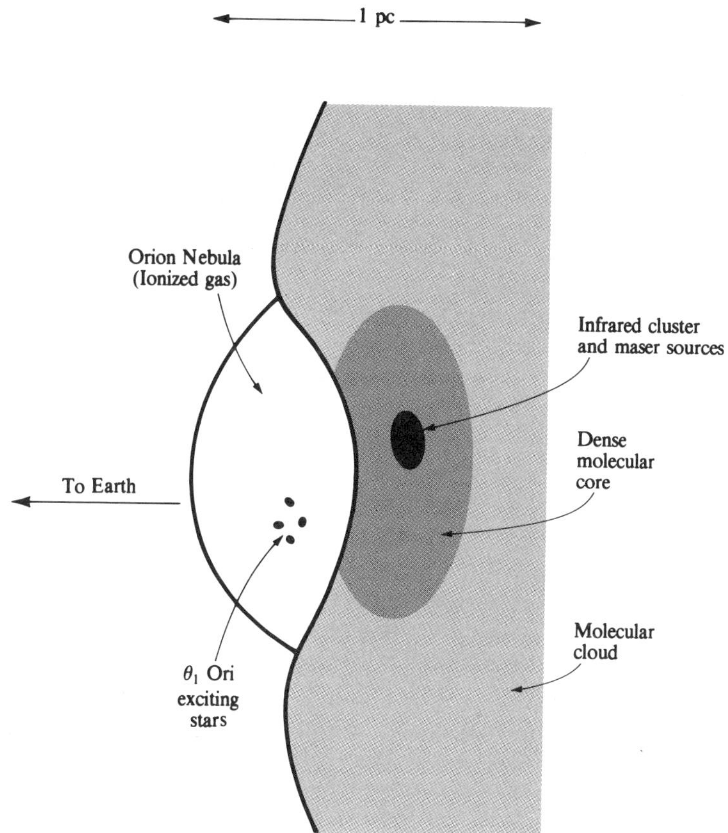

FIGURE 6.4
Schematic sketch of NGC 1976, the Orion Nebula, an ionized blister of gas expanding away from the Orion molecular cloud, into which the ionization front is moving.

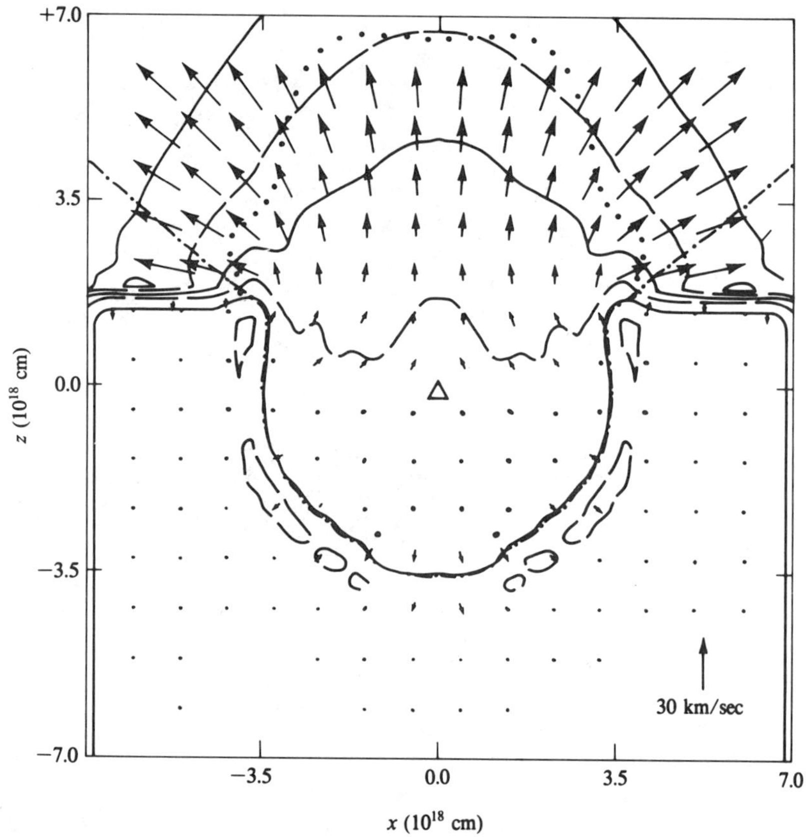

FIGURE 6.5
Cross section of dense molecular cloud (in lower part of figure) ionized by star (at triangle), expanding into low-density intercloud medium (in upper part of figure). The initial density in the cloud is 10^3 cm^{-3}, and the isodensity surfaces and velocity vectors are shown 9×10^4 yr after the star (with $T_* = 40{,}000°$ K) turns on.

there is symmetry only about the z axis through the star and perpendicular to the contact discontinuity between the two media. Thus a two-dimensional (r, θ) or (ρ, z) axisymmetric hydrodynamics code must be used to calculate the evolution as the ionization front eats into the dense cloud, and the ionized gas expands rapidly outward into the less dense medium.

The flow patterns for these simplified "champagne" or "blister" models are much more like the observed velocity fields (for instance, in NGC 1976) than are the simpler results calculated for spherically symmetric models. However, they still do not show the complicated "turbulent" type of velocity field exhibited by the observations such as Figure 6.2 and 6.3. This small-scale structure results from the small-scale density structure of the initially neutral region through which the ionization front is running. It can be mimicked by a more complicated model, with small, dense condensations, idealized as spherical, in the initial neutral medium. Such a model will produce velocity patterns centered at each condensation, and if there are enough of them of different density contrasts and randomly scattered through the medium, it is reasonable to suppose that the "turbulence," including the observed line doubling, can be reproduced. No calculations have been carried to this level of sophistication to date, however.

6.6 The Expansion of Planetary Nebulae

The expansion of planetary nebulae is better understood than the internal velocity distribution of H II regions, apparently largely because the planetary nebulae have more nearly symmetric and initially more nearly homogeneous structures and velocity fields. The earliest high-dispersion spectral studies of the planetary nebulae showed that, in several objects, the emission lines have the double "bowed" appearance shown in Figure 6.6. Later, more complete observational studies of many nebulae showed that in nearly all of them the emission lines are double at the center of the nebula; the line splitting is typically of order 50 km sec^{-1} between the two peaks, but decreases continuously to 0 km sec^{-1} at the apparent edge of the nebula. This can be understood on the basis of the approximately radial expansion of the nebula from the central star, with a typical expansion velocity of order 25 km sec^{-1}. Observations further show that there is a systematic variation of expansion velocity with degree of ionization: the ions of highest ionization have the lowest measured expansion velocity, while the ions of lowest ionization have the highest expansion velocity. Since the degree of ionization decreases outward from the star, this observation clearly shows that the expansion velocity increases outward. In this section this expansion will first be discussed theoretically, and then the theory will be compared in detail with the available observational results.

Planetary nebulae are hot ionized gas clouds, which, once formed, must

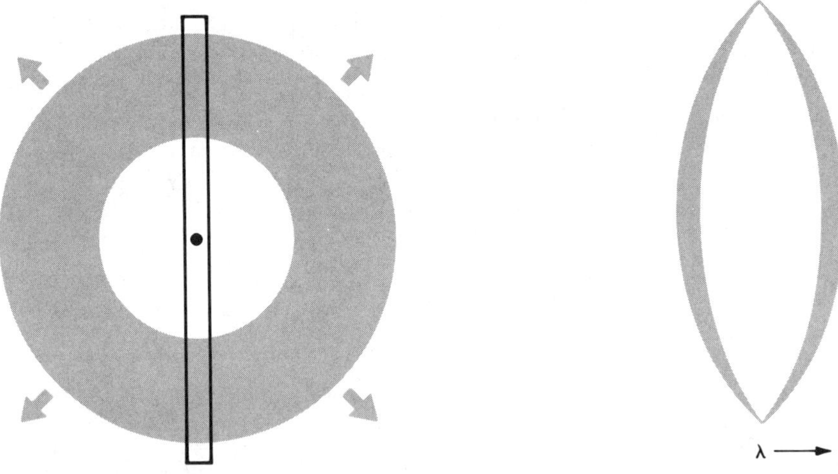

FIGURE 6.6
Diagram on left shows rectangular slit of spectrograph superimposed on expanding idealized, spherically symmetric planetary nebula. Resulting image of a spectral line emitted by the planetary, split by twice the velocity of expansion at the center and with splitting decreasing continuously to 0 at edges, is shown on right.

expand into the near vacuum that surrounds them. The expansion of a gas cloud into a vacuum is an old problem that was first treated by Riemann, and only the results will be given here. In the expansion of a finite homogeneous gas cloud, $\rho = \rho_0 =$ constant, released at $t = 0$ from the state of rest $v = 0$, and following the adiabatic equation (6.9), the edge of the gas cloud expands outward with a velocity given by

$$v_e = \frac{2}{\gamma - 1} c_\gamma, \tag{6.29}$$

while a rarefaction wave moves inward into the undisturbed gas with the adiabatic sound speed c_γ. Thus, at a later time t, the rarefaction wave has reached a radius

$$r_i = r_0 - c_\gamma t, \tag{6.30}$$

where r_0 is the initial radius of the gas cloud, while the outer edge has reached a point

$$r_e = r_0 + v_e t, \tag{6.31}$$

and all the gas between these two radii is moving outward with velocity increasing from 0 at r_i to v_e at r_e. For a spherical nebula, the inward-running rarefaction wave ultimately reaches the center of the nebula and is reflected, and the gas near the center is then further accelerated outward.

In an actual planetary nebula, the adiabatic equation (6.9) is not a very good approximation because, as was discussed previously, the heating and cooling are mainly by radiation, and the resulting flow is very nearly isothermal except at extremely low densities. The isothermal approximation corresponds to the limit $\gamma \to 1$ in equations (6.29) to (6.31), in which case the outer edge of the nebula expands with velocity $v_e \to \infty$, but the density within the rarefaction wave falls off exponentially, so the bulk of the gas has a velocity not much higher than the sound velocity.

To go beyond this description, it is necessary to integrate numerically the hydrodynamic equations of Section 6.2. This has been done for a few specific models of planetary nebulae, assuming complete spherical symmetry and an idealized radiative-cooling law, and one set of results is shown in Figure 6.7. Here the initial configuration was a spherical shell, with inner radius 2.4×10^{17} cm and outer radius 3.0×10^{17} cm, set into motion with $v =$ 20 km sec^{-1} at $t = 0$; the density and velocity fields are shown at several subsequent times. Notice that the expansion velocity at all times increases more or less linearly outward, a common result in all spherical expansion problems, since the central boundary condition ensures that $v = 0$ there, and the material at the outer edge generally has the highest expansion velocity. The main difficulty with these models is that the calculated density profiles shown in Figure 6.7 characteristically have the highest density near $r = 0$, while most observed planetaries, including nearly all those for which velocities of expansion have been measured, have a ring-shaped appearance in the sky; that is, they appear to be objects with lower densities near their centers. Likewise, Figure 6.7 shows that, at the edge of the calculated model expanding planetary nebula, the density decreases outward with a long tail, which would be observed as a diffuse outer edge, while most real planetaries have a sharper outer edge.

The forms of most planetary nebulae clearly are not spherically symmetric. Many appear to be objects of approximately cylindrical symmetry, seen in various orientations. Of these, many can be understood to have roughly spheroidal density distributions, and others appear to have toroidal structure. Naturally, the problem of calculating numerically the expansion of a spheroidal or toroidal object would be considerably more difficult, and in fact no such calculations are yet available.

However, if a spherically symmetric model is assumed, then in order to get the observed central "hole" of a typical planetary nebula, we must suppose that some additional repulsive force is exerted on the nebular gas from the central star, or from the center of the nebula. This is produced by a "stellar wind" of high-speed gas flowing out from the outer layers of the central star. The existence of such winds from planetary-nebula central stars with velocities of order 10^3 km sec^{-1}, was predicted on the basis of such considerations. Their existence has since been confirmed by ultraviolet spectra of the central stars, obtained with the *IUE* satellite, which show strong, broad absorption lines arising from the gas in the wind. The wind velocities observed in this way range from about 300 km sec^{-1} up through 4,000 km sec^{-1}, to possibly as high as 8,000 km sec^{-1}. A computed model, in which such a stellar wind was taken into account as a pressure exerted on the inner edge of the spherical shell of the planetary nebula, is shown in Figure 6.8. This model shows significantly better agreement with both the shape and the velocity field of a typical planetary nebula, and its success is a confirmation of the idea of a stellar wind. However, since the pressure taken into account in the calculation results from the stopping of the high-speed "wind" particles in the gas at the inner edge of the planetary nebula, these particles must deliver energy to the gas there, heating it and resulting in additional radiation. This radiation might be in the form of X-ray radiation, far-ultraviolet emission lines or optical forbidden lines of high stages of ionization, such as [Fe X] λ6375 or [Fe XI] λ7892. In most planetary nebulae, no signs of such radiation have yet been observed, but to detect it would probably require high-resolution measurements of individual points near the inner edge of the nebula. However, in one peculiar object, Abell 30, very strong collisionally excited ultraviolet lines, especially C IV λ1549, clearly indicate additional heating, well above photo-ionization, by this mechanism. The ultraviolet spectrum of the central star in Abell 30 has strong P Cygni profiles, indicating a stellar wind with a velocity of 4,000 km sec^{-1}. Similar, weaker examples of wind heating may be present but as yet still undetected in other planetaries.

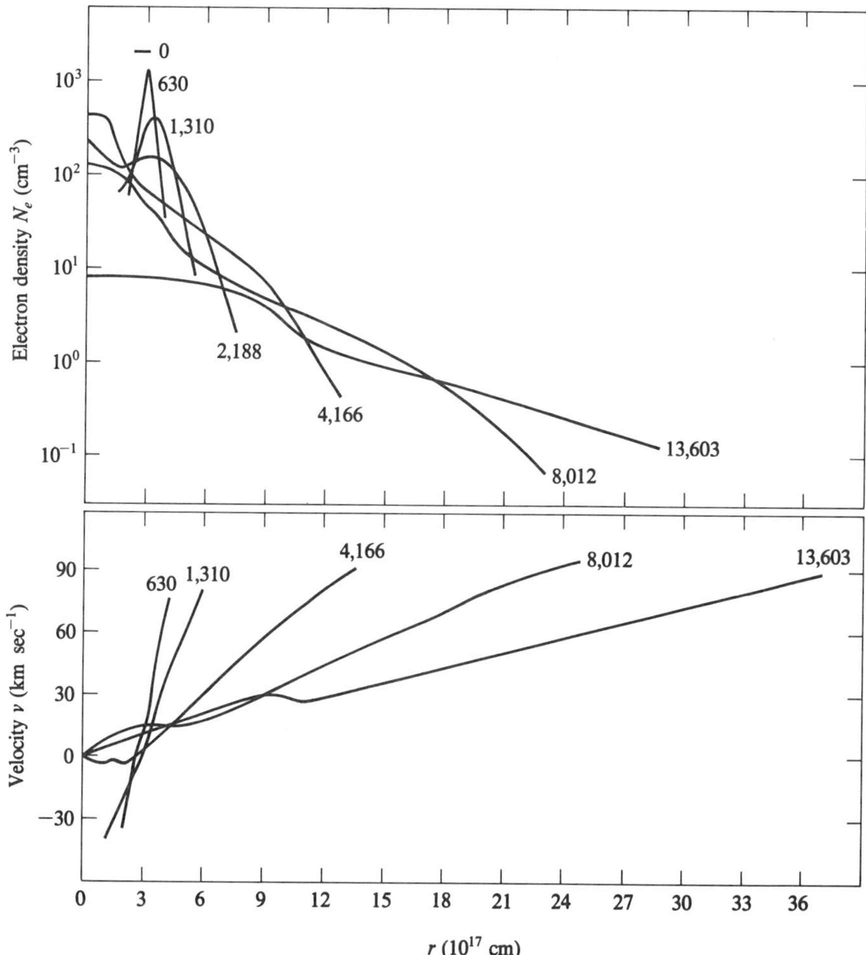

FIGURE 6.7
Top diagram shows calculated variation of electron density N_e with radius r at several times (given, in years, on curves) after expansion begins at $t = 0$ for model planetary nebula described in text. Bottom diagram shows calculated variation of expansion velocity v for same models. Initial homogeneous density distribution $N_e = 1.66 \times 10^3$ cm^{-3} is shown for $t = 0$; initial velocity is $v = 20$ km sec^{-1} at all r.

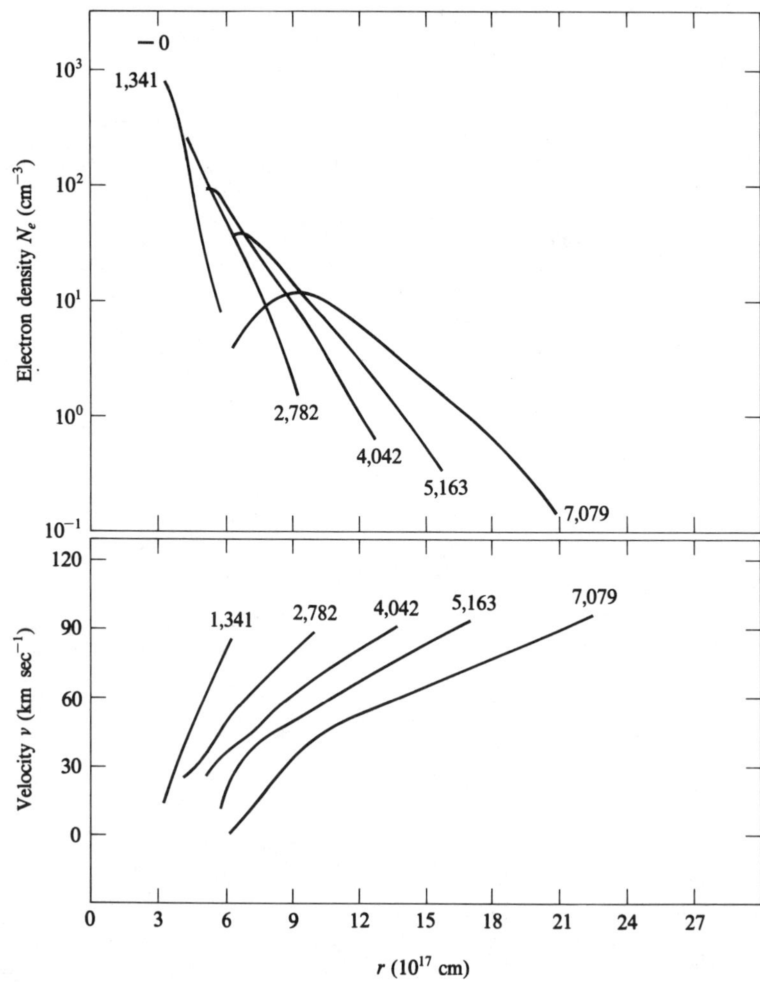

FIGURE 6.8
Top diagram shows calculated variation of electron density N_e with radius r at several times (given, in years, on curves) after expansion begins at $t = 0$ for model planetary nebula with stellar wind described in text. Bottom diagram shows calculated variation of expansion velocity v for same models. Initial conditions are the same as those for models described in Figure 6.7.

TABLE 6.1
Measured expansion velocities in planetary nebulae

Nebula	Velocity (km sec^{-1})					
	[O I]	H I	[O II]	[O III]	[Ne III]	[Ne V]
NGC 2392	–	–	53.0	52.6	57.0	0:
NGC 3242	–	20.4	–	19.8	19.5	–
NGC 6210	–	21.0	35.6	21.4	20.8	–
NGC 6572	16.0	–	16.8	–	–	–
NGC 7009	–	21.0	20.4	20.6	19.4	–
NGC 7027	22.8	21.2	23.6	20.4	22.4	19.1
NGC 7662	–	25.8	29.0	26.4	25.9	19.3
IC 418	25.0	17.4	0:	0:	0:	–

Let us turn next to brief examination of some of the observational data on the expansion velocities of planetary nebulae. Measured velocities (half the separation of the two peaks seen at the center of the nebula) for several ions in a number of fairly typical planetaries are listed in Table 6.1. In addition, in some planetary nebulae, small, diametrically opposite features ("bipolar jets") with considerably higher velocities, up to 200 km sec^{-1} in NGC 2392, are observed.

In addition, for some of the brighter planetary nebulae, spectrograms taken at a dispersion of about 4.1 Å mm^{-1} in the blue and about 6.5 Å mm^{-1} in the red, giving line profiles with a velocity resolution of about 5 km sec^{-1}, are available. For instance, the line profiles of NGC 7662 are shown in Figure 6.9. The double peaks, with wings extending over a total velocity range of 100 km sec^{-1}, are clearly shown, as well as the fact that the lines are asymmetric, a common feature in planetary nebulae. Though in NGC 7662 the peak with positive radial velocity is stronger, in other nebulae the reverse is true, and this asymmetry is clearly an effect of the departure from complete symmetry of the structure of the nebula itself.

The observed line profile results from the integration along the line of sight of radiation emitted by gas moving with different expansion velocities, further broadened by thermal Doppler motions, so that the observed profile $P(V)$ can be represented as an integral

$$P(V) = \text{const} \int_{-\infty}^{\infty} E(U) e^{-m(U-V)^2/2kT} \, dU, \tag{6.32}$$

where $E(U)$ is the distribution function of the emission coefficient in the line per unit radial velocity U for an ion of mass m in an assumed isothermal nebula of temperature T. In the lower part of Figure 6.9, synthetic line profiles calculated from this equation using an approximately triangular distribution function $E(U)$ (half of which is shown in the insert) can be seen to represent the observed profiles quite well. Though other forms of $E(U)$ also fit the observations just as well, all of them require a peak at approximately 25 km sec^{-1}, decreasing to nearly zero at approximately 10 km sec^{-1} and 40 km sec^{-1}. The expansion velocity in some nebulae is small enough that the ion with lowest atomic mass, H^+, has sufficiently large thermal Doppler broadening that a double peak is not seen; these are the nebulae measured to have "zero" expansion velocity in Table 6.1. However, in nebulae of this type for which line profiles have been measured, the same distribution function $E(U)$, which fits the resolved line profile of an ion of higher mass, such as [O III] $\lambda\lambda 4959$, 5007, also fits the unresolved H I profiles.

For some of the nearby planetary nebulae, the average velocity of expansion of 25 km sec^{-1} is large enough that the proper motion of expansion should be marginally detectable with a long-focal-length telescope over a baseline of order 50 yr. Though some measurements have been obtained, to date the results are inconclusive and contradictory. Part of the difficulty is to find sharp, well-defined features, whose positions can be accurately measured. To obtain results of this type that have been published to date, it was necessary to use first-epoch plates taken on blue-sensitive emulsions, which combine many different emission lines into the image. These plates unfortunately do not show many sharp features. It would be a very worthwhile long-term program to take systematically new first-epoch plates or images with a long-focus telescope using red-sensitive emulsions or CCDs and a filter to restrict the photograph of $H\alpha$ + [N II] $\lambda\lambda 6548$, 6583, a combination that does show many sharp features. Plates taken with the 200-inch Hale telescope by Minkowski and Baade in the 1950s could be repeated now with a baseline of over 30 years.

The main features of the expansion of planetary nebulae seem to be fairly well understood, but a good comparison of Doppler and proper-motion determinations of expansion velocity would give independent data on the distances of the nebulae measured.

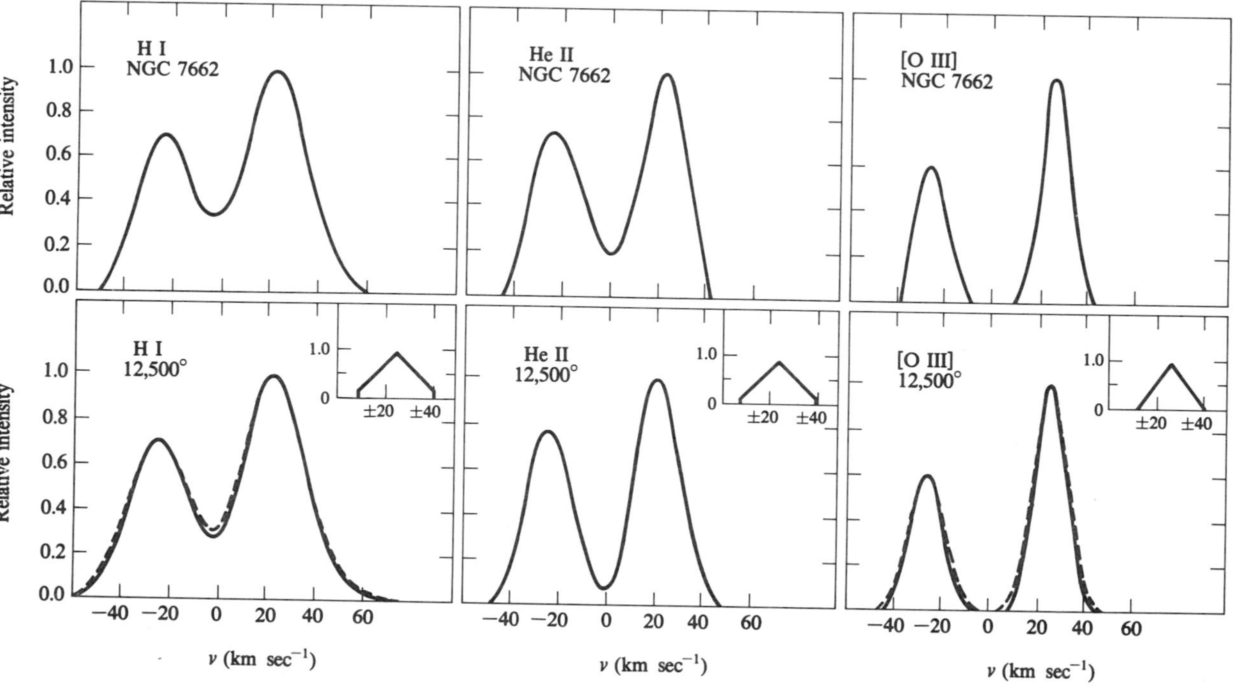

FIGURE 6.9
Top diagrams show observed emission-line profiles at center of planetary nebula NGC 7662. Bottom diagrams show calculated line profiles for simple model of expanding planetary nebula; distribution functions of emission coefficient in the line is shown in inserts in upper right-hand corner of each diagram. Dashed lines in H I and [O III] models show model corrected for very slight effects of instrumental broadening.

References

The importance of hydrodynamical studies of nebulae is clear, but little serious theoretical work on these problems was done previous to the Symposium on the Motions of Gaseous Masses of Cosmical Dimensions, sponsored by the International Union of Theoretical and Applied Mechanics (IUTAM) and the International Astronomical Union (IAU), held in Paris in 1949, the proceedings of which were published as

Burgers, J. M., and van de Hulst, H. C., eds. 1951. *Problems of Cosmical Aerodynamics.* Dayton, Ohio: Central Air Documents Office.

This symposium awakened the interest of astronomers, physicists, and gas dynamicists in studying nebular dynamics problems. Succeeding symposia continued to stimulate research in this field. The last of them,

Habing, H. J., ed. 1970. *Interstellar Gas Dynamics* (IAU Symposium No. 39). Dordrecht: Reidel.

contains many excellent reviews and original papers plus references to early publications on the subject of astrophysical hydrodynamics.

A short monograph on the subject is

Kaplan, S. A. 1966. *Interstellar Gas Dynamics,* ed. F. D. Kahn. London: Pergamon Press.

This book, and indeed any text on hydrodynamics, discusses the hydrodynamical equations in Section 6.2. A very useful book for a review of this material is

Courant, R., and Friedrich, K. O. 1948. *Supersonic Flow and Shock Waves.* New York: Interscience.

Another good book which contains much material on interstellar gas dynamics is

Dyson, J. E., and Williams, D. A. 1980. *Physics of the Interstellar Medium.* Manchester: Manchester University Press.

The classification of ionization fronts and shock fronts was exhaustively discussed by

Kahn, F. D. 1954. *Bull. Astr. Inst. Netherlands* **12,** 187.

Axford, W. I. 1961. *Phil. Trans. Roy. Soc. Lon. A.* **253,** 301.

Early numerical integrations of the dynamical evolution of an H II region were carried out by

Mathews, W. G. 1965, *Ap. J.* **142,** 1120.

Lasker, B. M. 1966, *Ap. J.* **143,** 700.

(The model shown in Figure 6.1 is taken from the latter reference.)

Specific numerical calculations of D-type ionization fronts advancing into neutral gas are given by

Mallik, D. C. V. 1975. *Ap. J.* **197,** 355.

(The numerical values quoted at the end of Section 6.3 were taken from it.)

Much of the material on dynamics of H II regions is summarized in

Mathews, W. G., and O'Dell, C. R. 1969. *Ann. Rev. Astr. Ap.* **7,** 67.

Spitzer, L. 1978. *Physical Processes in the Interstellar Medium.* New York: Wiley, chap. 12.

By far the most complete earlier observational material on the dynamics of an H II region was is the Coudé-spectrograph survey of NGC 1976 by

Wilson, O. C., Münch, G., Flather, E. M., and Coffeen, M. F. 1959. *Ap. J. Supp.* **4,** 199.

The results were discussed by

Münch, G. 1958. *Rev. Mod. Phys.* **30,** 1035.

(Figure 6.3 is taken from this reference.) Fabry-Perot observational material on internal velocities is included in the following papers:

Smith, M. G., and Weedman, D. W. 1970. *Ap. J.* **160,** 65.

Meaburn, J. 1971. *Astr. Ap.* **13,** 110.

Dopita, M. A. 1972. *Astr. Ap.* **17,** 165.

Dopita, M. A., Gibbons, A. H., and Meaburn, J. 1973. *Astr. Ap.* **22,** 33.

These four papers include measurements of M 8, M 16, M 17, and NGC 1976. More recent high-resolution spectroscopic and Fabry-Perot measurements, and discussions based on these data, which continue to confirm the existence of turbulence and the importance of the energy fed into the nebula at all scales in generating it, include

Roy, J. R., and Joncas, G. 1985. *Ap. J.* **288,** 142.

Roy, J. R., Arsenault, R., and Joncas, G. 1986. *Ap. J.* **300,** 624.

O'Dell, C. R. 1986. *Ap. J.* **304,** 767.

O'Dell, C. R., Townsley, L. K., and Castaneda, H. O. 1987. *Ap. J.* **317,** 676.

O'Dell, C. R., and Castaneda, H. O. 1987. *Ap. J.* **317,** 686.

The interpretation of NGC 1976 as an ionized blister at the edge of a dense molecular cloud was proposed by

Zuckerman, B. 1973. *Ap. J.* **183,** 863.

Figure 6.4 is based on a later discussion by

Werner, M. W. 1982. *Ann. N.Y. Acad. Sci.* **395,** 79.

A very complete summary of observational results in the Orion Nebula, including its structure at the edge of a giant molecular cloud is given in the monograph

Goudis, C. 1982. *The Orion Complex: A Case Study of Interstellar Matter.* Dordrecht: Reidel.

The blister or "champagne" models of H II regions are described, computed, and summarized in a series of papers, of which only three (and a recent review which gives references to the other papers of the series) are listed here:

Tenorio-Tagle, G. 1979. *Astr. Ap.* **7,** 59.

Bodenheimer, P., Tenorio-Tagle, G., and Yorke, H. W. 1979. *Ap. J.* **233,** 85.

Yorke, H. W., Tenorio-Tagle, G., and Bodenheimer, P. 1984. *Astr. Ap.* **128,** 325.

Yorke, H. W. 1986. *Ann. Rev. Astr. Ap.* **24,** 49.

The last of these papers is an excellent overall survey of the internal dynamics of H II regions, at a more advanced level than in this chapter. Figure 6.5 is based on numerical calculations described in the third of these papers.

The theory of the ionization of a dense globule is worked out by

Dyson, J. E. 1968. *Astrophys. Space Sci.* **1,** 388.

Complete numerical studies of the expansion of planetary nebulae are included in

Mathews, W. G. 1966. *Ap. J.* **143,** 173.

Hunter, J. H., and Sofia, S. 1971, *M.N.R.A.S.* **154,** 393.

(The calculated models shown in Figures 6.7 and 6.8 are taken from the first of these references.) A good general reference on gas dynamics is:

Stanyukovich, K. P. 1960. *Unsteady Motions of Continuous Media,* trans. J. G. Adashko, ed. M. Holt. London: Pergamon Press.

This work includes a good discussion of the analytic treatment of the expansion of a spherical gas cloud.

The first spectroscopic measurements of line splitting in planetary nebulae were made many years ago by

Campbell, W. W., and Moore, J. H. 1918. *Pub. Lick Obs.* **13,** 75.

However, they were unable to interpret the observation fully, and subsequently a very complete high-dispersion survey was made by Wilson:

Wilson, O. C. 1948. *Ap. J.* **108,** 201.

Wilson, O. C. 1950. *Ap. J.* **111,** 279

This survey includes extensive data and discussion in terms of expansion. (Table 6.1 is taken from the latter reference.) A very good summary of these radial-velocity measurements and their significance is:

Wilson, O. C. 1958. *Rev. Mod. Phys.* **30,** 1025.

Later, Fabry-Perot interferometer and CCD spectrograph measurements of many more planetary nebulae, including large, low-surface-brightness objects, are given by

Bohuski, T. J., and Smith, M. G. 1974. *Ap. J.* **193,** 197.

Robinson, G. J., Reay, N. K., and Atherton, P. D. 1982. *M.N.R.A.S.* **199,** 649.

Sabbadin, F., and Hamzaoglu, L. 1982. *Astr. Ap.* **110,** 105.

Higher-resolution measurements (line profiles) are available in

Osterbrock, D. E., Miller, J. S., and Weedman, D. W. 1966. *Ap. J.* **145,** 697.

(Figure 6.9 is taken from the preceding reference.) Higher-resolution measurements are also available in

Weedman, D. W. 1968. *Ap. J.* **153,** 49.

Osterbrock, D. E. 1970. *Ap. J.* **159,** 823.

Atherton, P. D., Hicks, T. R., Reay, N. K., Woswick, S. P., and Smith, W. H. 1978. *Astr. Ap.* **66,** 297. (NGC 6720).
Geiseking, F., Becker, I., and Solf, J. 1985. *Ap. J.* **295,** L17. (NGC 2392).
Solf, J., and Weinberger, R. 1984. *Astr. Ap.* **130,** 1984. (NGC 7026).

The analysis of Abell 30, showing the importance of heating at the surfaces of dense condensations by the stellar wind in the nebula, is by

Harrington, J. P., and Feibelman, W. A. 1984. *Ap. J.* **277,** 716.

The observations of proper motions of expansion of planetary nebulae are discussed by

Liller, M. H., Welther, B. L., and Liller, W. 1966. *Ap. J.* **144,** 280.
Liller, M. H., and Liller, W. 1968. *Planetary Nebulae* (IAU Symposium No. 34). Dordrecht: Reidel, p. 38.
Liller, W. 1978. *Planetary Nebula* (IAU Symposium No. 76). Dordrecht: Reidel, p. 35.

7

Interstellar Dust

7.1 Introduction

The discussion in the first six chapters of this book has concentrated entirely on the gas within H II regions and planetary nebulae, and in fact these objects are usually simply called gaseous nebulae. However, they really contain dust particles in addition to the gas, and the effects of this dust on the properties of the nebulae are by no means negligible. Therefore, this chapter will discuss the evidence for the existence of dust in nebulae, its effects on the observational data on nebulae and how the measurements can be corrected for these effects. The measurements of the radiation of both H II regions and planetary nebulae are then considered, and the dynamical effects that result from this dust are briefly discussed.

7.2 Interstellar Extinction

The most obvious effect of interstellar dust is its extinction of the light from distant stars and nebulae. This extinction in the ordinary optical region is due largely to scattering, but it is also partly due to absorption. (Nevertheless, the process is very often referred to as interstellar absorption.) It reduces the amount of light from a source shining through interstellar dust according to the equation

$$I_\lambda = I_{\lambda 0} e^{-\tau_\lambda}, \tag{7.1}$$

where $I_{\lambda 0}$ is the intensity that would be received at the Earth in the absence of interstellar extinction along the line of sight, I_λ is the intensity actually observed, and τ_λ is the optical depth at the wavelength observed. This equation also applies to stars, in which we observe the total flux, with πF_λ substituted for I_λ. The equation is correct when radiation is either absorbed or scattered out of the beam, but only if other radiation is not scattered into the beam. It is a good approximation for all stars and for nebulae that do not themselves contain interstellar dust, but it is incorrect if the dust is mixed with the gas in the nebula and scatters nebular light into the observed ray as well as out of it. (This point will be discussed in Section 7.3.) The interstellar extinction is thus specified by the values of τ_λ along the ray to the star or nebula in question.

The interstellar extinction has been derived for many stars by spectrophotometric measurements of pairs of stars selected because they have identical spectral types. The ratio of their brightnesses,

$$\begin{aligned} \frac{\pi F_\lambda(1)}{\pi F_\lambda(2)} &= \frac{\pi F_{0\lambda}(1)\ e^{-\tau_\lambda(1)}}{\pi F_{0\lambda}(2)\ e^{-\tau_\lambda(2)}} \\ &= \frac{D_2^2}{D_1^2}\ e^{-[\tau_\lambda(1)-\tau_\lambda(2)]}, \end{aligned} \qquad (7.2)$$

depends on the ratio of their distances D_2^2/D_1^2 and on the difference in the optical depths along the two rays. Interstellar extinction, of course, increases toward shorter wavelengths (in common terms, it reddens the light from a star); so by comparing a slightly reddened or nonreddened star with a reddened star, we can determine $\tau_\lambda(1)-\tau_\lambda(2) \approx \tau_\lambda(1)$, essentially the interstellar extinction along the path to the more reddened star. The logarithm of the ratio of the fluxes differs by an additive constant 2 ln D_2/D_1 that is independent of wavelength. The constant is not determined, because it depends on the distance of the reddened star, which is generally not independently known. However, for any kind of interstellar dust, or indeed for any kind of particles, $\tau_\lambda \to 0$ as $\lambda \to \infty$, and it is thus possible to determine the constant approximately by making measurements at sufficiently long wavelengths.

Such measurements have been made over the years for many stars, and from them we have a fairly good idea of the interstellar extinction. They show that in the ordinary optical region, the form of the wavelength dependence of the interstellar extinction is approximately the same for all stars, and only the amount of extinction varies, so that

$$\tau_\lambda = Cf(\lambda), \qquad (7.3)$$

where the constant factor C depends on the star, but the function $f(\lambda)$ is the same for all stars. This result implies physically that, to this same first approximation, the optical properties of the dust are similar everywhere in the observed region of interstellar space. Figure 7.1 shows this standard interstellar extinction expressed relative to the extinction at Hβ, and normalized so that $\tau_{H\gamma} - \tau_{H\alpha}$, the difference in optical depths between the two wavelengths, is 0.50. Notice that the extinction is plotted in terms of reciprocal wavelength (proportional to frequency) because it is nearly linear in this variable. Notice also that the extrapolation to $\lambda \to \infty$ establishes the zero point for the extinction shown on the right-hand scale.

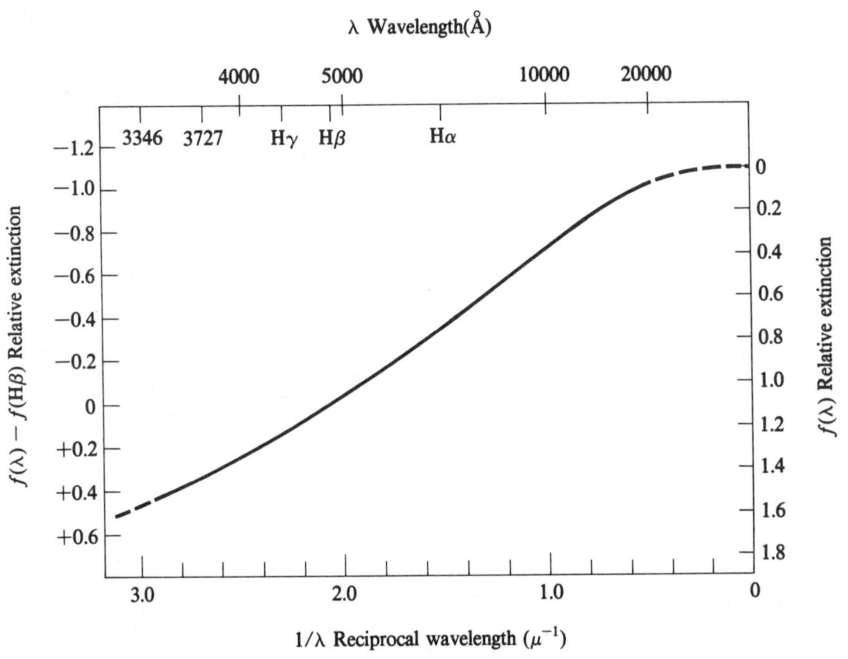

FIGURE 7.1
Standard interstellar extinction curve as a function of wavelength as described in text and listed in Table 7.1. Note that the left-hand scale gives extinction relative to extinction at Hβ; the right-hand scale shows total extinction and is chosen so that $\tau_\lambda \to 0$ as $\lambda \to \infty$.

Although the standard form of the interstellar extinction is a good first approximation, careful observations of different stars reveal variations in the wavelength dependence of the interstellar extinction along different light paths

TABLE 7.1
Standard interstellar extinction curve

λ(Å)	$1/\lambda(\mu^{-1})$	$f(\lambda) - f(H\beta)$	λ (Å)	$1/\lambda(\mu^{-1})$	$f(\lambda) - f(H\beta)$
∞	0.00	−1.09	4545	2.20	+0.09
20000	0.50	−1.02	λ_B	2.30	+0.15
12500	0.80	−0.86	H_γ	2.304	+0.15
10000	1.00	−0.72	4167	2.40	+0.20
8333	1.20	−0.56	Hδ	2.438	+0.22
7143	1.40	−0.43	4000	2.50	+0.25
Hα	1.524	−0.35	3846	2.60	+0.29
6250	1.60	−0.29	3727	2.683	+0.33
5556	1.80	−0.16	3571	2.80	+0.39
λ_V	1.83	−0.14	3346	2.989	(+0.46)
5000	2.00	−0.04	3333	3.00	(+0.46)
Hβ	2.057	0.00			

in the Galaxy. The most extreme deviations have been measured for the stars of θ^1 Ori, the Trapezium stars exciting NGC 1976, and in Figure 7.2 their average extinction is compared with the average extinction for a large number of stars in Cygnus, Cepheus, Perseus, and Monoceros; this average is essentially the same as the standard extinction curve. It can be seen that the differences are small but not completely insignificant, particularly at the extreme wavelengths shown. Measurements of other stars show that similar deviations from the standard interstellar extinction curve of Figure 7.1 tend to be largest for stars in H II regions. There also seem to be regional variations of the extinction (with galactic longitude), but these variations are even smaller than the differences between the extinction within and without H II regions.

However, observations made from satellites reveal that there are larger differences in interstellar extinction between different stars in the ultraviolet spectral region, particularly at the shortest measured wavelengths. This is shown graphically in Figure 7.3, in which a mean extinction curve is plotted extending down to $\lambda 1100$, as is one of several abnormal extinction curves, this one for σ Sco, normalized to the same difference in extinction between $\lambda 4348$ and $\lambda 5465$, the effective wavelengths of the B and V filters. The broad dip (actually a "bump" or increase in extinction) centered near 4.6 μ^{-1} or 2175 Å is due to graphite or other particles rich in carbon, such as hydrogenated amorphous carbon particles. The differences in the overall extinction curve show that dust particles of at least two, and probably three or more, compo-

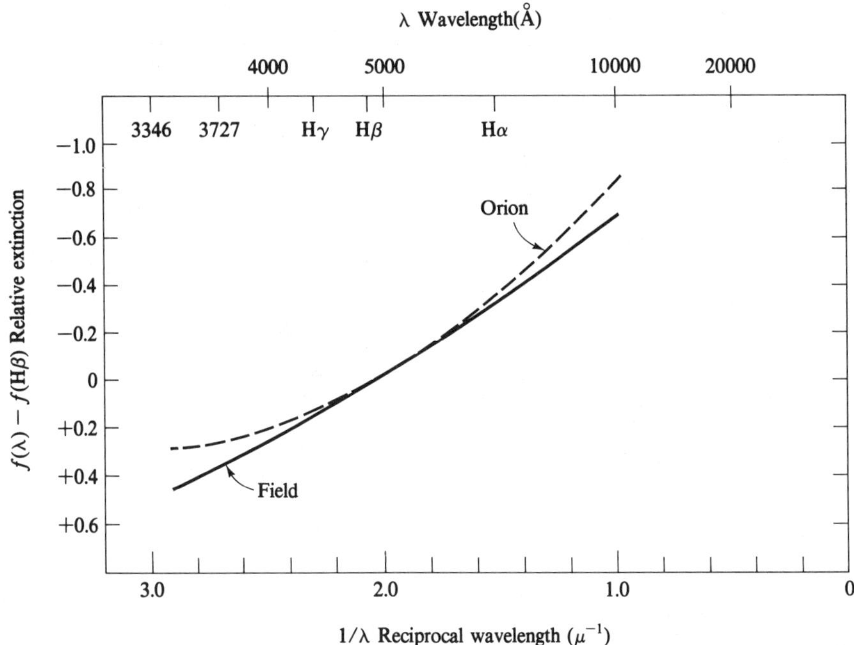

FIGURE 7.2
Average extinction for θ^1 Ori, the Trapezium stars, in NGC 1976, compared with average extinction for stars in Cygnus, Cepheus, Perseus and Monoceros.

sitions are responsible for the extinction. One may be graphite; good matches to the observed extinctions can be obtained by combining mixtures of graphite, olivine (Fe, Mg_2) SiO_4, and silicon carbide (SiC) particles in various proportions. These compositions are also indicated, or at least favored, by other considerations, as we will see. The best average extinction based on optical plus ultraviolet measurements is listed in Table 7.2. In the optical region, it differs only slightly from Table 7.1, but it is based on more and later measurements, and so should be used.

Interstellar extinction naturally makes the observed ratio of intensities of two nebular emission lines $I_{\lambda_1}/I_{\lambda_2}$ differ from their ratio as emitted in the nebula $I_{\lambda_1 0}/I_{\lambda_2 0}$:

$$\frac{I_{\lambda_1}}{I_{\lambda_2}} = \frac{I_{\lambda_1 0}}{I_{\lambda_2 0}}\, e^{-(\tau_{\lambda_1} - \tau_{\lambda_2})}. \tag{7.4}$$

The observations must be corrected for this effect before they can be discussed physically. As a first approximation, the interstellar extinction can ordinarily

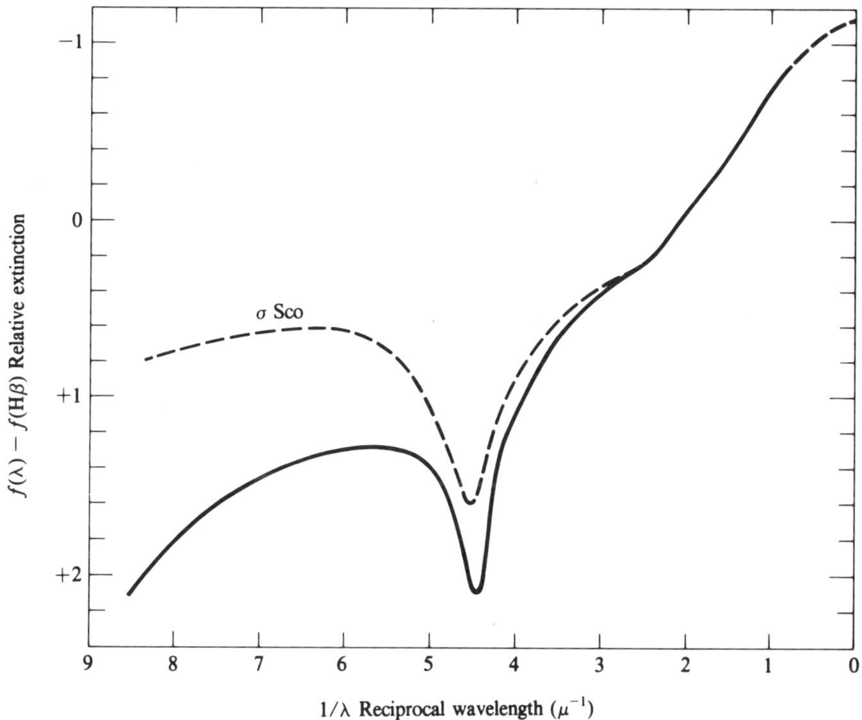

FIGURE 7.3
Average interstellar extinction curve extending into the ultraviolet spectral region, as described in text and listed in Table 7.2 (solid line). Also, abnormal extinction curve of σ Sco (dashed line).

be assumed to have the average form, unless the amount of extinction is very large, so this ratio can also be written

$$\frac{I_{\lambda_1}}{I_{\lambda_2}} = \frac{I_{\lambda_1 0}}{I_{\lambda_2 0}} e^{-C[f(\lambda_1)-f(\lambda_2)]}. \tag{7.5}$$

Note that only the difference in optical depths at the two wavelengths enters this equation; so the correction depends on the form of the interstellar extinction curve and on the amount of extinction, but not on the more-uncertain extrapolation to infinite wavelength. To find the amount of correction, the principle is to use the measured ratio of strengths of two lines for which the

TABLE 7.2
Average interstellar extinction curve

λ(Å)	$1/\lambda(\mu^{-1})$	$f(\lambda)-f(\mathrm{H}\beta)$	λ (Å)	$1/\lambda(\mu^{-1})$	$f(\lambda)-f(\mathrm{H}\beta)$
∞	0.00	−1.13	Hδ	2.438	+0.19
34000	0.29	−1.08	4000	2.50	+0.21
22000	0.45	−1.02	3727	2.683	+0.28
12500	0.80	−0.86	3440	2.91	+0.36
9000	1.11	−0.67	3346	2.989	+0.40
7000	1.43	−0.42	3000	3.33	+0.54
Hα	1.524	−0.37	2740	3.65	+0.76
5494	1.82	−0.18	2500	4.00	+1.09
Hβ	2.057	0.00	1909	5.238	+1.32
4405	2.27	+0.12	1549	6.456	+1.35
Hγ	2.304	+0.13	1216	8.224	+1.93

relative intensities, as emitted in the nebula, are known independently; thus in equation (7.4) only $\tau_{\lambda_1}-\tau_{\lambda_2}$, or equivalently in equation (7.5) only C, is unknown and can be solved for. Once C is determined, the average reddening curve (which is listed in Table 7.2) gives the optical depths at all wavelengths.

The ideal line ratio to determine the amount of extinction would be one that is completely independent of physical conditions and that is easy to measure in all nebulae. Such an ideal pair of lines does not exist in nature, but various approximations to it do exist and can be used to get a good estimate of the interstellar extinction of a nebula.

The best lines would be a pair with the same upper level, whose intensity ratio would therefore depend only on the ratio of their transition probabilities. An observable case close to this ideal is [S II], in which $^4S\ -\ ^2P$ $\lambda\lambda$4069, 4076 may be compared with $^2D-\ ^2P$ $\lambda\lambda$10287, 10320, 10336, 10370. Here both multiplets arise from a double upper term rather than from a single level, and the relative populations in the two levels depend slightly on electron density, so the calculated ratio of intensities of the entire multiplets varies between the limits $0.55 \geq I(^2D-^2P)/I(^4S-^2P) \geq 0.51$ over the range of densities $0 \leq N_e \leq 10^7$ cm^{-3}. Although this [S II] ratio has been used to determine the interstellar extinction in a few galaxies with emission lines, the planetary nebula NGC 7027, and the supernova remnant NGC 1952, the lines involved are all relatively weak. Furthermore, the measurement of $I(^2D-^2P)$

is difficult because of contamination due to infrared OH atmospheric emission lines, and also because the sensitivities of CCDs and other infrared detectors are relatively low and dropping rapidly near $\lambda 1.03\ \mu$.

A somewhat easier observational method for determining the amount of interstellar extinction of a nebula is to compare an H I Paschen line with a Balmer line from the same upper set of levels. For example, we could compare $P\delta$ $\lambda 10049$ with Hϵ $\lambda 3970$, both of which arise from the excited terms with principal quantum number $n = 7$. Since several different upper terms are involved, $7\ ^2S$, $7\ ^2P$, ...$7\ ^2F$, the relative strengths depend slightly on excitation conditions, but as Table 4.4 shows, the variation in Paschen-to-Balmer ratios is quite small over the whole range of temperatures expected in gaseous nebulae. This Paschen-to-Balmer ratio method is, in principle, excellent, but it has the same problems as the [S II] method, because of contamination by infrared night-sky emission, plus the relative insensitivity of photomultipliers and of CCDs at wavelengths longward of 1μ.

Hence the method most frequently used in practice to determine the interstellar extinction is to measure the ratios of two or more H I Balmer lines, for instance, Hα/Hβ and Hβ/Hγ. Though the upper levels are not the same for the two lines, the relative insensitivity of the line ratios to temperature, shown in Table 4.2, means that the interstellar extinction can be determined with relatively high precision even though the temperature is only roughly estimated. The Balmer lines are strong and occur in the part of the spectrum that is ordinarily observed, so this method is now the one most often used. The fact that different pairs of Balmer lines (usually Hα/Hβ and Hβ/Hγ) give the same result tends to confirm observationally the recombination theory outlined in Chapter 4.

It can be seen from equation (7.3) that the normalization of the function $f(\lambda)$ that gives the form of the wavelength dependence of the interstellar extinction is arbitrary. The normalization we have adopted is convenient for nebular work, as is the idea of tabulating $f(\lambda) - f(\mathrm{H}\beta)$, so that it is simple to correct all emission-line ratios involving Hβ, the usual nebular standard reference line. Then, working with logarithms, it is often convenient to write

$$\begin{aligned} \frac{I_\lambda}{I_{\mathrm{H}\beta}} &= \frac{I_{\lambda 0}}{I_{\mathrm{H}\beta 0}}\ 10^{-0.434(\tau_\lambda - \tau_{\mathrm{H}\beta})} \\ &= \frac{I_{\lambda 0}}{I_{\mathrm{H}\beta 0}}\ 10^{-c[f(\lambda) - f(H\beta)]} \end{aligned} \tag{7.6}$$

and to use $c = 0.434C$ as a measure of the amount of extinction. Nebulae are observed to differ greatly in the amount of extinction; for instance, $c \approx 0.02$

for NGC 6853, while the most heavily reddened planetary nebula for which optical observations have been published to date is probably NGC 6369, with $c \approx 2.3$.

Naturally, the nebulae with the strongest interstellar extinction are too faint to observe in the optical region, though they can be measured in the radio-frequency region. This suggests still another way to measure the amount of interstellar extinction, namely, to compare the intensity of the radio-frequency continuum (at a frequency at which the nebula itself is optically thin) to that of an optical H I recombination line. This is the same principle as that used in comparing two optical lines, except that one of the lines is effectively at infinite wavelength in this method. The intrinsic ratio of intensities $j_\nu / j_{H\beta}$ can be calculated explicitly for any assumed temperature using equations (4.22) and (4.30) and Table 4.4. It depends on the ratio $N_+ \langle Z^2 \rangle / N_p$, because the free-free emission contains contributions from all ions, but since $N_{He}/N_H \approx 0.10$ and the abundances of all the other elements are smaller, this quantity,

$$\frac{N_+ \langle Z^2 \rangle}{N_p} \approx 1 + \frac{N_{He+}}{N_p} + 4 \frac{N_{He++}}{N_p}, \qquad (7.7)$$

is rather well determined. The temperature dependence of $j_\nu / j_{H\beta}$ is low, approximately as $T^{1/3}$, but nevertheless is considerably stronger than that of an optical recombination-line ratio such as $H\alpha / H\beta$.

Table 7.3 lists values of c determined for several planetaries from the most accurate optical and radio measurements. The probable error of each method is approximately 0.1 in c, and it can be seen that the methods agree fairly well for these accurately measured planetaries. The theory behind these determinations is quite straightforward, and the expected uncertainty because of the range T in nebulae is relatively small, so this method, in principle, provides a good absolute determination of the interstellar extinction at the measured optical wavelength; that is, it should determine the extrapolation $\lambda \to \infty$ of the interstellar extinction curve quite accurately.

7.3 Dust within H II Regions

Dust is certainly present within H II regions, as can clearly be seen on direct photographs. Many nebulae show "absorption" features that cut down the nebular emission and starlight from beyond the nebula. Very dense small features of this kind are often called globules; others at the edges of nebulae are known as elephant-trunk or comet-tail structures. Two examples are shown in Figures 7.4 and 7.5. Many of these absorption features appear to be almost

TABLE 7.3
Interstellar extinction for planetary nebulae

Nebula	c (Balmer-line method)	c (Radio-frequency-Hβ method)
NGC 650/1	0.14	0.15
NGC 6572	0.37	0.41
NGC 6720	0.16	0.12
NGC 6803	0.56	0.73
NGC 6853	0.00	0.02
NGC 7009	0.14	0.17
NGC 7027	1.21	1.50
IC 418	0.26	0.32

completely dark; this indicates not only that they have a large optical depth at the wavelength of observation (perhaps $\tau \geq 4$ if the surface brightness observed in the globule is a small percentage of that observed just outside it), but also that they are on the near side of the nebula, so that very little nebular emission arises between the globule and the observer. A few large-absorption features that are not so close to the near side of the nebula can be seen on photographs; they are features in which the surface brightness is smaller than in the surrounding nebula, but not zero. There must be many more absorption structures with smaller optical depths, or located deeper in the nebula, that are not noticed on ordinary photographs. It is difficult to study these absorption features quantitatively, except to estimate their optical depths, from which the amount of dust can be estimated if its optical properties are known. If, in addition, the gas-to-dust ratio is known, the total mass in the structure can be estimated. We shall return to a consideration of these questions after examining the scattered-light observations of the dust.

The dust particles scatter the continuous radiation of the stars immersed in nebulae, resulting in an observable nebular continuum. Measurements of this continuum must be made with a scanner or interference-filter system designed to avoid the strong nebular-line radiation, and photographs taken in the continuum, such as Figure 7.6, also require filters that avoid strong nebular lines. Measurements of an H recombination line such as Hβ are made at the

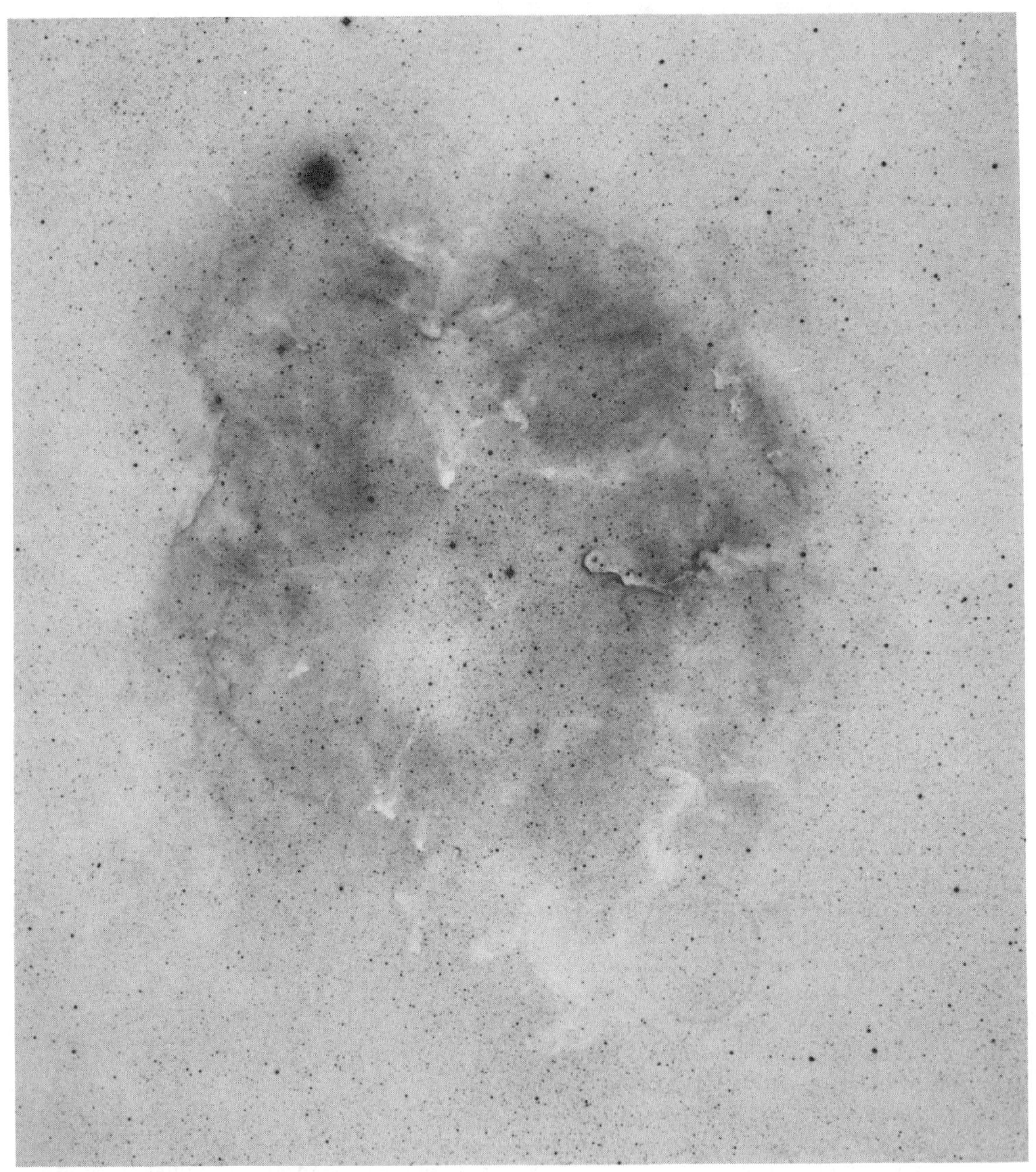

FIGURE 7.4

IC 1396, the H II region in Cepheus, taken with 48-inch Schmidt telescope, red filter and 103a-E plate, emphasizing Hα, [N II]. Note the overlying interstellar extinction, particularly the large globule (or comet-tail structure), to the west (*right*). (*Palomar Observatory photograph.*)

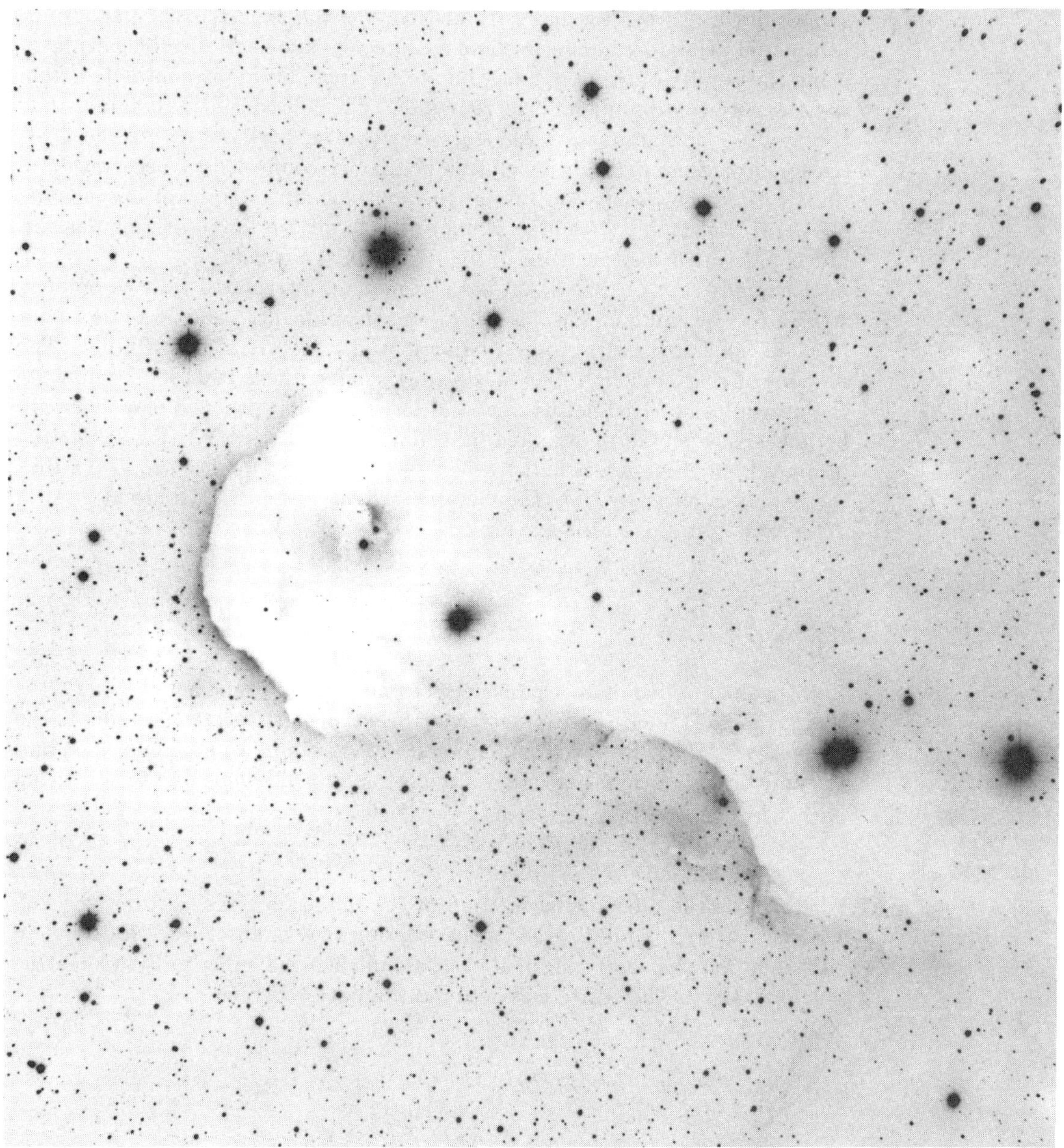

FIGURE 7.5

Large globule (or comet-tail structure) in IC 1396, taken with 5-m Hale telescope, RG-2 filter and 103a-E plate, emphasizing Hα, [N II]. The very sharp reduction in star density shows that the globule, particularly near its east (*left*) end, is practically opaque. Note the bright edge between the H II region and the globule; this is the ionization front progressing into the dense globule. (*Palomar Observatory photograph.*)

same time, and from the intensity of that line the expected nebular atomic continuum caused by bound-free and free-free emission can then be calculated using the results of Section 4.3. The atomic contribution is subtracted from the observed continuum, and the remainder, which is considerably larger than the atomic continuum in most observed nebulae, must represent the dust-scattered continuum. This conclusion is directly confirmed by observation of the He II λ4686 absorption line in the continuous spectrum of one nebula, NGC 1976. This line, of course, cannot arise in absorption in the nebular gas, but is present in the spectrum of the O star in the nebula.

Generally, the observational data cannot be interpreted in a completely straightforward and unique way because of the difficulties caused by the complicated (and unknown) geometry and spatial structure of real nebulae. The amount of scattered light depends strongly upon these factors. To indicate the principles involved, let us treat the very simplified problem of a spherical, homogeneous nebula illuminated by a single central star. Writing L_ν for the luminosity of the star per unit frequency interval, and further supposing that the nebula is optically thin, the flux of starlight within the nebula at a point distance r from the star is given by

$$\pi F_\nu = \frac{L_\nu}{4\pi r^2}. \tag{7.8}$$

If N_D is the number of dust particles per unit volume in the nebula, and C_λ is their average extinction cross section at the wavelength λ corresponding to the frequency ν, then the extinction cross section per unit volume is $N_D C_\lambda$, and the emission coefficient per unit volume per unit solid angle due to scattering is

$$j_\nu = \frac{A_\lambda N_D C_\lambda \pi F_\nu}{4\pi} = \frac{A_\lambda N_D C_\lambda L_\nu}{16\pi^2 r^2}, \tag{7.9}$$

where A_λ is the albedo, the fraction of the radiation removed from the flux that is scattered, while $1 - A_\lambda$ is the fraction that is absorbed. Note that in this equation the scattering has been assumed to be spherically symmetric. The intensity of the scattered continuum radiation is then

$$\begin{aligned} I_\nu(b) &= \int j_\nu \, ds \\ &= \frac{A_\lambda N_D C_\lambda L_\nu}{8\pi^2} \cdot \frac{1}{b} \cos^{-1} \frac{b}{r_0} \end{aligned} \tag{7.10}$$

for a ray with a minimum distance b from the central star in a spherical homogeneous nebula of radius r_0.

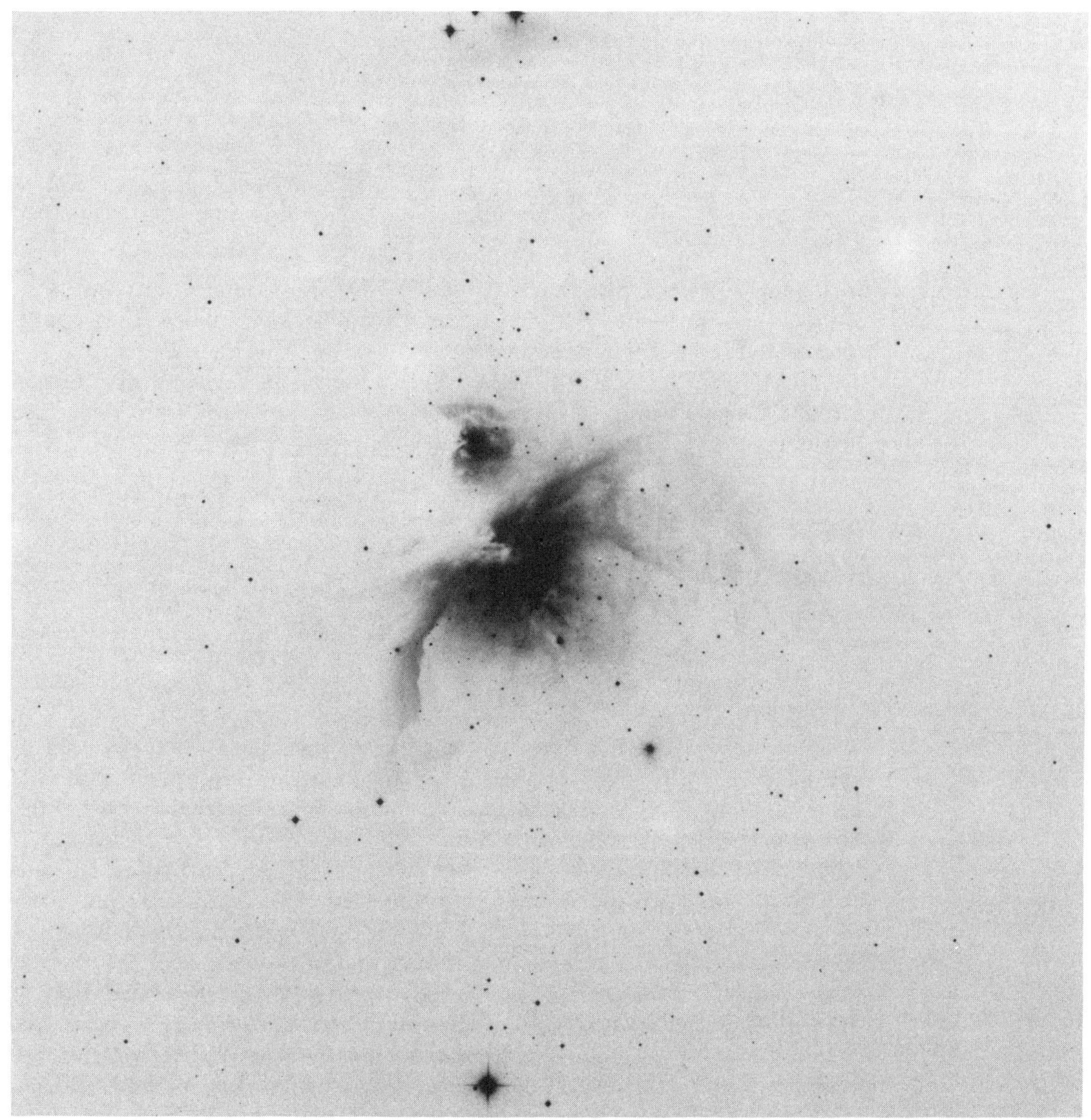

FIGURE 7.6

NGC 1976, the Orion Nebula, taken with 48-inch Schmidt telescope, using Wratten 15 filter and 103a-J plate, in the continuum $\lambda\lambda$5100-5500. Compare this photograph with the Frontispiece (particularly the 40-second exposure), which shows the same nebula in Hα, [N II]. Many differences are apparent. (*Palomar Observatory photograph.*)

This may be compared with the Hβ surface brightness observed from the same nebula, which may, however, be assumed to have a possibly different Strömgren radius r_1 limiting the ionized gas,

$$I_{\mathrm{H}\beta}(b) = \int j_{\mathrm{H}\beta}\, ds$$
$$= \frac{1}{4\pi} N_p N_e \alpha_{\mathrm{H}\beta}^{eff} h\nu_{\mathrm{H}\beta} 2\sqrt{r_1^2 - b^2}. \tag{7.11}$$

In Figure 7.7, these two surface-brightness distributions are compared with observational data for NGC 6514, the most nearly symmetric H II region illuminated by a single dominant central star for which measurements are available. It can be seen that the model is a reasonable representation of this nebula. Then dividing (7.10) and (7.11), we may write the ratio of surface brightness in Hβ to surface brightness in the continuum as

$$\frac{I_{\mathrm{H}\beta}(b)}{I_\nu(b)} = \left[\frac{N_p N_e \alpha_{\mathrm{H}\beta}^{eff} h\nu_{\mathrm{H}\beta}}{A_\lambda N_D C_\lambda}\right]$$
$$\times \left(\frac{4\pi D^2}{L_\nu}\right)\left(\frac{r_0 r_1}{D^2}\right)\left[\frac{(b/r_0)\sqrt{1-(b/r_1)^2}}{\cos^{-1}(b/r_0)}\right]. \tag{7.12}$$

In equation (7.12) we have inserted D, the distance from the nebula to the observer. Notice that the first factor (in square brackets) involves atomic properties and properties of the dust, the second factor is the reciprocal of the flux from the star observed at the Earth, the third factor is the product of the angular radii of the nebula in the continuum (r_0/D) and in Hβ (r_1/D), and the fourth factor gives the angular dependence of the surface brightnesses, expressed in dimensionless ratios. Thus the first factor can be determined from measurements of surface brightnesses and of the flux from the star, to the accuracy with which the model fits the data. If the electron density is determined either from the Hβ surface-brightness measurements themselves or from [O III] or [S II] line ratio measurements, the ratio $N_p/A_\lambda N_D C_\lambda$, which is proportional to the ratio of densities of gas to dust, is determined; note that this quantity is proportional to the reciprocal of the poorly known electron density. Table 7.4 lists numerical values of ratios found in this way from continuum observations of several H II regions. For NGC 1976, a model in which the average electron density decreases outward, with a range from $N_e \approx 2 \times 10^3$ cm^{-3} in the inner part to $N_e \approx 2 \times 10$ cm^{-3} in the outer part, was used.

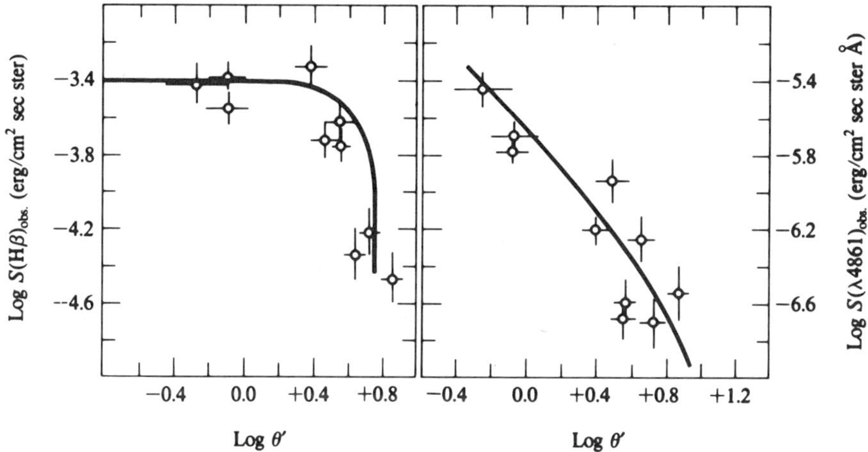

FIGURE 7.7
Diagram on left shows Hβ surface brightness as a function of angular distance from the central star in H II region NGC 6514. Diagram on right shows continuum surface brightness near Hβ (corrected for atomic continuum) as a function of angular distance from the same star.

TABLE 7.4
Gas-to-dust ratios in H II *regions*

Nebula	Assumed N_e (cm^{-3})	$N_p/A_\lambda N_D C_{\lambda 4861}$ (cm^{-2})
NGC 1976 (inner)	model	1.4×10^{22}
NGC 1976 (outer)	model	5×10^{20}
NGC 6514	1.3×10^{2}	4×10^{20}
NGC 6523	4.4×10	2×10^{21}
NGC 6611	5.5×10	2×10^{21}
Field	–	2×10^{21}

The size distribution of the interstellar particles can, of course, be found from the detailed study of their extinction properties. This is a long and complicated subject in itself, approached from measurements of the continuous spectra of stars with different amounts of reddening as discussed in Section 7.2, and it will not be discussed in detail here. It is clear that, for any model of the composition and spectral size of interstellar particles, an extinction curve

$$\tau_\lambda = <N_D> C_\lambda s = Cf(\lambda) \tag{7.13}$$

is predicted, and comparison of a predicted curve with observational data, such as Figure 7.1, 7.2, or 7.3, shows how well the model represents the actual properties of the particles. It is obvious that the larger the wavelength range for which observational data are available, the more specifically the properties of the interstellar particles can be determined. The product τ_λ is well determined observationally, much better than its factors $< N_D >$ and C_λ. Knowledge of the abundances of the elements in interstellar matter, and physical descriptions of the processes by which the particles are formed and destroyed, also provide information on the properties of the particles.

All these methods, however, still do not tell us exactly what kinds of dust particles are present in interstellar matter, because different mixtures of particle distributions can be matched to the same observational data to a fairly good approximation. Until twenty years ago, the particles were generally believed to be dielectric, "dirty-ice" particles, consisting mainly of frozen H_2O, CH_4, and NH_3, with smaller amounts of such impurities as Fe and Mg. However, measurements of ultraviolet extinction of stars from satellites have provided important new information on the nature of interstellar particles. These measurements have shown that, in addition to the dielectric particles, there are also other kinds of interstellar particles, most probably graphite or hydrogenated amorphous carbon, silicate, and silicon carbide particles. In addition, infrared extinction measurements made with high wavelength resolution have shown that the H_2O (ice) absorption band $\lambda 3.1\ \mu$ expected in dirty-ice particles is not present in the spectra of many heavily reddened stars, and where observed it is much weaker than predicted by the earlier ideas.

A very rough mean value of the radius of the particles effective in extinction in the optical region is $a \approx 1.5 \times 10^{-5}$ cm. Particles of this size, which is comparable with the wavelength of light, are the easiest to observe, and the information on extremely large particles is rather incomplete. If we now adopt as a very crude mean $A_\lambda \approx 1$, $C_{\lambda 4861} \approx \pi a^2 \approx 7 \times 10^{-10}$ cm^2, we find that $N_p/A_\lambda N_D C_{\lambda 4861} = 2 \times 10^{21}$ cm^{-2}, a typical value from Table 7.4, corresponds to $N_p/N_D \approx 1.5 \times 10^{12}$. If we further assume that the density within the dust particles is $\rho \approx 1$ gm cm^{-3}, we find that the ratio of masses $N_p m_H/N_D m_D \approx 1.5 \times 10^2$. The relative gas-to-dust ratio found in this way within H II regions is, in most cases, quite similar to that in an average region of interstellar space, the "field" of Table 7.4; and the ratio of masses is also comparable to the ratio of the mass of hydrogen to the mass of heavy elements in typical astronomical objects. Since most of the mass of the dust particles consists of heavy elements – because neither H nor He can be in solid form at the very low density of interstellar space, and only compounds like C (graphite), (Fe, Mg_2) SiO_4 (silicates), SiC (silicon carbide), H_2O, NH_3, and CH_4 are expected in the particles – it appears that a fairly large fraction of the heavy-element content of interstellar space is locked up in dust; so the material in H II regions does not significantly differ from the typical interstellar material in this respect. Therefore, the abundance of an element such as C derived from observations of the gas in a nebula is a lower limit to its actual abundance, because significant quantities may be present in the form of dust. Note from Table 7.4 that dust is considerably less abundant in the inner part of NGC 1976; as we will see in the next section, this probably results from the presence of the hot stars there.

Of course, the scattering of stellar continuous radiation within the nebula shows that the emission-line radiation emitted by the gas must also be scattered. In fact, observations show that much of the line radiation observed in the faint outer parts of NGC 1976 actually consists of scattered photons that were originally emitted in the bright central parts of the nebula. If the albedo of the dust were $A_\lambda = 1$ at all wavelengths, this scattering would not affect the total emission-line flux from the whole nebula, because every photon generated within it would escape, although the scattering would transfer the apparent source of photons within the nebula. In reality, of course, $A_\lambda \neq 1$ (although it is relatively high), and some emission-line photons are destroyed by dust within the nebula. Therefore, the procedure for correcting observed nebular emission line intensities for interstellar extinction described in Section 7.2 is not completely correct, because it is based on stellar measurements in which radiation scattered by dust along the line of sight does not reach the observer. However, numerical calculations of model nebulae using the best available information on the properties of dust show that corrections determined in the way described are approximately correct and give very nearly the right *relative* emission-line intensities, because the wavelength dependence of the extinction, however it occurs, is relatively smooth. Hence, the observational procedure, which amounts to adopting an amount of extinction that correctly fits the observational data to theoretically known relative-line strengths near both ends of the observed wavelength range, cannot be too far off anywhere within that range. Naturally, the longer the range of wavelengths over which these corrections are applied, the larger the error may be.

Finally, let us estimate the amount of dust within a globule with a radius 0.05 pc that appears quite opaque, so that it has an optical depth $\tau_{H\beta} \geq 4$ along its diameter. Many actual examples with similar properties are known to exist in observed H II regions. Supposing that the dust in the globule has the same properties as the dust in the ionized part of the nebula, we easily see that $N_D \geq 2 \times 10^{-8}$ cm^{-3}; further supposing that the gas-to-dust ratio is the same, $N_H \geq 2 \times 10^4$ cm^{-3}. Thus the observed extinction indicates quite high gas densities in globules of this type.

7.4 Infrared Emission

Dust is also observed in H II regions by its infrared thermal emission. The measurements are relatively recent, depending as they do upon sensitive infrared detectors and the sophisticated observational techniques necessary to use them effectively. Absorption and emission in the Earth's atmosphere become increasingly important at longer wavelengths, but there are windows through which observations can be made from the ground out to just beyond $\lambda \approx 20\ \mu$. Most of the still-longer wavelength measurements have been made from high-altitude balloons and especially from airplanes, though a few observations have been made from mountain-top observatories through partial windows out to 350 μ. Most recently a very complete survey extending from 10 μ to 100 μ was made with the Infrared Astronomical Satellite (IRAS). In the infrared, subtraction of the sky emission is always very important, and this is accomplished by switching the observing beam back and forth rapidly between the "object" being measured and the nearby "blank sky." This scheme is highly effective for measurements of stars and other objects of small angular size, but it is clear that nebulae with angular sizes comparable with or larger than the angular separation of the object and reference beams are not detected by this method. Most of the early observations were taken with broad-band filters, but more recent higher-resolution measurements have shown spectral features in the infrared radiation from dust. All the observations immediately show that in H II regions the infrared radiation is far greater than the free-free and bound-free continuous radiation predicted from the observed Hβ on radio-frequency intensities.

Let us first examine the available observational data on NGC 1976, the best studied of the H II regions. There are several infrared "point" sources in this nebula, at least one of which (the "Becklin-Neugebauer object") evidently is a highly luminous, heavily reddened star. It is in the dense molecular cloud, well behind the ionized nebula, seen within it only in projection. In addition, two extended peaks of intensity are measured at 10 μ and 20 μ; one centered approximately on the Trapezium (nearest the stars θ^1 Ori C and D), and the other centered approximately on the Becklin-Neugebauer object about $1'$ northwest of the Trapezium. Both these peaks (the first is known as the Trapezium infrared nebula, the second as the Kleinmann-Low nebula) have angular sizes of order $30''$ to $1'$, and are only the brightest and smallest regions of a larger complex of infrared emission. At much longer wavelengths, such as 100 μ and 350 μ, the Kleinmann-Low nebula remains a bright feature, but the Trapezium nebula is scarcely distinguishable from the background approximately $2'$ east or west.

The measured nebular infrared continuous radiation, of order 10^2 to 10^3 as large as the expected free-free and bound-free continua, can only arise by radiation from dust. To a first crude approximation, the dust emits a blackbody spectrum, so measurements at two wavelengths approximately determine its temperature. For instance, in the infrared Trapezium nebula, the color temperature determined from the measured fluxes at 11.6 μ and 20 μ is $T_c \approx$ 220°K; this must approximately represent the temperature of dust particles. They are heated to this temperature by the absorption of ultraviolet and optical radiation from the Trapezium stars, and possibly also from the nearby nebular gas that is ionized by these same stars. Likewise, the dust observed as the Kleinmann-Low nebula is heated by absorption of shorter wavelength radiation emitted by or ultimately due to the Becklin-Neugebauer star within it. However, the measured intensity of the Trapezium nebula at 11.6 μ is only about $10^{-3} B_\nu(T_c)$; this indicates that it has an effective optical depth of only about $\tau_{11.6\mu} \approx 10^{-3}$. These infrared measurements are the basis for the picture of NGC 1976 as an ionized region at the edge of a giant dust cloud described in Chapter 6. The Kleinmann-Low nebula is a dense region within the cloud but near its ionized surface.

Furthermore, the description and calculations of blackbody spectra are somewhat simplified, for measurements with better frequency resolution show that the continuous spectrum does not accurately fit $I_\nu = \text{const } B_\nu(T)$, for any T, or even a range of temperature, but rather has a relatively sharp peak at $\lambda \approx 10$ μ, similar to the sharp peak observed in the infrared emission of many cool stars, such as the M2 Ia star μ Cep. A more accurate wavelength of the peak is $\lambda 9.8$ μ, and the full width at half maximum is about 2.5 μ. In late-type supergiants, this feature is attributed to circumstellar silicate particles, which have a band near this position; the presence of the feature in the nebular infrared spectrum reveals the presence of similar particles in the nebula. Somewhat narrower features are also observed in NGC 1976, other H II regions including NGC 6523, and several planetary nebulae, at $\lambda\lambda 3.28$, 3.4, 6.2, 7.7, 8.6, and 11.3 μ. They are too broad to be emission lines of ions, and are most probably the result of infrared fluorescence from vibrationally excited, polycyclic aromatic hydrocarbon (PAH) molecules consisting of 20 to 50 atoms, such as C_{24} H_{12}, or more generally, hydrogenated amorphous carbon particles. These large molecules or small particles are excited by ultraviolet and optical radiation, and then decay to excited vibrational levels that emit photons in the 3.28 μ and other bands in decaying to the ground level.

Though the most complete observational data is available for NGC 1976, similar results, though not so detailed, are available for NGC 6523, in which the Hourglass region is a local peak of infrared nebular emission, and for several other H II regions with such infrared peaks. These peaks are clearly regions of high dust density close to high-luminosity stars, which produce the energy that is absorbed and reradiated by the solid particles.

One interesting result originally obtained from measurements of a wide band in the far infrared ($\lambda \approx 400\ \mu$, the band is actually approximately 45-750 μ) is that the measured infrared flux is roughly proportional to the measured radio-frequency flux, as shown in Figure 7.8. Since the radio-frequency flux from a nebula is proportional to the number of recombinations within the nebula, the infrared emission is also roughly proportional to the number of recombinations, that is, to the number of ionizations, or the number of ionizing photons absorbed in the nebula. A plausible interpretation might be that since every ionization by a stellar photon in an optically thick nebula leads ultimately to a recombination and the emission of a Lα photon, or of two photons in the $2\ ^2S \rightarrow 1^2S$ continuum, as explained in Chapters 2 and 4; and since the Lα photons are scattered many times by resonance scattering before they can escape, then perhaps every Lα photon is absorbed by dust in the nebula and its energy is re-emitted as infrared radiation. According to this interpretation, the ratio of total infrared flux to radio-frequency flux would be

$$\frac{j_{\mathrm{IR}}}{j_\nu} = \frac{N_p N_e(\alpha_B - \alpha^{eff}_{2\ ^2S})h\nu_{L\alpha}}{4\pi j_\nu}, \tag{7.14}$$

where the radio-frequency emission coefficient j_ν is given by equations (4.22) and (4.30). Here α_B, $\alpha^{\mathrm{eff}}_{2\ ^2S}$, and j_ν depend only weakly on T, and their ratio depends on it even more weakly; j_{IR} and j_ν have the same density dependence, so this ratio is quite well determined.

For $\nu = 1.54 \times 10^{10}$ Hz, the radio frequency used in Figure 7.8 and a representative $T = 7500°$ K, the calculated ratio from equation (7.14) is $j_{\mathrm{IR}}/j_\nu = 1.3 \times 10^{15}$ Hz, but the line drawn through the data corresponds to $j_{\mathrm{IR}}/j_\nu = 7.5 \times 10^{15}$ Hz, approximately five times larger. The conclusion is that the infrared emission is larger than can be accounted for by absorption of Lα alone; in addition to Lα, some of the stellar radiation with $\nu < \nu_0$ and probably also some of the ionizing radiation with $\nu \geq \nu_0$ must be absorbed by the dust. This conclusion has been confirmed by several subsequent studies. Clearly some of the ionizing photons are destroyed in this way; so the Zanstra method of determining the effective temperature of the ionizing stars must be corrected for this effect. Detailed calculations of model nebulae, taking the absorption and scattering by dust particles into account, have been made. The main uncertainty is in the composition and optical properties of the solid particles.

Though the optical continuum measurements had shown dust to be present in H II regions, there was no previous clear evidence of dust in planetary nebulae before the infrared observations were made. However, these measurements revealed that in many planetaries there is an infrared continuum that is from 10 to 100 times stronger in the 5 μ to 18 μ region than the extrapolated free-free and bound-free continua. Later airborne observations at longer wavelengths out to 100 μ show that the infrared continua of most planetaries peak at about 30 μ. NGC 7027, the planetary nebula in which the infrared

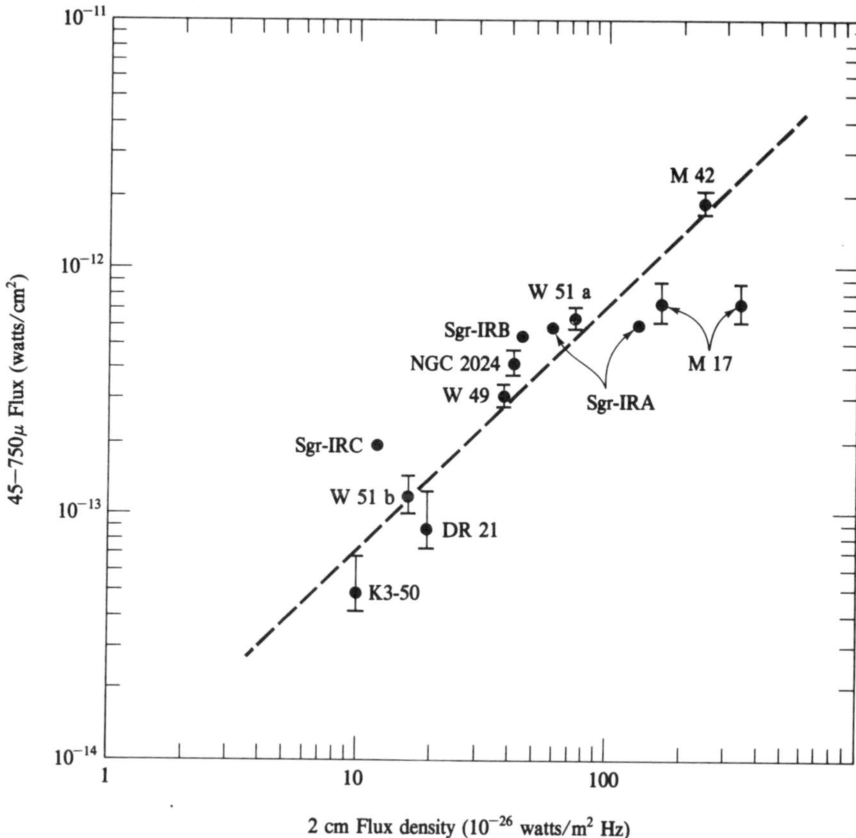

FIGURE 7.8
Measured far infrared (45-750 μ) flux and radio-frequency flux for several H II regions, showing proportionality between the two. The dashed line is drawn to fit the data on the average; it corresponds to more infrared emission than can be accounted for from absorption of all the Lα radiation calculated from the radio-frequency emission.

continuum was first discovered, is still the brightest throughout this spectral region.

As in the H II regions, the infrared continuum radiation of planetaries is due to dust, heated by absorption of stellar radiation and nebular resonance-line radiation, particularly Lα. In NGC 7027, for which the most detailed infrared spectrum is available, the far-infrared continuum corresponds approximately to the radiation from dust particles at $T \approx 90°$ K, but there is an excess over this at shorter infrared wavelengths, including a range of temperatures to higher values. In addition, the broad emission features between 3.28

and 11.3 μ are observed, probably indicating the presence of hydrogenated amorphous carbon particles, as in H II regions. The total optical depth of the dust in most planetary nebulae is small, and it therefore does not affect the Zanstra temperature determinations as much as in H II regions.

7.5 Survival of Dust Particles in an Ionized Nebula

The observational data obtained from the nebular-scattered optical continuum and thermal infrared continuum show that dust particles exist in H II regions and planetary nebulae. At least in H II regions, their optical properties, and the ratio of amounts of dust to gas, are approximately the same as in the general interstellar medium. Three questions then naturally arise: How are the dust particles initially formed? How long do they survive? How are they ultimately destroyed in nebulae? Much research effort has been expended on these questions, with results that can only be very briefly summarized here.

Let us begin by examining the formation of dust particles. All theoretical and experimental investigations indicate that though dust particles, once formed, can grow by accretion of individual atoms from the interstellar gas, they cannot initially form by atomic collisions at even the highest densities in gaseous nebulae. Thus in planetary nebulae, where the gaseous shell has undoubtedly been ejected by the central star, the dust must have been present in the atmosphere of the star or must have formed during the earliest stages of the process, at the high densities that occurred close to the star. Infrared measurements show that many cool giant and supergiant stars have dust shells around them, so there is observational evidence that this process does occur. Very probably the original dust particles in H II regions were formed in the same way.

Dust particles in a nebula are immersed in a harsh environment containing both ionized gas, with $T \approx 10{,}000°$ K, and high-energy photons. Collisions of the ions and photons with a dust particle tend to knock atoms or molecules out of its surface and thus tend to destroy it. We must examine very briefly some of the problems connected with the survival of dust particles in a nebula.

First of all, let us consider the electrical charge on a dust grain in a nebula. This charge results from the competition among photoejection of electrons from the solid particle by the ultraviolet photons absorbed by the grain (which tends to make the charge more positive) and captures of positive ions and electrons from the nebular gas (which tend to make the charge more positive and negative, respectively). It is straightforward to write the equilibrium equation for the charge on a grain. The rate of increase of the charge Ze due to photoejection of electrons can be written

$$\left(\frac{dZ}{dt}\right)_{pe} = \pi a^2 \int_{\nu_K}^{\infty} \frac{4\pi J_\nu}{h\nu} \phi_\nu \, d\nu, \tag{7.15}$$

where ϕ_ν is the photodetachment probability $(0 \leq \phi_\nu \leq 1)$ for a photon that strikes the geometrical cross section of the particle. If the dust particle is electrically neutral or has a negative charge, the effective threshold $\nu_K = \nu_c$, the threshold of the material; but if the particle is positively charged, the lowest energy photoelectrons cannot escape, so in general, the threshold is

$$\nu_K = \left\{ \begin{array}{ll} \nu_c + \dfrac{Ze^2}{ah} & Z > 0 \\ \nu_c & Z \leq 0 \end{array} \right\}, \tag{7.16}$$

where $-Ze^2/a$ is the potential energy of an electron at the surface of the particle. The rate of increase of the charge due to capture of electrons is

$$\left(\frac{dZ}{dt}\right)_{ce} = -\pi a^2 N_e \sqrt{\frac{8kT}{\pi m}} \xi_e Y_e, \tag{7.17}$$

where ξ_e is the electron-sticking probablility $(0 \leq \xi_e \leq 1)$, and the factor due to the attraction or repulsion of the charge on the particle is

$$Y_e = \left\{ \begin{array}{ll} 1 + \dfrac{Ze^2}{akT} & Z > 0 \\ e^{Ze^2/akT} & Z \leq 0 \end{array} \right\} \tag{7.18}$$

The rate of increase of the charge caused by capture of protons is, completely analogously,

$$\left(\frac{dZ}{dt}\right)_{cp} = \pi a^2 N_p \sqrt{\frac{8kT}{\pi m_{\rm H}}} \xi_p Y_p, \tag{7.19}$$

with

$$Y_p = \left\{ \begin{array}{ll} e^{-Ze^2/akT} & Z > 0 \\ 1 - \dfrac{Ze^2}{akT} & Z \leq 0 \end{array} \right\} \tag{7.20}$$

Thus the charge on a particle can be found from the solution of the equation

$$\frac{dZ}{dt} = \left(\frac{dZ}{dt}\right)_{pe} + \left(\frac{dZ}{dt}\right)_{ce} + \left(\frac{dZ}{dt}\right)_{cp} = 0, \tag{7.21}$$

in which the area of the particle cancels out, but the dependence on a through the surface potential remains. Equation (7.21) can be solved numerically for any model of nebula for which the density and the radiation field are known, but the main difficulty is that the properties of the dust particles are only poorly known. Still, for any apparently reasonable values of the parameters, we find that, in the inner part of an ionized nebula, photoejection dominates and the particles are positively charged; but in the outer parts, where the ultraviolet flux is smaller, photoejection is not important and the particles are negatively charged because more electrons, with their higher thermal velocities, strike the particle. As a specific example, for a dirty-ice particle with $a = 3 \times 10^{-5}$ cm, $\xi_e \approx \xi_p \approx 1$, and $\phi_\nu \approx 0.2$ for $h\nu > h\nu_c \approx$ 12 eV, the calculated result is that for a representative O5 star with $L \approx 5 \times 10^5 L_\odot$, $T_* \approx 50{,}000°$ K, in a nebula with $N_e \approx N_p \approx 16$ cm^{-3}, $Z \approx 380$ at a distance $r = 3.8$ pc from the star, $Z \approx 0$ at $r = 8.5$ pc, and $Z \to -360$ as $J_\nu \to 0$.

Knowledge of this charge is important for estimating the rate of sputtering, that is, the knocking out of atoms from the particle by energetic positive ions. The threshold for the process is estimated to be about 2 eV, which is somewhat larger than the mean thermal energy of a proton in a nebula, but of the same order of magnitude as the Coulomb energy at the surface of the particle, $Ze^2/a \approx 0.005\ Z$ eV. Therefore, in the inner part of a nebula, where, say, $Z \approx 400$, the positive charge of the particle raises the threshold significantly, and decreases the sputtering rate, but in the outer part of the nebula, the negative charge of the particle increases the sputtering rate. The efficiency or yield, expressed as the probability that a molecule will be knocked out of the particle per incident-fast proton, is quite uncertain, but according to the best available estimate, $\sim 10^{-3}$, the lifetime of a particle against sputtering is approximately $10^{15}\ a/N_p$ yr in the inner part (with $Z \approx 400$), $2 \times 10^{13}\ a/N_p$ yr where $Z = 0$, and $3 \times 10^{12}\ a/N_p$ yr in the outer part (with $Z \approx -400$). Taking a representative size $a = 3.0 \times 10^{-5}$ cm and $N_p \approx 16$ cm^{-3}, the lifetime of the dust against sputtering is long in comparison with the lifetime of the H II region itself, except possibly in the very outermost parts of the nebula.

Photons tend to destroy a dust particle by heating it to a temperature at which molecules vaporize from the surface. The temperature of the particle, T_D, is given by the equilibrium between the energy absorbed, mostly ultraviolet photons, and the energy emitted, mostly infrared photons:

$$\int_0^\infty 4\pi J_\nu(1 - A_\lambda)C_\lambda d\nu = \int_0^\infty 4\pi B_\nu(T_D)(1 - A_\lambda)C_\lambda\, d\nu. \tag{7.22}$$

On the left-hand or absorption side of the equation, since $\lambda \ll a$, the absorption cross section is essentially the same as the geometrical cross section, $(1 - A_\lambda)C_\lambda \approx \pi a^2$; but on the right-hand or emission side, where $\lambda \gg a$, $(1 - A_\lambda)C_\lambda \approx \pi a^2 \epsilon(2\pi a/\lambda)$ with $\epsilon \approx 0.1$ for dielectric particles containing a small contamination of such elements as Fe and Mg. As a result,

$$T_D \propto \left(\frac{L}{4\pi r^2 a}\right)^{1/5}, \tag{7.23}$$

and for a representative particle with $a = 3.0 \times 10^{-5}$ cm, $T_D \approx 100°$ K at $r =$ 3 pc from the star. The evaporation temperatures of the main constituents of a dielectric particle are $T_v \approx 20°$ K for CH_4, $T_v \approx 60°$ K for NH_3, and $T_v \approx 100°$ K for H_2O, suggesting that CH_4 cannot be held by dust particles anywhere in the nebula, that NH_3 vaporizes everywhere except in the outer parts, and that H_2O evaporates only in the innermost parts. On the other hand, graphite, silicate and silicon carbide particles have much higher evaporation temperatures ($T_v \approx 10^3$ °K) and are not appreciably affected.

Another possibly important destructive mechanism for dust particles is "optical erosion," which might better be called photon sputtering, the knocking out of atoms or molecules immediately following absorption of a photon. The process can result either from photoionization of a bonding electron of a molecule in the surface of the particle, or from excitation of a bonding electron to an antibonding level. Very little experimental data is available on this process, and the calculations depend on many unchecked assumptions, but they seem to show that dirty-ice particles would be destroyed by this process in a typical H II region in a relatively short time, of order 10^2 to 10^3 yr, and therefore should not exist at all in nebulae. Again, however, silicate or graphite grains are quite safe for times of order 10^7 yr.

The overall conclusion is that, for classical dielectric particles, sputtering is not an important mechanism except possibly in the outer parts of a nebula; vaporization is important in the inner parts; and optical erosion is probably very important but is only poorly understood. Carbon and silicate particles, on the other hand, are more tightly bound and more likely to survive than dirty-ice particles. This conclusion, of course, agrees with the observational certainty that dust particles do exist in H II regions and planetary nebulae.

7.6 Dynamical Effects of Dust in Nebulae

Dust particles in a nebula are subjected to radiation pressure from the central star. However, the coupling between the dust and gas is very strong, so the dust particles do not move through the gas to any appreciable extent, but rather transmit the central repulsive force of radiation pressure to the entire nebula. Let us look at this a little more quantitatively. The radiation force on a dust particle is

$$F_{\rm rad} = \pi a^2 \int_0^\infty \frac{\pi F_\nu}{c} P_\nu \, d\nu$$

$$= \pi a^2 \int_0^\infty \frac{L_\nu}{4\pi r^2 c} P_\nu \, d\nu \approx \frac{a^2 L}{4r^2 c}, \tag{7.24}$$

where P_ν is the efficiency of the particle for radiation pressure. Since most of the radiation from the hot stars in the nebula has $\lambda \ll a$, $P_\nu \approx 1$ for classical dirty-ice particles. However, this is not true for very small graphite or silicate particles. Here only the radiation force of the central star has been taken into account; the diffuse radiation field is more nearly isotropic and can, to a first approximation, be neglected in considering the motion due to radiation pressure. The force tends to accelerate the particle through the gas, but its velocity is limited by the drag on the particle produced by its interaction with the nebular gas. If the particle is electrically neutral, this drag results from direct collisions of the ions with the grain, and the resulting force is

$$F_{coll} = \frac{4}{3} N_p \pi a^2 \left(\frac{8kTm_{\rm H}}{\pi} \right)^{1/2} w, \tag{7.25}$$

where w is the velocity of the particle relative to the gas, assumed to be small in comparison with the mean thermal velocity. Thus the particle is accelerated until the two forces are equal, and reaches a thermal velocity

$$w_t = \frac{3L}{16\pi r^2 c N_p} \left(\frac{\pi}{8kTm_{\rm H}} \right)^{1/2}, \tag{7.26}$$

which is independent of the particle size. As an example, for a particle at a distance of 3.3 pc from the O star we have been considering, $w_t = 10$ km sec^{-1}, and the time required for a relative motion of 1 pc with respect to the surrounding gas is about 10^5 yr.

However, for charged particles, the Coulomb force increases the interaction between the positive ions and the particle significantly, and the drag on a charged particle has an additional term,

$$F_{Coul} \approx \frac{2N_p Z^2 m_{\rm H}}{T^{3/2}} w, \tag{7.27}$$

with T expressed in $^\circ$ K. Comparison of equation (7.27) with equation (7.25) shows that Coulomb effects dominate if $|Z| \geq 50$. Since, in most regions of the nebula the particles have a charge greater than this, the terminal velocity is even smaller, and the motion of the particle with respect to the gas is smaller yet. Under these conditions the dust particles are essentially frozen to the gas, the radiation pressure on the particles is communicated to the nebular material, and the equation of motion therefore contains an extra term on the right-hand side; so equation (6.1) becomes

$$\rho \frac{D\mathbf{v}}{Dt} = -\nabla p - \rho \nabla \phi + N_D \frac{a^2 L}{4r^2 c} \mathbf{e}_r, \tag{7.28}$$

Substitution of typical values, including the observationally determined gas-to-dust ratio, shows that the accelerations produced can be appreciable, and the radiation-pressure effects should therefore be taken into account in the calculation of a model of an evolving H II region. An approximate calculation of this type has shown that, with reasonable amounts of dust, old nebulae will tend to develop a central "hole" that has been swept clear of gas by the radiation pressure transmitted through the dust. An example of a real nebula to which this model may apply is NGC 2237, shown in Figure 7.9. Likewise radiation pressure or dust is probably important in the early stages of formation of planetary nebulae from red giant stars, as further discussed in Chapter 8.

The observational data clearly show that dust does exist in nebulae, but unfortunately its optical and physical properties are still not accurately known, so the specific calculations that have been carried out to date must be considered schematic and indicative rather than definitive. Astronomical measurements over the entire spectrum from ultraviolet to infrared, combined with laboratory data and physical theory applicable to small, cold, very "impure" particles, may be expected to lead to further progress.

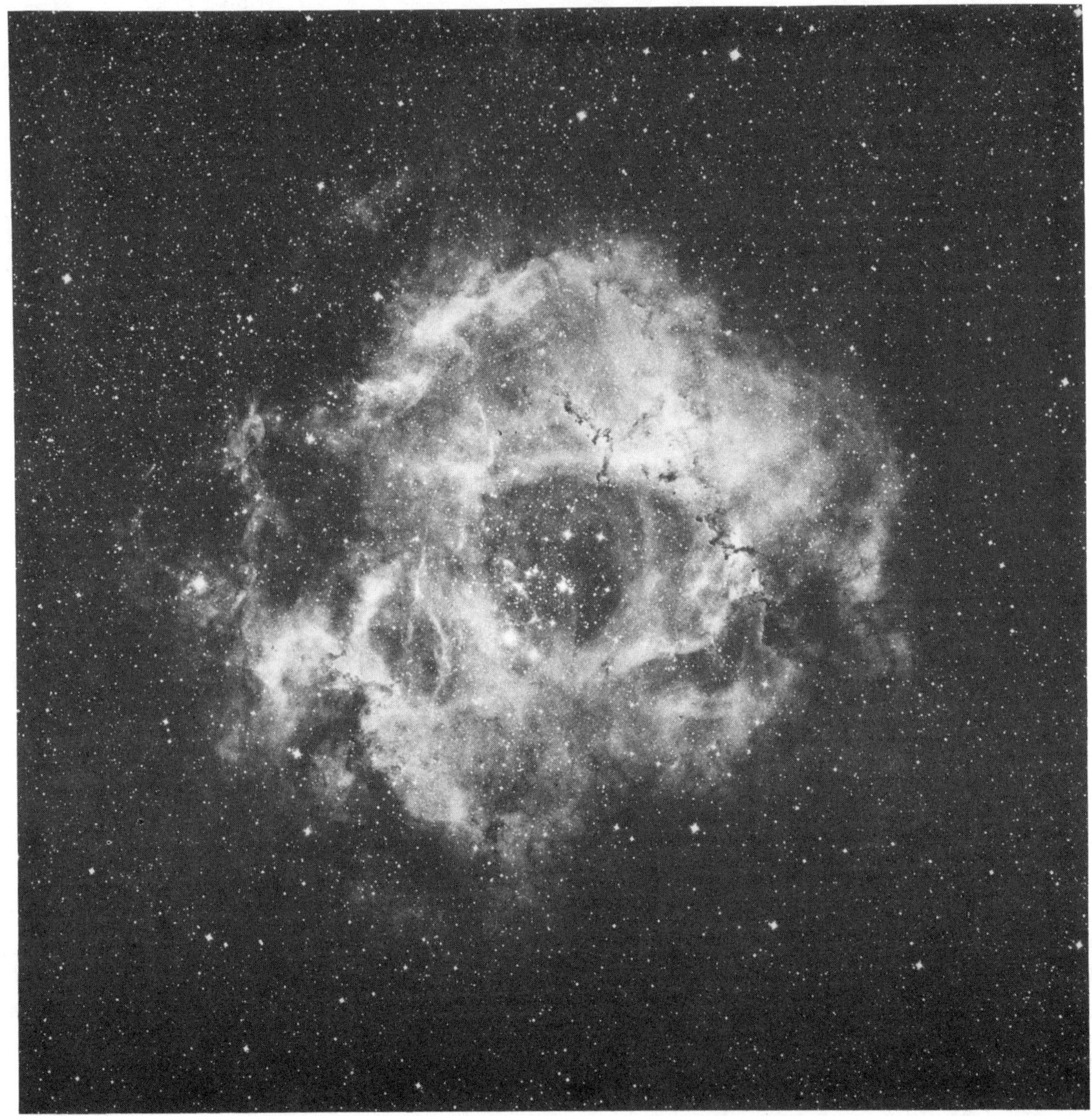

FIGURE 7.9

NGC 2237, an H II region in Monoceros, taken with 24-Schmidt telescope, red filter and 103a-E plate, emphasizing Hα, [N II]. The central hole may have been swept clear of gas by radiation pressure on the dust from the central star. (*National Optical Astronomy Observatories photograph.*)

References

Interstellar extinction is a subject with a long history, most of which is not directly related to the study of gaseous nebulae. Several useful summaries of the available data going back to an early review which contains the standard or Whitford interstellar extinction of Figure 7.1 and Table 7.1 are:

Whitford, A. E. 1958. *A. J.* **63,** 201.

Lynds, B. T., and Wickramasinghe, N. C. 1968. *Ann. Rev. Astr. Ap.* **6,** 215.

Martin, P. G. 1978. *Cosmic Dust.* Oxford: Clarendon Press.

Savage, B. D., and Mathis, J. S. 1979. *Ann. Rev. Astr. Ap.* **17,** 73.

(The average interstellar extinction curve of Figure 7.2 and Table 7.2, which extends into the satellite ultraviolet spectral region, is taken from this last reference.)

Another, somewhat different interstellar extinction curve is given, with an analytic interpolation formula to it, by

Seaton, M. J. 1979. *M.N.R.A.S.* **187,** 73P.

Two papers on regional differences of extinction are

Whiteoak, J. B. 1966. *Ap. J.* **144,** 305.

Anderson, C. M. 1970. *Ap. J.* **160,** 507.

(The data for Figure 7.3 are taken from the first of these references.)

The [S II] method of measuring the extinction of the light of a nebula is discussed by

Allen, D. A. 1979. *M.N.R.A.S.* **186,** 1P.

A list of nebular line ratios that may in principle be used for extinction determinations is given by

Draine, B. T., and Bahcall, J. N. 1981. *Ap. J.* **250,** 579.

The Paschen/Balmer and Balmer-line ratio methods are described parenthetically in many references chiefly devoted to observational data on planetary nebulae, for instance,

Miller, J. S. and Mathews, W. G. 1972, *Ap. J.* **172,** 593.

This paper contains accurate measurements of NGC 7027, and shows that calculated Balmer-line and Balmer-continuum ratios, modified by interstellar extinction, agree accurately with the measurements. These authors give convenient interpolation formulas for fitting the standard interstellar extinction curve. The Balmer radio-frequency method of determining the extinction is discussed for instance, in

Cahn, J. H. 1976. *A. J.* **81,** 407.

The results of Table 7.3 are taken from this reference.

The presence of a nebular continuum in diffuse nebulae, resulting from dust scattering of stellar radiation, is an old concept. The fact that it is the dominant contributor to the observed continuum was first quantitatively proved by

Wurm, K., and Rosino, L. 1956. *Mitteilungen Hamburg Sternwarte Bergedorf,* **10,** Nr. 103.

They compared photographs of NGC 1976 (similar to the Frontispiece) taken with narrow-band filters, which isolated individual spectral lines with photos of a region in the continuum free of emission lines (similar to Figure 7.6), and showed that the appearance of the nebula in the continuum is different from its appearance in any spectral lines, and that the continuum cannot have an atomic origin and therefore (by implication) must arise from dust. The He II $\lambda 4686$ absorption line in the continuum of NGC 1976 was observed by

Peimbert, M., and Goldsmith, D. W. 1972. *Astr. Ap.* **19,** 398.

Quantitative measurements of the continuum surface brightnesses in several nebulae, and comparisions with $H\beta$ surface brightnesses, were made by

O'Dell, C. R., and Hubbard, W. B. 1965. *Ap. J.* **142,** 591.
O'Dell, C. R., Hubbard, W. B., and Peimbert, M. 1966. *Ap. J.* **143,** 743.
Lynds, B. T., Canzian, B. J., and O'Neil, E. J. 1985. *Ap. J.* **288,** 164.

The analysis given in the text is based on the second of these references. (Figure 7.7 and Table 7.4 are taken from it.) The simplifying assumptions made in the analysis are not necessary, and more realistic models have been calculated by

Mathis, J. S. 1972, *Ap. J.* **176,** 651.

The classical optical work on interstellar dust, its extinction, and its scattering properties, are very well summarized in the Martin 1978 reference above. The deduced properties of interstellar particles, including numerical values for C_λ, A_λ, and a used in the text, are suggested by it. Additional discussion of the types of particles required to fit the observational data on extinction is contained in

Mathis, J. H., Rumpl, W., and Nordsieck, K. H. 1977. *Ap. J.* **217,** 425.

The effect of nebular scattering on the correction of measured line-intensity ratios for extinction is discussed by

Mathis, J. S. 1983. *Ap. J.* **267,** 119.
Caplan, J., and Deharveng, L. 1986. *Astr. Ap.* **155,** 297.

Globules were named and discussed by

Bok, B. J., and Reilly, E. F. 1947. *Ap. J.* **105,** 255.

A good summary of their properties is given in

Bok, B. J., Cordwell, C. S., and Cromwell, R. H. 1971. *Dark Nebulae, Globules, and Protostars,* ed. B. T. Lynds. Tucson: University of Arizona Press, p. 33.

Good reviews of the infrared work on nebulae (including planetaries) are

Neugebauer, G., Becklin, E., and Hyland, A. R. 1971, *Ann. Rev. Astr. Ap.* **9,** 67.

Wynn-Williams, C. G., and Cruikshank, D. P. eds. 1981. *Infrared Astronomy* (IAU Symposium No. 96). Dordrecht: Reidel.

Barlow, M. J. 1983. *Planetary Nebulae* (IAU Symposium No. 103). Dordrecht: Reidel, p. 105.

A survey of a large number of infrared sources, including many H II regions, is included in

Hoffman, W. L., Frederick, C. L., and Emery, R. J. 1971, *Ap. J.* **170,** L89.

The rough proportionality of the wide-band infrared fluxes of H II regions to their free-free radio-frequency fluxes was discovered by

Harper, D. A., and Low, F. J. 1971. *Ap. J.* **165,** L9.

(Figure 7.8 is taken from this reference.)

Very good theoretical treatments of the effects of dust within a nebula or the ionizing, continuum, and line photons, and the consequent infrared emission are contained in

Panagia, N. 1974. *Ap. J.* **192,** 221.

Natta, A., and Panagia, N. 1976. *Astr. Ap.* **50,** 191.

Perinotto, M., and Picchio, G. 1978. *Astr. Ap.* **68,** 275.

Tielens, A. G. G. M., and de Jong, T. 1979. *Astr. Ap.* **75,** 326.

Good reviews on infrared radiation from dust in planetary nebulae are:

Rank, D. M. 1978. *Planetary Nebulae* (IAU Symposium No. 76). Dordrecht: Reidel, p. 103.

Barlow, M. J. 1983. *Planetary Nebulae* (IAU Symposium No. 103). Dordrecht: Reidel, p. 105.

Problems of survival of dust particles in ionized nebulae are reviewed by

Salpeter, E. E. 1977. *Ann. Rev. Astr. Ap.* **15,** 267.

Spitzer, L. 1978. *Physical Processes in the Interstellar Medium.* New York: Wiley, Chap. 9.

Draine, B. T. 1979. *Ap. J.* **230,** 106.

Draine, B. T., and Salpeter, E. E. 1979. *Ap. J.* **231,** 77 and 438.

The model expanding nebula (taking radiation pressure into account), which ultimately develops a central cavity, was calculated by

Mathews, W. G. 1967. *Ap. J.* **147,** 965.

8

H II Regions in the Galactic Context

8.1 Introduction

In the first five chapters of this book, we examined the equilibrium processes in gaseous nebulae and compared the models calculated on the basis of these ideas with observed H II regions and planetary nebulae. In Chapter 6 and 7 the basic ideas of the internal dynamics of nebulae, and of the properties and consequences of the interstellar dust in nebulae, were discussed, worked out, and compared with observational data. Thus we have a fairly good basis for understanding most of the properties of the nebulae themselves. In this chapter we consider the H II regions in the wider context of galaxies. The discussion includes the distributions of these regions, both in our own Galaxy and in other galaxies so far as they are known, and the regions' galactic kinematics. Then we examine the stars in H II regions, and what is known about their formation and H II region formation, and we sketch the evolution of H II regions on the basis of the ideas presented in the earlier chapters. The chapter concludes with a brief discussion of the molecules in H II regions, as revealed by radio-frequency, microwave, and infrared spectroscopy.

8.2 Distribution of H II Regions in Other Galaxies

H II regions can be recognized on direct photographs of other galaxies taken in the radiation of strong nebular emission lines. The best spectral region for this purpose is the red, centered around Hα λ6563 and [N II] $\lambda\lambda$6548, 6583. Most pictures of nebulae in this and other astronomical books were taken in this way, using various red filters – often red Plexiglass ($\lambda > 6000$), which was used in the National Geographic Society–Palomar Observatory Sky Survey, or Schott RG 2 glass ($\lambda > 6300$), often used with large telescopes – together with 103a-E plates ($\lambda < 6700$). It is possible to use quite narrow-band interference filters for the maximum rejection of unwanted continuum radiation, and comparison with a narrow-band photograph in the nearby continuum permits nearly complete discrimination between H II regions and continuum sources, which are apparently mostly luminous stars and star clusters. A photographic subtraction, combining a negative of one plate with a positive of the other, can be used to compare the two exposures, though difficulties are caused by the nonlinearity of the photographic process. CCDs, with their much higher sensitivity, permit the use of very narrow interference filters, and have the additional advantages of digital readout and complete linearity.

Many external galaxies have been surveyed for H II regions in these ways (for example, NGC 1232, which is shown in Figure 8.1). In such studies the entire galaxy can be observed (except for the effects of interstellar extinction), and all parts of it are very nearly the same distance from the observer, in contrast to the situation in our own Galaxy, where the more distant parts are nearly completely inaccessible to optical observation. These photographic surveys show that essentially all the nearby, well-studied spiral and irregular galaxies contain many H II regions. On the other hand, elliptical and S0 galaxies typically do not contain H II regions; a few S0s have some H II regions, but many less than typical later-type spirals.

In spiral galaxies the H II regions are strikingly concentrated along the spiral arms, and in fact are the main objects seen defining the spiral arms in many of the published photographs of galaxies. Often there are no H II regions in the inner parts of the spiral galaxies, but the spiral arms can be seen as concentrated regions of interstellar extinction – "dust" in the terminology used in galactic structure. Evidently, in these regions there is interstellar matter, but no O stars to ionize it and make it observable as H II regions. Different galaxies have different amounts of dust and different densities of H II regions along the spiral arms, but the concentration of H II regions along relatively narrow spiral arms and spurs is a general feature of spiral galaxies.

In irregular galaxies the distribution of H II regions is less well-organized. In some of the galaxies classified as irregular, such as the Large Magellanic Cloud, features resembling spiral arms can be traced in the distribution of H II regions, but in other irregular galaxies, such as the Small Magellanic Cloud, the distribution of H II regions is often far less symmetric; one or more areas may contain many H II regions, but other areas may be essentially devoid of H II regions. Some of these galaxies contain H II emission spread through much of their volume; these are often called "extragalactic H II regions," or "blue compact dwarf galaxies."

From the direct photographs, it is clear that spiral galaxies are highly flattened, disk systems and that the H II regions are strongly concentrated, not only to the spiral arms, but also to the galactic plane in which the arms lie. Photographs or maps from which the effects of the inclination of the plane of the galaxy to the line of sight have been removed have very symmetric structures that can result only if the H II regions are nearly in a plane.

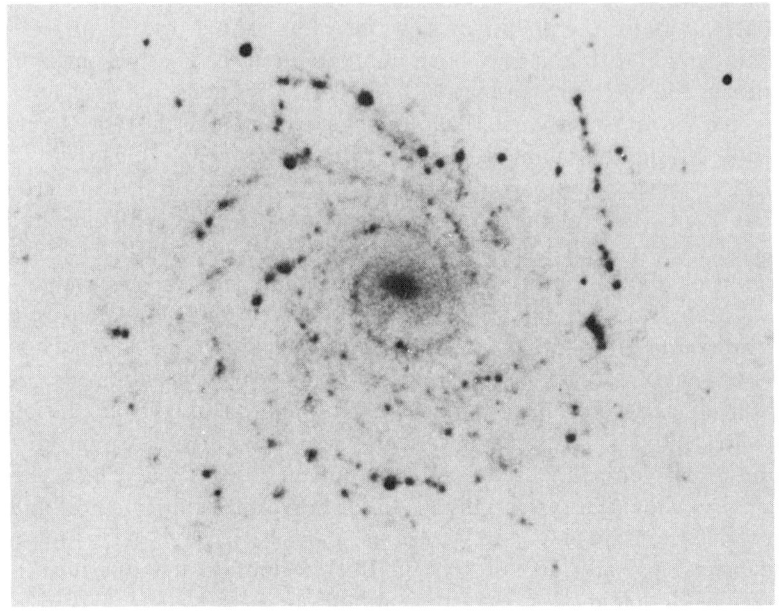

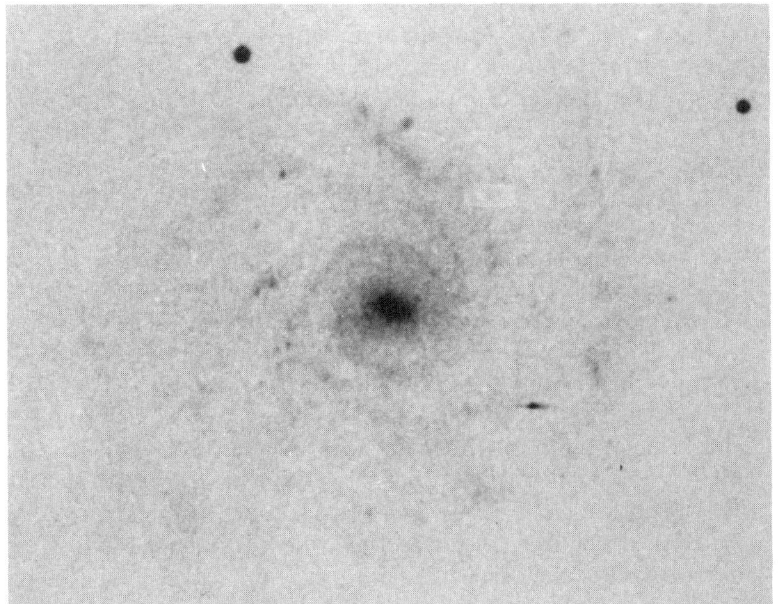

FIGURE 8.1

NGC 1232, a nearly face-on spiral galaxy. Top photograph, taken using Hα, [N II] filter, emphasizes H II region; bottom photograph, taken using red continuum filter, suppresses emission nebulae. Exposure times were chosen so that stars appear similar in two photographs. Note how the H II regions lie along the spiral arms in the top photograph. Original plates were taken with an image-tube camera on the Kitt Peak 2.1-m telescope. *(National Optical Astronomy Observatories photographs.)*

Because the light of H II regions is largely concentrated into a relative few spectral lines, they are favorite objects for spectroscopic determinations of radial velocities in external galaxies. For many of the larger nearby galaxies, extensive lists of radial velocities of H II regions are available, chiefly from Hα and [N II]; but in earlier work this information was obtained from [O II] $\lambda3727$, Hβ and [O III] $\lambda\lambda4959$, 5007. These measurements show that, at a specific point in a galaxy, the dispersion of the radial velocities of the H II regions is quite small, and that, when corrected for effects of projection, the velocities show the familiar galactic rotation pattern. Indeed, much of the available information on galactic rotation and the masses of spiral galaxies has been derived from these radial-velocity measurements of H II regions. The most convincing evidence for the existence of dark matter in galaxies, the "missing mass," has come from rotation curves obtained from H II regions in this way, and from similar H I $\lambda21$ cm radial-velocity measurements.

The spectra of the brighter H II regions in many external galaxies have been studied and are quite similar to the spectra of H II regions in our Galaxy. The best quantitative information on helium abundances in other galaxies comes from measurements of He I and H I recombination lines in H II regions, and the results shown in Table 8.1 indicate that there are true He abundance differences among the nearby observed galaxies. The estimated errors

TABLE 8.1
Helium abundance in galaxies

Galaxy	H II region	N_{He^+}/N_p	$N_{\mathrm{He}}/N_{\mathrm{H}}$
Our Galaxy	NGC 1976	–	0.101
	η Car	–	0.102
LMC	30 Dor I	0.083	0.085
	30 Dor II	0.082	0.085
	NGC 1714 I	0.079	0.082
	NGC 2079 I	0.081	0.085
	IC 2111 I	0.075	0.079
	IC 2111 II	0.080	0.087
	Mean	–	0.084
SMC	NGC 346 I	0.077	0.077
	NGC 346 II	0.078	0.078
	NGC 356 I	0.076	0.077
	NGC 456 I	0.077	0.079
	Mean	–	0.078

of the mean values of $N_{\rm He}/N_{\rm H}$ are approximately $\pm$ 0.005. In both Magellanic Clouds He is relatively less abundant than in our Galaxy; in addition, He is probably less abundant in the Small Magellanic Cloud than in the Large.

Furthermore, there are real differences in the heavy-element contents measured in H II regions in other galaxies. The results listed in Table 8.2 show a correlation of He abundance with the abundances of O, N, and Ne. (Tables 8.1 and 8.2, taken from different sources, list slightly different He abundances for the Small Magellanic Cloud.) The first three objects in the table are low-luminosity H II region galaxies. The natural interpretation of these observed relationships among the abundances is that nuclear burning in stars followed by return of the processed material to interstellar gas has enriched the helium and heavy elements in the different galaxies by different amounts. These measurements of elemental abundances in H II regions thus allow us to study the integrated effects of stellar evolution in galaxies. Of course, a danger is that the measured heavy-element abundances in the gas may be affected by significant amounts of matter locked up as solids in the dust particles, but this does not seem to be the major critical effect, at least for O and Ne.

TABLE 8.2
Element abundances in galaxies

Galaxy	$\frac{N_{\rm He}}{N_{\rm H}}$	$\frac{N_{\rm O}}{N_{\rm H}}$	$\frac{N_{\rm N}}{N_{\rm H}}$	$\frac{N_{\rm Ne}}{N_{\rm H}}$	$\frac{N_{\rm S}}{N_{\rm H}}$
I Zw 18	6.3×10^{-1}	1.8×10^{-5}	–	3.6×10^{-6}	–
II Zw 40	7.8×10^{-1}	1.5×10^{-4}	7.2×10^{-6}	2.2×10^{-5}	4.0×10^{-5}
II Zw 70	7.8×10^{-1}	1.2×10^{-4}	6.2×10^{-6}	2.6×10^{-5}	8.3×10^{-6}
<SMC>	8.0×10^{-1}	1.0×10^{-4}	2.8×10^{-6}	2.5×10^{-5}	1.5×10^{-5}
<LMC>	8.4×10^{-1}	2.4×10^{-4}	9.7×10^{-6}	5.6×10^{-5}	2.6×10^{-5}
NGC 1976	10.0×10^{-1}	3.6×10^{-4}	3.1×10^{-5}	6.6×10^{-5}	2.4×10^{-5}
<Galaxy>	11.7×10^{-1}	4.0×10^{-4}	3.9×10^{-5}	1.3×10^{-4}	1.8×10^{-5}

The three low-luminosity irregular galaxies at the beginning of Table 8.2, with the lowest helium and heavy-element abundances, are presumably relatively recently formed objects; so in them not much processed material has yet been returned to the interstellar gas. Objects with zero heavy-element content would presumably then give the He abundance in the early universe, before any star formation, consequent processing, and recycling to interstellar gas has occurred. No such objects have been observed to date, though

I Zw 18 is a close approximation to them. The best single heavy element to study is O, since its abundance is relatively straightforward to measure in H II regions. It is probably not strongly depleted from the gas by being in dust particles, and is chiefly formed in massive stars directly by He-burning reactions. Extrapolating the correlation from the measurements upon which Table 8.2 is based to zero O abundance gives a primordial abundance ratio He/H $= 0.069 \pm 0.006$ by number of atoms, or in the form used by researchers in cosmology, a primordial mass fraction of He $Y = 0.216 \pm 0.015$. A more recently measured set of abundances in H II regions in our Galaxy, and in low-luminosity irregular galaxies, is shown in Figure 8.2. The best straight line through these points gives the primordial He fraction by mass $Y = 0.232 \pm 0.013$; the exact value is highly significant in testing and evaluating current big-bang models of the early universe. Slight observational errors, and probably more importantly, details of the theoretical calculations, are thus of great importance. In particular, collisional excitation from the metastable $2\,^3$S level of He I, ignored in the reduction of the measurements to the abundances of Tables 8.1 and 8.2, may in reality be significant enough to reduce some or all of the derived He abundances. Actual nebulae certainly have much more complicated structure than the idealized models described in this book, and whether the derived He abundances are correct to within a few percent is not obvious.

H II regions in nearby galaxies such as M 33 and M 101 may also be used to measure possible variations of the abundances with distance from the center. Direct observations show that especially in Sc galaxies the [O III] ($\lambda4959 + \lambda5007$)/H$\beta$ intensity ratio increases outward, but the [N II]($\lambda6548 + \lambda6583$)/H$\beta$ ratio decreases outward. Combining these ratios with electron temperatures determined from [O III] ($\lambda4959 + \lambda5007$)/$\lambda4363$, we find that in most of the observed galaxies the abundance ratios O/H and N/H both decrease outward. The reasons can easily be seen. Since O is the most abundant and thus most important element for collisional cooling, its abundance decrease causes an outward increase of the temperature in the H II regions, essentially until increased radiation in the high-energy ($h\nu > kT$) $\lambda\lambda4959$, 5007 lines compensates for decreased radiation in the low-energy fine-structure lines. In the somewhat lower energy [N II] $\lambda\lambda6583$, 6548 lines, this temperature increase does not completely compensate for the abundance decrease.

To derive quantitative abundances also requires the use of highly schematized ionization-correction factors, derived essentially from the [O II]/[O III] optical line ratios by assuming that the O^+ and O^{++} zones have the same T. A difficulty is that at low nebular temperatures [O III] $\lambda4363$ is often too weak to measure. Hence it is conventional to adopt empirical relationships derived from the H II regions for which T has been

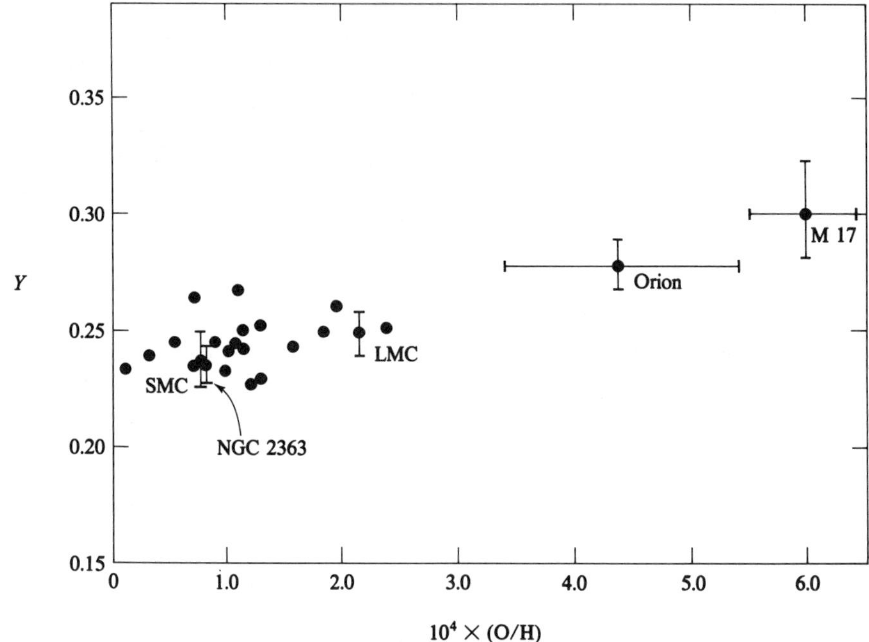

FIGURE 8.2
Fraction of mass that is helium, Y, plotted against relative oxygen abundance O/H (by numbers) from H II regions in our Galaxy, and in irregular and blue compact dwarf galaxies.

measured, for instance between combined ratios such as ([O II] $\lambda3727$ + [O III] $\lambda\lambda4959$, 5007)/Hβ and O/H abundance ratios, and to interpolate or extrapolate them to other regions or even other galaxies in which $\lambda4363$ was not observed. These methods are subject to uncertain corrections for extinction, but seem to work well enough to confirm many of the general ideas of nucleogenesis in stars and of mass return to interstellar space.

A better method is to use simplified models of H II regions, with as many adjustable parameters as can be determined from the available line ratios. Some good steps have been taken in this direction. Generally the mixture of heavy elements is taken as fixed, and only the overall amount of these elements, taken all together, relative to H and He, is varied. A more sophisticated procedure is to allow the abundance of N, a "secondary element" in terms of nucleogenesis, to vary with respect to the abundances of the "primary elements," O and C. In either case, the heavy-element content in the stellar-atmosphere models that supply the ionizing photons should be varied also, to match that in the model nebulae. The distribution function of luminosities and effective temperatures of the stellar-model atmospheres that occur in real clusters is generally modeled by assuming a single effective temperature; this is not a bad assumption because the great bulk of the ionizing photons come from the hottest (O5) stars. Typical resulting values are $T_* \approx 45,000^\circ$ K. Alternatively, the relative numbers of early type-stars (the "initial mass function") may be taken as fixed, corresponding to a fixed T_*. The other important variable is then the ionization parameter

$$\Gamma = \frac{Q(\mathrm{H}^0)}{4\pi < r^2 > c < N_\mathrm{H} >} = \frac{< N_\mathrm{ph} >}{< N_\mathrm{H} >}, \tag{8.1}$$

where $Q(\mathrm{H}^0)$ is the number of ionizing photons emitted per unit time by the central source, as defined by equation (2.19); so $Q(\mathrm{H}^0)/4\pi<r^2>$ would be the flux at a typical point in the nebula, in the absence of absorption, and N_ph the corresponding mean density of ionizing photons. Other definitions of the "ionization parameter" are used by various authors, but they all involve in some way or other the dimensionless form adopted here.

One interesting result is that spectroscopic studies of H II regions in external galaxies indicate systematic changes in the ionization as a function of the heavy-element abundances. Observationally, as the O/H abundance ratio increases (and other heavy element abundances also increase) the ionic ratio $\mathrm{O}^{++}/\mathrm{O}^{+}$ decreases, because of the increased importance of the absorption at the ionization limits ν_T of O^+ and Ne^+ (see Table 2.7), resulting in less ionizing radiation from the star for these same ions in the nebula. The increased importance of these absorption edges is caused by the increased abundances of Ne and O, and by an apparent tendency (that is not understood) for heavy-element-rich H II regions to have cooler ionizing stars. For a range of T_* around 37,000° K, and for O/H abundance ratios greater than in NGC 1976 or the Sun, the resulting stellar radiation field is strong above the ionization potential of He^0, $h\nu > h\nu_2 = 24.6$ eV, but is quite weak above the ionization potential of O^+, $h\nu > 35.1$ eV. As a result, O^{++} is confined to a small volume near the ionizing stars, and O^+ coexists with He^+ throughout most of the H II region. This ionization structure, which typically occurs in H II regions near the centers of spiral galaxies, differs from that of the H II regions in our Galaxy near the Sun, which typically have larger O^{++} volumes in their H^+, He^+ regions. The available calculations suggest that differences in abundance alone cannot cause the entire observed effect; it is necessary to assume also that the ionizing stars in the heavy-element-rich H II regions are somewhat cooler, as stated above.

Present-day computers enable us to calculate whole families of simplified model H II regions quickly and relatively inexpensively. Very promising first steps have been taken in interpreting the observational data in this way, and no doubt there will be rapid future progress in deducing further abundance information from even more sophisticated models, and measurements of even more lines, including those in the ultraviolet and infrared spectral regions.

The ionization parameter essentially determines the nature of a gaseous nebula or ionized region. If Γ is large, each nebular ion is strongly affected by the direct nebular radiation. The ionization is high, and the diffuse radiation field is comparatively unimportant. On the other hand, for small Γ the fractions of H^0 and He^0 in the nebula are relatively larger, these neutral atoms absorb comparatively more of the ionizing photons, leaving fewer for the heavy ions, and the average degree of ionization is lower, even for very hot "central" stars. The "nebula" hardly has a well-defined "edge." This is the situation in large regions of our Galaxy, far from the nearest O star but ionized by photons that have escaped from matter-bounded H II regions. Even though $< N_H >$ may be relatively small, $< r^2 >$ is so large that Γ is small, and only a low-ionization spectrum is observed. Diffuse emission-line radiation from other galaxies, as well as our own, results not only from the classical, bright, well studied H II regions with large ionization parameters, but also from these extended, faint regions of low ionization parameters.

8.3 Distribution of H II Regions in Our Galaxy

The analogy with observed external galaxies, of course, strongly suggests that the H II regions in our own Galaxy are also concentrated to spiral arms. There is no doubt that H II regions are strongly concentrated to the Galactic plane because, except for the very nearest, they are all close to the Galactic equator in the sky. However, our location in the system and the strong concentration of interstellar dust to the Galactic plane make it difficult to survey much of the Galaxy optically for H II regions and to determine their distances accurately. The surveys are carried out photographically in the light of Hα plus [N II] $\lambda\lambda$6548, 6583. Many early surveys were made with smaller cameras, but all were supplanted by the National Geographic Society–Palomar Observatory Sky Survey, made with the 48-inch Schmidt telescope, which (including its southern extension) covers all of the sky north of declination $-33°$. In this survey, the red plates were taken with a red Plexiglass filter and 103a-E plates, isolating a spectral region approximately 6000 Å $< \lambda <$ 6700 Å, as described in Section 8.2. The Whiteoak extension to $-45°$, taken with a wider spectral range (5400 Å $< \lambda <$ 6700 Å, defined by an amber Plexiglass filter) is also available. Even better surveys are now available for the southern hemisphere, made with the 1-m European Southern Observatory Schmidt telescope at La Silla, Chile, and the 1.2-m United Kingdom Schmidt, at Siding Springs, Australia. Both have focal lengths and hence plate scales very similar to the Palomar Schmidt. A "blue" survey (λ3700 - λ5000) has been completed with the ESO Schmidt, and a "J" or "blue-green" survey (λ3800 - λ5300) with the UK Schmidt. Both cover the entire sky south of $\delta = -18°$. A red survey ($\lambda6300$ - $\lambda6800$, and thus ideal for discovering H II regions) with the ESO Schmidt is approaching completion for the same area of the sky, and another with a somewhat wider passband ($\lambda5900$ - $\lambda6900$) is underway for the equatorial region ($-18° < \delta < +3°$) with the UK Schmidt. A second Palomar Survey of the northern sky, using the 48-inch Schmidt telescope with an improved corrector plate and finer-grain photographic plates was begun in 1985. It will cover the same area as the original survey in three wavelength bands blue, red and infrared.

The only accurate method of finding the distance of H II regions is the method of spectroscopic parallaxes, that is, spectral classifications of the stars involved to find their absolute magnitudes and hence their distances. To determine the distance accurately requires highly accurate spectral classification, accurate color measurements to determine the interstellar extinction, and careful calibration of the relationships between spectral type, absolute magnitude, and intrinsic color.

Part of the problem in finding the distances of H II regions is simply in identifying the exciting star or stars whose distance, when measured, will give the distance of the nebula. Although for most of the nearby bright H II regions the exciting O stars can easily be recognized from available spectral surveys, there are some, even among the nearest, in which the exciting star has not yet been identified with certainty. For instance, in NGC 7000, the North America Nebula, shown in Figure 8.3, the exciting star is probably the sixth magnitude O5 star HD 199579 near the middle of the photograph, but some workers have thought that it may be simply a projected foreground object, and that other exciting stars may be hidden behind the dense obscuring cloud that runs across the H II region and divides North America from the Pelican. In more distant H II regions, the problem often is that the exciting star or stars cannot be identified among the many foreground and background stars projected on the nebula because there are no adequate spectral surveys of OB stars fainter than about twelfth magnitude. In fact, with a CCD spectrograph and direct camera or photometer, we can determine the spectral types and colors of quite faint O stars once they have been found, and thus measure their distances, but the main obstacle is the lack of spectral surveys which are necessary to isolate the O stars to be observed in detail.

The distances of H II regions and OB star aggregates, plus those of a few very distant single stars, all determined in this way, are plotted in Figure 8.4, which shows that sections of three spiral arms are well-delineated from the optical measurements. However, the range of observation is small, and the distances of most of the H II regions plotted are less than 2 kpc from the Sun, while the distance to the Galactic center is much larger. Notice also that the extreme southern-hemisphere part of the Galactic plane, which cannot be observed from the northern observatories and was therefore much less completely studied, appears on this map as a large blank sector between Galactic longitudes approximately 210° and 330°.

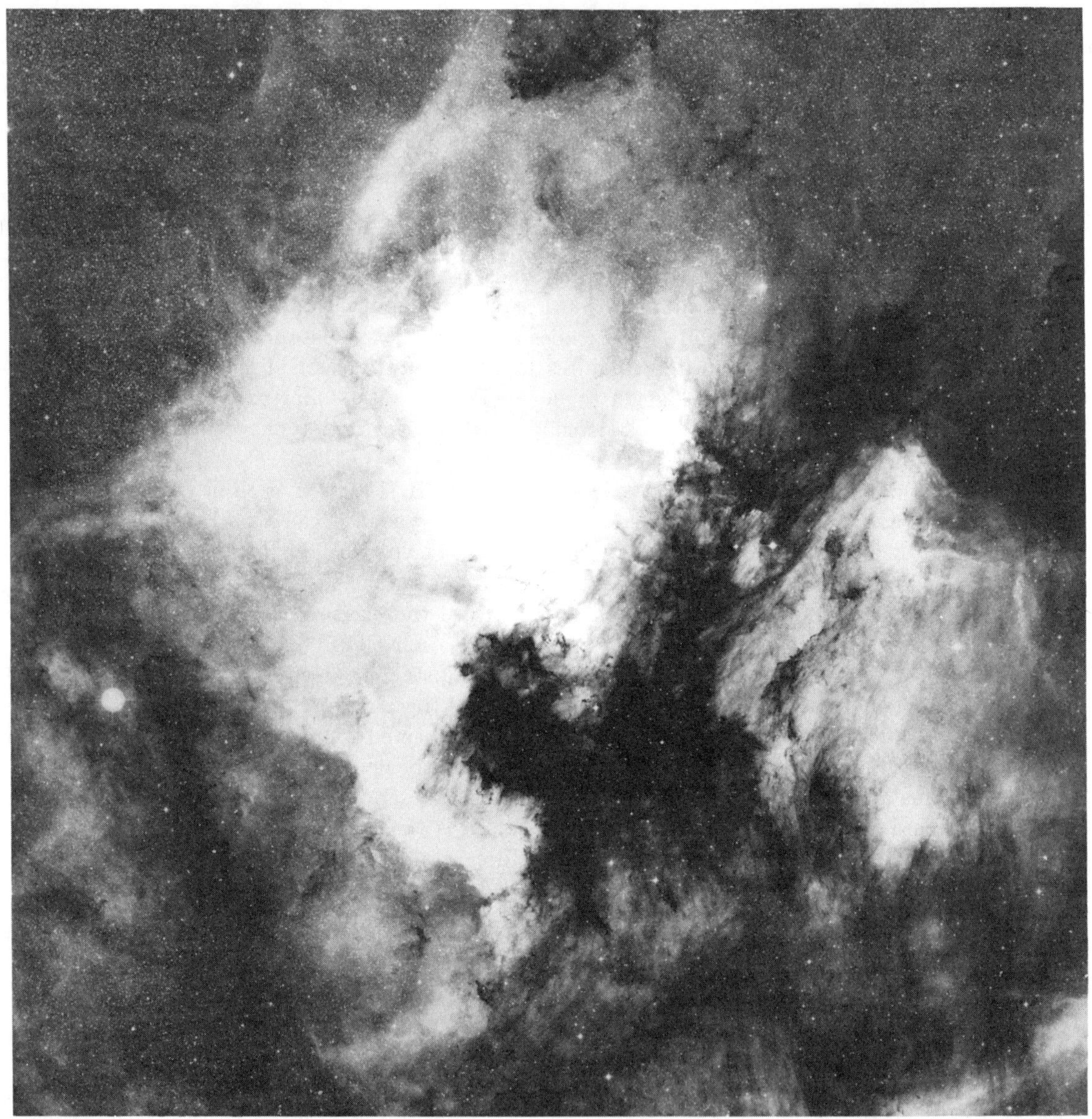

FIGURE 8.3
NGC 7000, North America Nebula (*to left*), and smaller IC 5067, Pelican Nebula (*to right*). Both are apparently parts of one large H II region, separated in the sky by foreground extinction of a dense interstellar cloud. HD 199579, the O5 star referred to in the text, is approximately 2 inches down from top edge, 2-3/4 inches in from right edge. Original plate, taken with the Palomar 48-inch Schmidt, with red filter, emphasizing Hα, [N II] $\lambda\lambda$6548, 6583. (*Palomar Observatory photograph.*)

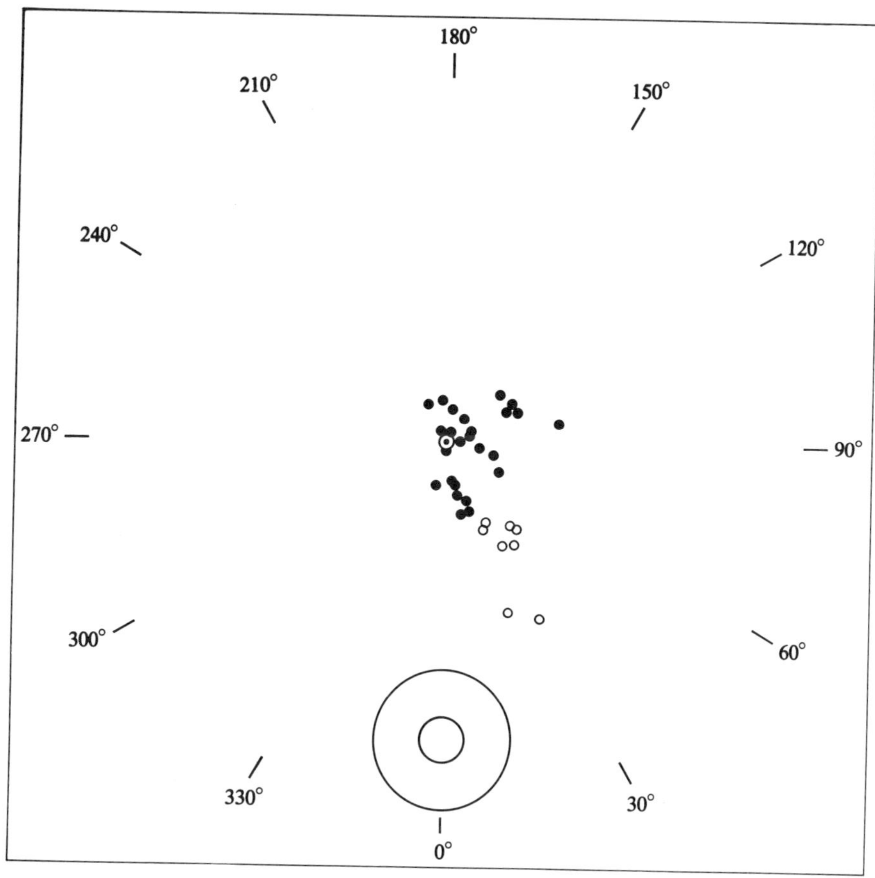

FIGURE 8.4
Northern-hemisphere H II regions plotted as solid circles in the plane of the Galaxy. The point ⊙ indicates the Sun, and eight distant OB stars are plotted as open circles. Sections of three spiral arms can be seen: the outer (Perseus) arm, the local (Orion) arm, and the inner (Sagittarius) arm. Galactic longitudes are indicated, with 0° toward the Galactic center, which is shown as two concentric circles at an adopted distance of 10 kpc from the sun.

A later, more complete map which includes the southern-hemisphere results, Figure 8.5, shows that, around Galactic longitude 290°, a spiral arm apparently runs off with the opposite inclination to the northern spiral arms. Part of the explanation is that there is a clear region in this direction, with very little interstellar extinction, so that it is possible to observe quite distant stars and nebulae in the longitude range 280° to 300°. If the distances of these stars are not accurately known, relatively small percentage errors appear as rather large displacements, and this probably contributes to the apparent general spreading out of the spiral arm in this direction.

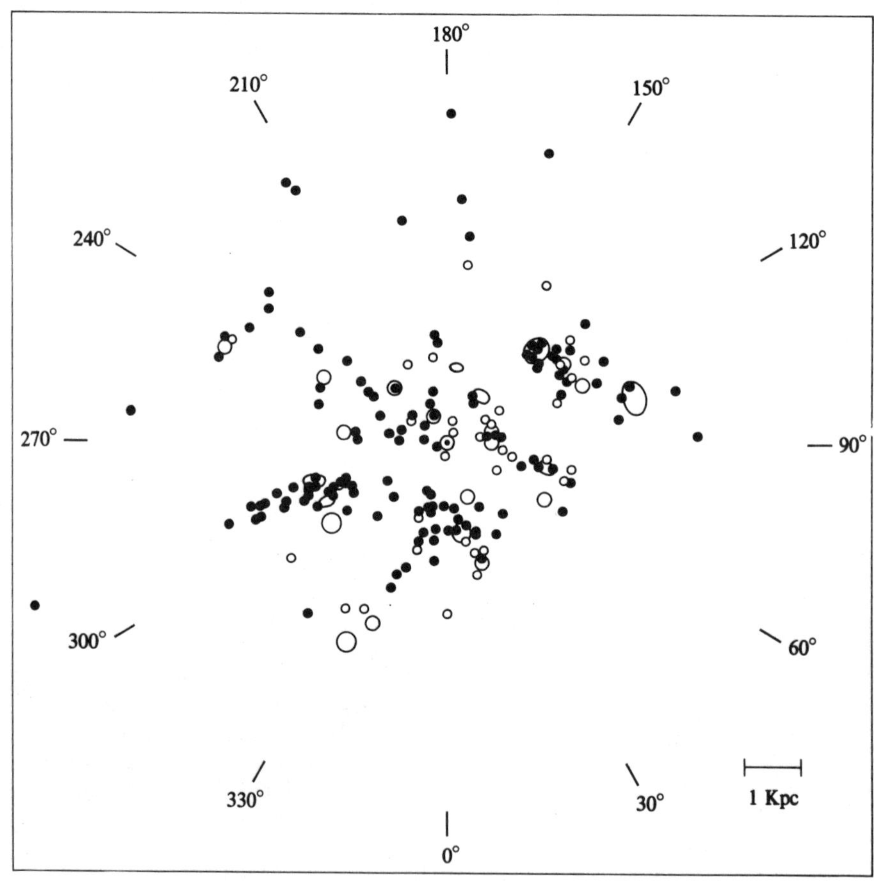

FIGURE 8.5
Distribution in Galactic plane of young stars ○, and young galactic clusters (B2-B3) •. The position of the Sun is shown as ⊙.

Optical studies of more distant H II regions in our Galaxy are rendered impossible by interstellar extinction. However, this is not a problem in the radio-frequency region, and the radio-continuum and recombination-line measurements have detected very distant H II regions. Unfortunately, further study of these optically invisible nebulae is difficult; there is no direct way to measure their distances, since stars cannot be identified in them. The only method for finding the distance is then to use the observed radial velocity of the H II region, together with a model of the variation of Galactic rotation velocity with distance. Such models, of course, have been widely used for the 21-cm H I observations. However, careful comparison of distances derived in this way with distances derived from spectroscopic classification of the identified stars in relatively nearby H II regions shows that the models are too simplified. Although the velocity measurements give distance results that are better than having no information whatever, they are not highly reliable.

Figure 8.6, a map of the entire Galaxy, shows H II regions, most of which were observed by their radio-frequency recombination-line radiation, but a few by their Hα emission. Notice that approximately 100 nebulae are included, which are the most luminous H II regions observed in our Galaxy and are more or less comparable with the 100 brightest H II regions that show on an Hα photograph of an external galaxy such as M 31. They are, on the average, significantly more luminous than the 30 or so H II regions, most within 2 kpc of the Sun, that are plotted in Figure 8.4. Note that the Galactic rotation model does not uniquely determine the distance of H II regions that are closer to the center of the Galaxy than the Sun, since the same measured radial velocity corresponds to two possible positions equally distant from the Galactic center, one closer to the Sun, on the Sun's side of the perpendicular from the center of the Galaxy to the line from the Sun through the nebula, and the other further from the Sun, on the other side of this perpendicular. In Figure 8.6 these ambiguities are resolved for the H II regions observed in the radio-frequency region only by additional information from radio absorption lines if available, or by continuity arguments in the absence of other information. This ambiguity does not exist for H II regions more distant from the Galactic center than the Sun. However, for all H II regions, any errors in the kinematic model that links position in the Galaxy and velocity, and also any dispersion about this relationship, leads to errors in the derived distance. Note that according to Figure 8.6, the line of H II regions in which the Sun lies is not a major spiral arm, but is a "spur" or branch between arms.

Even with only rough distances, H II regions may be used to study abundance gradients in our own Galaxy. Although only the nearer H II regions can be observed in the optical spectral region, measurements of the mean temperatures within the nebulae from the radio-frequency recombination lines may be used to supplement them. The result found from the measurements

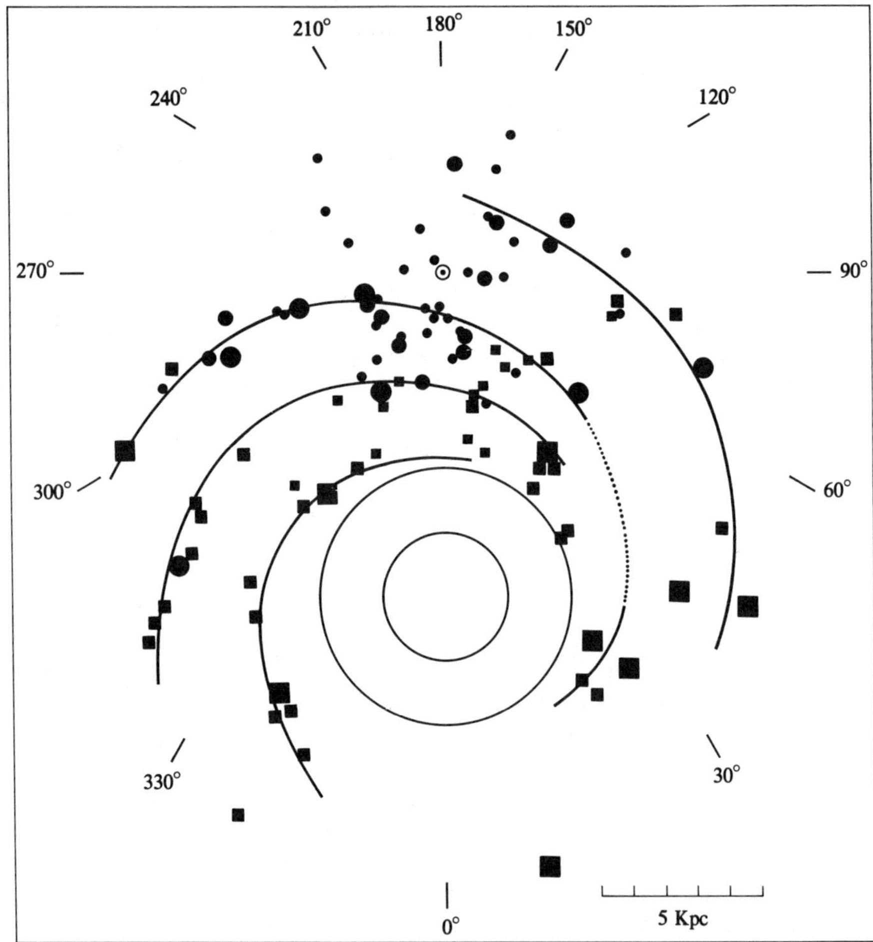

FIGURE 8.6
Distribution in Galactic plane of H II regions observed in optical (circles) or radio (squares) spectral regions. The size of the symbol indicates the luminosity of the H II region. The deduced positions of the spiral arms are drawn as solid lines.

to date, as mentioned in Chapter 5, is that the mean temperature increases outward, and the relative abundances of O and N decrease outward, as in other spiral galaxies.

It is also possible to get some information from the radio observations on the total amount of ionized gas in the Galaxy. This is most easily done from measurements of the radio-frequency continuum, which is the strongest radio-frequency emission from H II regions. Measurements must be made at two well-separated frequencies, at both of which the Galaxy is optically thin, in order to separate the nonthermal synchrotron radiation (which increases toward lower frequency) from the thermal free-free radiation (which has a flux nearly independent of frequency). The free-free surface brightness in any direction gives the integral

$$T_{b\nu} = 8.24 \times 10^{-2}\, T^{-0.35} \mu^{-2.1} \int N_+ N_e \, ds \tag{8.2}$$

(see equations 4.37 and 4.32). The mean effective length is defined by a Galactic model, so the measurement gives the mean-square ion density averaged along the entire ray. The integral is, of course, proportional to the number of recombinations along the ray, and hence to the number of ionization processes; so it gives directly the number of ionizing photons absorbed in H II regions and in the lower-density gas between the nebulae. To derive the amount of ionized gas requires an estimate of its distribution along the ray, that is, of the clumpiness or filling factor. With an estimated density in the emitting regions $N_e \approx 5$ cm^{-3}, corresponding to a filling factor ranging from 0.1 to 0.01 depending on distance from the Galactic center, such observations indicate a total mass of ionized gas in the Galactic plane of order $4 \times 10^7\ M_\odot$. This is only a small fraction of the total amount of gas determined from 21-cm H I observations, $5 \times 10^9\ M_\odot$, which, in turn, is itself only a small fraction of the mass of the Galaxy out to 20 kpc radius, $2 \times 10^{11}\ M_\odot$. The total mass of the Galaxy, including dark matter out to $\sim$ 200 kpc, is probably about an order of magnitude greater, and the fraction of this mass that is ionized correspondingly smaller.

8.4 Stars in H II Regions

The existence of an H II region requires that there be one or more O stars within a region of reasonably high interstellar gas density, because only these stars emit enough photons with $h\nu > h\nu_0 = 13.6$ eV to ionize enough gas to make the nebula an easily observable H II region. Often these O stars are in early-type clusters containing relatively large numbers of B stars; for instance, the O stars that ionize NGC 6524 belong to the cluster NGC 6530, which contains at least 23 known B stars. It is known from theoretical stellar-evolution calculations that O stars have maximum lifetimes of a few $\times\ 10^6$ yr before they exhaust their central available H fuel and become supergiants; thus the clusters in which they occur must be recently formed star clusters. The luminosity function of such a cluster therefore provides a good estimate of the initial luminosity function of a group of recently formed stars.

In more distant galaxies, measuring the Hα emission-line luminosity as a function of position is a method of determining the current star-formation rate. Actually the Hα luminosity gives directly the number of ionizing photons emitted by the O stars and absorbed locally, but for an assumed initial luminosity function, this gives the number of stars of each luminosity and spectral type. Since the high-luminosity O stars exhaust their H and die so quickly, their present number divided by their calculated lifetimes *is* their present formation rate, on a galactic time scale. The initial luminosity function can be transformed into an initial mass function; assuming that these functions, derived in our Galaxy close to the Sun, apply in all galaxies gives the star-formation rates in them down to the faintest luminosities or smallest masses. Of course there are many possible errors in this chain of reasoning, and these must always be kept in mind when evaluating the results that come from it. Most importantly, because dust absorbs some of the potential ionizing photons, and some of the emitted Hα photons, this method in practice provides only a *lower limit* to the star-formation rate.

One type of low-luminosity star of which many examples have been found in nearby H II regions is the T Tauri stars, which are pre-main-sequence G and K stars that vary irregularly in light and have H and Ca II emission lines in their spectra. Only the nearest H II regions can be surveyed for these stars because of their intrinsically low luminosities, but many of them have been found in NGC 1976. On the other hand, T Tauri stars can also exist in regions of high-density interstellar gas that is not ionized; for instance, many of these stars are also found in the Taurus-Auriga dark cloud. Thus they are recently formed stars, not necessarily directly connected with the formation of O stars.

An H II region first forms when an O star "turns on" in a region of high interstellar-gas density. The star must have formed from interstellar matter, and observational evidence shows that a high density of interstellar matter is strongly correlated with star formation. Radio observations, in particular, have led to the discovery of many small, dense, "compact H II regions," with $N_e \sim 10^4\ \text{cm}^{-3}$, in nebulae that are optically invisible because of high interstellar extinction. Massive-star formation seems to occur in the dense cores of giant molecular clouds.

Once a condensation has become sufficiently dense to be self-gravitating, probably because of turbulent motions, it contracts, heating up and radiating photons by drawing on the gravitational energy source. Once the star becomes hot enough at its center for nuclear reactions to begin, it quickly stabilizes on the main sequence. Many nebulae probably form as a result of density increases, perhaps in the collision of two or more lower-density interstellar clouds, so that in the resulting high-density condensation, star formation rapidly begins. For instance, observations show that NGC 1976 has a very steep density gradient, with the highest density quite near but not exactly coincident with the Trapezium, which includes the ionizing stars θ^1 Ori C and θ^1 Ori D. Infrared measurements show that more star formation is going on inside the dense clouds of which NGC 1976 is an ionized edge.

After the O star or stars in a condensation stabilize on the main sequence, an R-type ionization front rapidly runs out into the gas at a rate determined by the rate of emission of ionizing photons by the star(s). Ultimately, the velocity of the ionization front reaches the R-critical velocity. At this stage the front becomes D-critical, and a shock wave breaks off and runs ahead of it, compressing the gas. The nebula continues to expand, and may develop a central local density minimum as a result of radiation pressure exerted on the dust particles in the nebula, or because of the ram pressure of stellar winds. Ultimately, the O star exhausts its nuclear-energy sources and becomes a supernova. In any case, the nebula expanding away from the central star has drawn kinetic energy from the radiation field of the star, and this kinetic energy is ultimately shared with the surrounding interstellar gas. From the number of O stars known to exist, it is possible to show that a significant fraction of the interstellar turbulent energy may be derived from the photoionization input of O stars, communicated through H II regions, though there are many observational uncertainties in such a picture.

8.5 Molecules in H II Regions

Within the past decades, many interstellar molecular emission lines have been detected in the infrared and radio-frequency regions. The first interstellar molecule detected by its radio-frequency lines, OH, has been observed in many H II regions. The transition is between the two components of the ground $^2\Pi_{3/2}$ level that are split by Λ-type doubling; each component is further split by hyperfine interactions, so that there is a total of four lines with frequencies 1612, 1665, 1667, and 1720 MHz. A typical observed line in an H II region has a profile that may be divided into several components with different radial velocities, and the relative strengths of these components often vary in times as short as a few months. Many of the individual components have narrow line profiles, nearly complete circular polarization or strong linear polarization, and high brightness temperature (sometimes $T_{b\nu} \geq 10^{12}$ °K). All of these characteristics are strong evidence for maser activity resulting from nonthermal population inversions of the individual molecular levels. Further, many but not all of the OH sources are also observed by their H_2O radio-frequency emission lines. Some of the OH radiation comes from extended regions in H II regions, but much of it, and all of the H_2O radiation, comes from very small, bright sources within the H II regions. Studies made with very long base-line interferometers show that the OH emission usually occurs in clusters of small sources, the clusters having sizes typically of 1″, while the individual sources within the cluster have diameters of order 0″.005 to 0″.5. Maser activity in H_2O and SiO is also observed in H II region–giant molecular cloud complexes such as Orion.

In the H II regions that have been studied to date, the OH masers tend to occur in areas of strong interstellar extinction. Since OH molecules would be rapidly dissociated in the strong ultraviolet radiation field within an H II region, the sources must be small, very dense condensations that are optically thick to ionizing radiation, so that their interiors are shielded by the surface layers of gas and dust. In these regions of high dust density, molecules are abundant. The molecules are excited, "pumping" the masers by collisions with other molecules and atoms, by infrared-radiation shock fronts, and presumably also by resonance fluorescence due to ultraviolet radiation with $h\nu < h\nu_0 = 13.6$ eV, which penetrates into the clouds.

In addition to OH and H_2O, the molecules CO, CN, CS, HCN, H_2CO, CH_3OH, and more than fifty others have all been detected in NGC 1976, which is the best studied H II region, and in other H II regions as well. Some of these molecules are undoubtedly concentrated in or are escaping from small, dense condensations in the nebula itself, but most of them are in the dense, dusty, dark molecular clouds into which the ionizing radiation cannot penetrate. The physical ideas of the study of molecular-line radiation are very similar to those used in treating the atomic and ionic lines, but the details, including especially the energy levels, the temperatures, and the level of ionization, are different enough so that we will not study them further in this book.

References

The distribution of H II regions "like beads on a string" along the spiral arms of M 31, and the photographic survey by which he discovered this distribution, are clearly described by

Baade, W. 1951. *Pub. Obs. Univ. Michigan* **10,** 7.

The detailed results of this survey, including coordinates of the H II regions and maps of their distribution projected on the sky and in the deduced plane of M 31 itself, are given in

Baade, W., and Arp, H. 1964. *Ap. J.* **139,** 1027.

Arp, H. 1964. *Ap. J.* **139,** 1045.

A photographic survey of a large number of other galaxies for H II regions is described and published in detail in

Hodge, P. W. 1982. *A. J.* **87,** 1341.

Hodge, P. W., and Kennicutt, R. C. 1983. *An Atlas of H II Regions in 125 Galaxies.* New York: American Institute of Physics.

Hodge, P. W., and Kennicutt, R. C. 1983. *A. J.* **88,** 296.

Three papers illustrating how improvements in detection sensitivity have led to improved spectroscopic measurements of radial velocities of H II regions in M 31 are

Babcock, H. W. 1939. *Lick Obs. Bull.* **19,** 41.
Mayall, N. U. 1951. *Pub. Obs. Univ. Michigan* **10,** 19.
Rubin, V. C., and Ford, W. K. 1970. *Ap. J.* **159,** 379.

Three papers summarizing a large amount of information on the rotation curves of spiral galaxies, as derived from measurements of the velocities of H II regions, are

Rubin, V. C., Ford, W. K., and Thonnard, N. 1980. *Ap. J.* **238,** 471.
Rubin, V. C., Ford, W. K., Thonnard, N., and Burstein, D. 1982. *Ap. J.* **261,** 439.
Rubin, V. C., Burstein, D., Ford, W. K., and Thonnard, N. 1985. *Ap. J.* **289,** 81.

Early papers on the spectra of H II regions in nearby galaxies are

Aller, L. H. 1942. *Ap. J.* **95,** 52.
Searle, L. 1971. *Ap. J.* **168,** 327.

The first paper gives measurements of the emission-line spectra of H II regions in M 33; the second paper discusses abundance variations with position of H II regions in M 33, M 51, and M 101. More recent references on the abundances of elements from measurements of H II regions in other galaxies include

Peimbert, M. 1975. *Ann. Rev. Astr. Ap.* **13,** 113.
Peimbert, M., and Torres-Peimbert, S. 1974. *Ap. J.* **193,** 327.
Peimbert, M., and Torres-Peimbert, S. 1976. *Ap. J.* **203,** 581.
Dufour, R. J., Talbot, R. J., Jensen, E. B., and Shields, G. A. 1980. *Ap. J.* **236,** 119.
French, H. B. 1980. *Ap. J.* **240,** 41.
Pagel, B. E. J., and Edmunds, M. G. 1981. *Ann. Rev. Astr. Ap.* **19,** 77.
Kunth, D. 1986. *P.A.S.P.* **98,** 984.
McCall, M. L., Rybski, P. M., and Shields, G. A. 1985. *Ap. J. Suppl.* **57,** 1.
Dopita, M. A., and Evans, I. N. 1986. *Ap. J.* **307,** 431.
Evans, I. N. 1986. *Ap. J.* **309,** 544.

Table 8.1 is based on data from the first three of these references, and Table 8.2 on the fifth, which deals with low-luminosity irregular galaxies. The sixth is a very good review of work in this field up to 1980; Figure 8.2 is based on the seventh.

The effects of abundance variations in the stars on the ionization in the nebulae, through heavy element, O, and Ne absorption edges, is discussed by

Balick, B., and Sneden, C. 1986. *Ap. J.* **208,** 336.
Shields, G. A., and Tinsley, B. M. 1976. *Ap. J.* **203,** 66.
Shields, G. A., and Searle, L. 1978. *Ap. J.* **222,** 821.

The uncertainties in the observational data and their interpretation are discussed in great detail in

Davidson, K., and Kinman, T. D. 1985. *Ap. J. Suppl.* **58,** 321.

The importance of primordial helium abundance in testing cosmological models is discussed in

Boesgaard, A. M., and Steigman, G. 1985. *Ann. Rev. Astr. Ap.* **23,** 319.

The National Geographic Society—Palomar Observatory Sky Survey is available in the form of photographic prints of the individual fields. Brief descriptions of it have been published in

Minkowski, R. L., and Abell, G. O. 1963. *Basic Astronomical Data,* ed. K. A. Strand. Chicago: University of Chicago Press, p. 481.

Lund, J. M., and Dixon, R. S. 1973. *P.A.S.P.* **85,** 230.

The ESO and United Kingdom Schmidt telescope surveys (the latter known as the SERC survey) are available as photographic films of the individual fields. They are described in the first three papers, by L. Woltjer, R. J. West, and R. D. Cannon, in the book

Capaccioli, M. 1984. *Astronomy with Schmidt-type Telescopes.* Dordrecht: Reidel.

Wide-angle photographs showing the brightest H II regions close to the Galactic equator are reproduced in

Osterbrock, D. E., and Sharpless, S. 1952. *Ap. J.* **115,** 89.

Mapping the spiral arms by optical determinations of the distance of OB stars in H II regions is discussed in

Morgan, W. W., Sharpless, S., and Osterbrock, D. E. 1952. *A. J.* **57,** 3.

Morgan, W. W., Whitford, A. E., and Code, A. D. 1953. *Ap. J.* **118,** 318.

(Figure 8.4 is adapted from this reference.)

Morgan, W. W., Code, A. D., and Whitford, A. E. 1965, *Ap. J. Supp.* **2,** 41.

Humphreys, R. M. 1979. *The Large-Scale Characteristics of the Galaxy* (IAU Symposium No. 84). Dordrecht: Reidel, p. 93.

(Figure 8.5 is adapted from this reference.)

The problem of finding very faint OB stars from objective prism or other spectroscopic surveys is very well described by

Morgan, W. W. 1951. *Pub. Obs. Univ. Michigan* **10,** 33.

Morgan, W. W., Meinel, A. B., and Johnson, H. M. 1954. *Ap. J.* **120,** 506.

Schulte, D. H. 1956. *Ap. J.* **123,** 250.

Orsatti, A. M., and Muzzio, J. C. 1980. *A. J.* **85,** 265.

Forti, J. C., and Orsatti, A. M. 1981. *A. J.* **86,** 209.

Distances of nebulae, determined from spectroscopic classification of their involved stars, have been compared with determinations based on measured velocities and a standard model of Galactic rotational velocities, by

Miller, J. S. 1968. *Ap. J.* **151,** 473.

Georgelin, Y. P., and Georgelin, Y. M. 1971. *Astr. Ap.* **12,** 482.

Other problems associated with attempting to find distances from velocities plus a rotational model are reviewed by

Burton, W. B. 1976. *Ann. Rev. Astr. Ap.* **14,** 275.

The combined optical and radio approach, with references to much of the earlier work, is given by

Georgelin, Y. M., and Georgelin, Y. P. 1976. *Astr. Ap.* **49,** 57.

(Figure 8.6 is adapted from the latter reference, but note that all three of Figures 8.4, 8.5, and 8.6 are based on an assumed distance of 10 kpc to the center of our Galaxy, while 8 kpc is the value more usually adopted today.) The continuum survey from which the entire amount of ionized gas in the Galaxy is derived is

Westerhout, G. 1958. *Bull. Ast. Inst. Netherlands* **14,** 261.

Compact H II regions are described in

Habing, H. J., and Israel, F. P. 1979. *Ann. Rev. Astr. Ap.* **17,** 345.

The discovery of the OH molecule in interstellar space by its radio-frequency absorption lines is reported in

Weinreb, S., Barrett, A. H., Meeks, M. L., and Henry, J. C. 1963. *Nature* **200,** 829.

Two later symposium volumes, each of which contains several papers on molecules in H II regions, are

Solomon, P. G., and Edmunds, M. G. 1979. *Giant Molecular Clouds in the Galaxy.* Oxford: Pergamon.

Andrew, B. H. 1986. *Interstellar Molecules* (IAU Symposium No. 87). Dordrecht: Reidel.

Star formation is a very large subject in itself. Three good reviews are the paper by Habing and Israel mentioned in connection with compact H II regions above, and

Strom, S. E., Strom, K. M., and Grasdalen, G. L. 1975. *Ann. Rev. Astr. Ap.* **13,** 187.

Shu, F., Adams, F. C., and Lizano, C. 1987. *Ann. Rev. Astr. Ap.* **25,** 23.

Good summaries of results on NGC 1976, the Orion Nebula, including the stars in it, are published in

Glassgold, A. E., Huggins, P. J., and Shucking, E. L. 1982. *Symposium on the Orion Nebula to Honor Henry Draper Ann. N. Y. Acad. of Sci.* **395**.

Goudis, C. 1981. *The Orion Complex: A Case Study of Interstellar Matter.* Dordrecht: Reidel.

9

Planetary Nebulae

9.1 Introduction

The previous chapters have summarized the ideas and methods of nebular research, first treating nebulae from a static point of view, then adding the effects of motions and of dust particles. In the first seven chapters, many references have been made to actual planetary nebulae, but only a relatively few of the known results about them have been discussed. This chapter will complete the discussion of our basic ideas about planetary nebulae. First their space distribution in the Galaxy and their Galactic kinematics are summarized; then what is known about the evolution of the nebulae and of their central stars, including ideas on the origin of planetary nebulae, is discussed. This leads naturally to a discussion of planetaries with unusual abundances of the elements and to the rate of mass return of interstellar gas to the Galaxy from planetary nebulae and its significance in Galactic evolution. Finally, the chapter summarizes what is known about planetary nebulae in other galaxies, and gives a brief introduction to molecules in planetary nebulae.

9.2 Space Distribution and Kinematics of Planetary Nebulae

Except for the brightest classical planetary nebulae, which were identified by their finite angular sizes, planetary nebulae are discovered photographically by objective-prism surveys or by direct photography in a narrow spectral region around a strong emission line or lines, such as [O III] $\lambda\lambda 4959, 5007$, or H$\alpha$ and [N II] $\lambda\lambda 6548, 6583$. An objective-prism survey tends to discover small, bright, high-surface-brightness objects, while direct photography tends to discover nebulae with large angular sizes, even though they have low surface brightness. These surveys, of course, penetrate only the nearer regions of the Galaxy, because of the interstellar extinction by dust concentrated to the

Galactic plane. A total of about 1,500 planetary nebulae are known, and their angular distribution on the sky, as shown in Figure 9.1, exhibits fairly strong concentration to the Galactic plane, but not so strong as H II regions, and strong concentration to the center of the Galaxy. It must be remembered that in this map the concentration to the Galactic equator and to the Galactic center would undoubtedly be more extreme if it were not for the interstellar extinction, which preferentially suppresses the more distant planetaries.

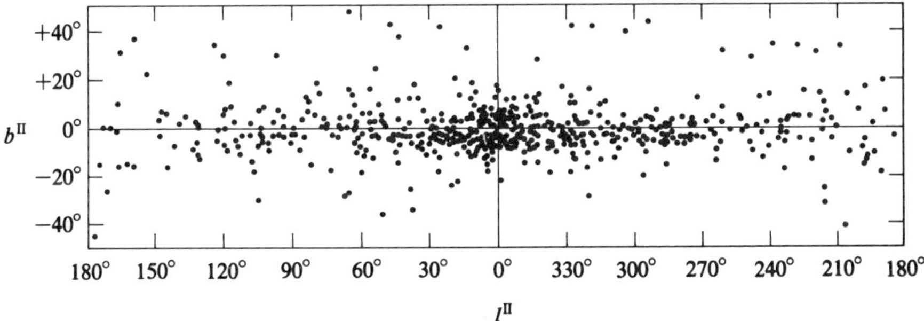

FIGURE 9.1
Distribution of planetary nebulae in the sky, plotted in Galactic coordinates. Note concentration to Galactic equator and to Galactic center.

The only directly measured trigonometric parallax of a planetary nebula that is reasonably well determined is $0\farcs042 \pm 0\farcs011$ for NGC 7293. The estimate of the probable error, which depends on the internal consistency of the individual measurements at one observatory, is a lower limit to the actual uncertainty, as can be seen from comparison of parallaxes of other planetary nebulae measured at two or more observatories. One planetary nebula, NGC 246, has a late-type main-sequence companion star, and the spectroscopic parallax of this star implies a distance of between 360 and 480 pc. Several other planetaries also have companions, or are in galactic clusters whose distances are similarly known. In a few nebulae, comparison of the tangential proper motion of expansion with the measured radial expansion velocity gives distance estimates, but these are very uncertain, because (as was discussed in Chapter 6) the velocity of expansion varies with position in the nebula, and in some nebulae the apparent motion of the outer boundary may be the motion of an ionization front rather than the mass motion, which is measured by the Doppler effect. Measurements of proper motions of about 35 planetary-nebula central stars are available, which give a statistical mean parallax for the group.

Another method, which can be applied to individual nebulae, is to measure their interstellar reddening or extinction by the methods described in

Chapter 7, and to map out the increase of extinction with distance along the same line of sight by measurements of O and B stars near it in the sky. Their distances can be found from their absolute magnitudes determined by spectroscopic classification, and their apparent magnitudes and reddening from UBV photometry. Then from a plot of extinction versus distance along this line, the distance of the planetary can be found. The main problems are the paucity of early-type stars, particularly at high Galactic latitude, and the extreme patchiness of interstellar matter. However, this and the other methods mentioned above provide all the direct information there is on the distances of planetary nebulae. It is not sufficient for a study of the space distribution of these objects; so we must use other, less-direct, and correspondingly still less-accurate methods of distance estimation.

The basic assumption of the indirect method, commonly known as the Shklovsky distance method, is that all planetary-nebula shells are completely ionized and have approximately the same mass, so that as they expand, their mean electron densities decrease and their radii r_N increase according to the law

$$\frac{4\pi}{3}\,\epsilon\, r_N^3\, N_e = \text{const.} \tag{9.1}$$

Here ϵ is the filling factor defined in Section 5.11. Hence measurement of the electron density of any planetary nebula allows us to determine its radius, and if its angular radius is then measured, its distance directly follows. The electron density cannot be directly measured except in planetaries with [O II], [S II], [Ar IV], or similar pairs of lines, but it is possible instead to measure the mean Hβ surface brightness (or intensity) of the nebula $I_{\mathrm{H}\beta}$;

$$I_{\mathrm{H}\beta} \propto \epsilon N_e N_p r_N \propto \epsilon N_e^2 r_N \propto \epsilon^{2/3} N_e^{5/3} \propto \epsilon^{-1} r_N^{-5}. \tag{9.2}$$

There is a check of this method, in that the expected relation between Hβ surface brightness and N_e is verified for those planetary nebulae in which the [O II] lines have been measured. Solving for the distance D of the nebula,

$$\begin{aligned} D &= \frac{r_N}{\phi} \propto \frac{\epsilon^{-1/5}(I_{\mathrm{H}\beta})^{-1/5}}{\phi} \\ &\propto \epsilon^{-1/5}(\pi F_{\mathrm{H}\beta})^{-1/5}\phi^{-3/5} \qquad (9.3) \\ &= K\epsilon^{-1/5}(\pi F_{\mathrm{H}\beta})^{-1/5}\phi^{-3/5}, \end{aligned}$$

where ϕ is the angular radius of the nebula, and $\pi F_{\mathrm{H}\beta}$ is its measured flux in Hβ (corrected for interstellar extinction) at the Earth, so that $I_{\mathrm{H}\beta} \propto \pi F_{\mathrm{H}\beta}\phi^{-2}$.

For instance, if we adopt a spherical planetary-nebula model in which the filling factor ϵ and densities N_p, N_e are constant, its mass is

$$M_N = \frac{4\pi}{3} N_p r_N^3 (1 + 4y) m_H \epsilon, \tag{9.4}$$

where y is the abundance ratio $N_{\rm He}/N_{\rm H}$. If we write the electron density $N_e = N_p\ (1 + xy)$, so that x gives the fractional ionization of $\rm He^+$ to $\rm He^{++}$, the expression for the flux from the nebula

$$\pi F_{\rm H\beta} = \frac{(4\pi/3) N_p N_e r_N^3 \epsilon \alpha_{\rm H\beta}^{eff} h\nu_{\rm H\beta}}{4\pi D^2} \tag{9.5}$$

can be expressed in terms of $r_N = \phi D$ and solved for

$$D = \left[\frac{3}{16\pi^2} \frac{M_N^2}{m_H^2} \frac{(1 + xy)}{(1 + 4y)^2} \epsilon \alpha_{\rm H\beta}^{eff} h\nu_{\rm H\beta}\right]^{1/5} \times (\pi F_{\rm H\beta})^{-1/5} \phi^{-3/5}. \tag{9.6}$$

This equation is then used to determine the distance of a planetary nebula from measured values of its flux and angular size, the first factor in square brackets being the constant K to be determined from the nebulae of known distance discussed previously. Alternatively, any other recombination line (such as Hα) or the radio-frequency continuum could be used; the equations are always similar to equation (9.6) and only the numerical value of the constant is different.

Unfortunately, for the reasons given above, the calibration is rather poorly determined. A study based on the most complete list of well-determined proper motions gave a coefficient $K = 108$ in equation (9.3), with D in pc, $\pi F_{\rm H\beta}$ in erg cm^{-2} sec^{-1}, and ϕ in seconds of arc. It corresponds to a total nebular mass $M_N \approx 0.5\ M_\odot$. A considerably smaller value $K = 75$, corresponding to $M_N \approx 0.2\ M_\odot$ seems to give a better overall fit to all the independent distance-determination methods. The nebular mass is poorly determined because, as equation (9.6) shows, the distance depends only weakly upon it.

One check on the distances is that, since planetary nebulae are expected to expand with constant velocity, the number within each range of radius between r_N and $r_N + dr_N$ should be proportional to dr_N. This test is in fact approximately fulfilled by the planetaries within a standard volume near the Sun, corrected for incompleteness, in the range of radii 0.1 pc $\leq r_N \leq$ 0.7 pc. The larger nebulae with lower density and correspondingly lower surface brightness are more difficult to discover, and objects with $r_N > 0.7$ pc are essentially undetectable by the survey methods that have been used to date. On the other hand, below a definite lower limit $r_N \leq r_1$ set by the ultraviolet luminosity of the central star, the nebula is so dense that it is not completely ionized, and consequently its true ionized mass is smaller than assumed under the constant-mass hypothesis. Thus its true distance is smaller than that calculated using a constant coefficient in equation (9.6), and we therefore should expect an apparent underabundance of nebulae with calculated radius $r_N \leq r_1$, and a corresponding excess of nebulae with $r_N \geq r_1$. This effect does occur in the statistics of observed planetary nebula sizes, with $r_1 \approx 0.07$ pc, which gives us some confidence in the indirect photometric distance method.

However, the derived distances of the planetary nebulae can only be approximately correct statistically, because direct photographs show that their forms, internal structures, and ionization all have considerable ranges. The central assumption of a uniform-density spherical model is too idealized to represent real nebulae accurately. Many planetary nebulae appear to be more nearly axisymmetrically toroidal than spherical, for example, NGC 6720, shown in Figure 9.2. Detailed studies of some of these nebulae show that they can be modeled in terms of axisymmetric objects with spheroidal density distributions, with density maxima at intermediate distances between their inner and outer edges. The statistical arguments can never eliminate the possibility that a small fraction of the planetary nebulae (say, 10 or 20 percent) differ completely in nature from other planetaries, but have a similar appearance in the sky. Nevertheless, the indirect Shklovsky distance method is the only method we have for measuring the distances of many planetaries and drawing statistical conclusions about their space distribution and evolution.

The radial velocities of many planetary nebulae have also been measured and exhibit a relatively high velocity dispersion. The measured radial velocities are plotted against Galactic longitude in Figure 9.3. It can be seen that the radial velocities of the planetary nebulae in the direction $l \approx 90°$ tend to be negative, and in the direction $l \approx 270°$ they tend to be positive, which shows that the planetary nebulae are "high-velocity" objects. That is, they belong to a system that actually has a considerably smaller rotational velocity about the Galactic center than the sun, so they appear to us to be moving, on the average, in the direction opposite the sun's Galactic rotation. Figure 9.3 also shows the high dispersion of velocities in the direction of the Galactic center. On the basis of the relative motion of the system of planetaries with respect to the local circular velocity, planetary nebulae are generally classified as Old Population I objects, but are not such outstanding high-velocity objects as Extreme Population II. More detailed investigations, mentioned in the next section, show that there is a range of ages among planetary nebulae, connected with the range of masses of their progenitor stars.

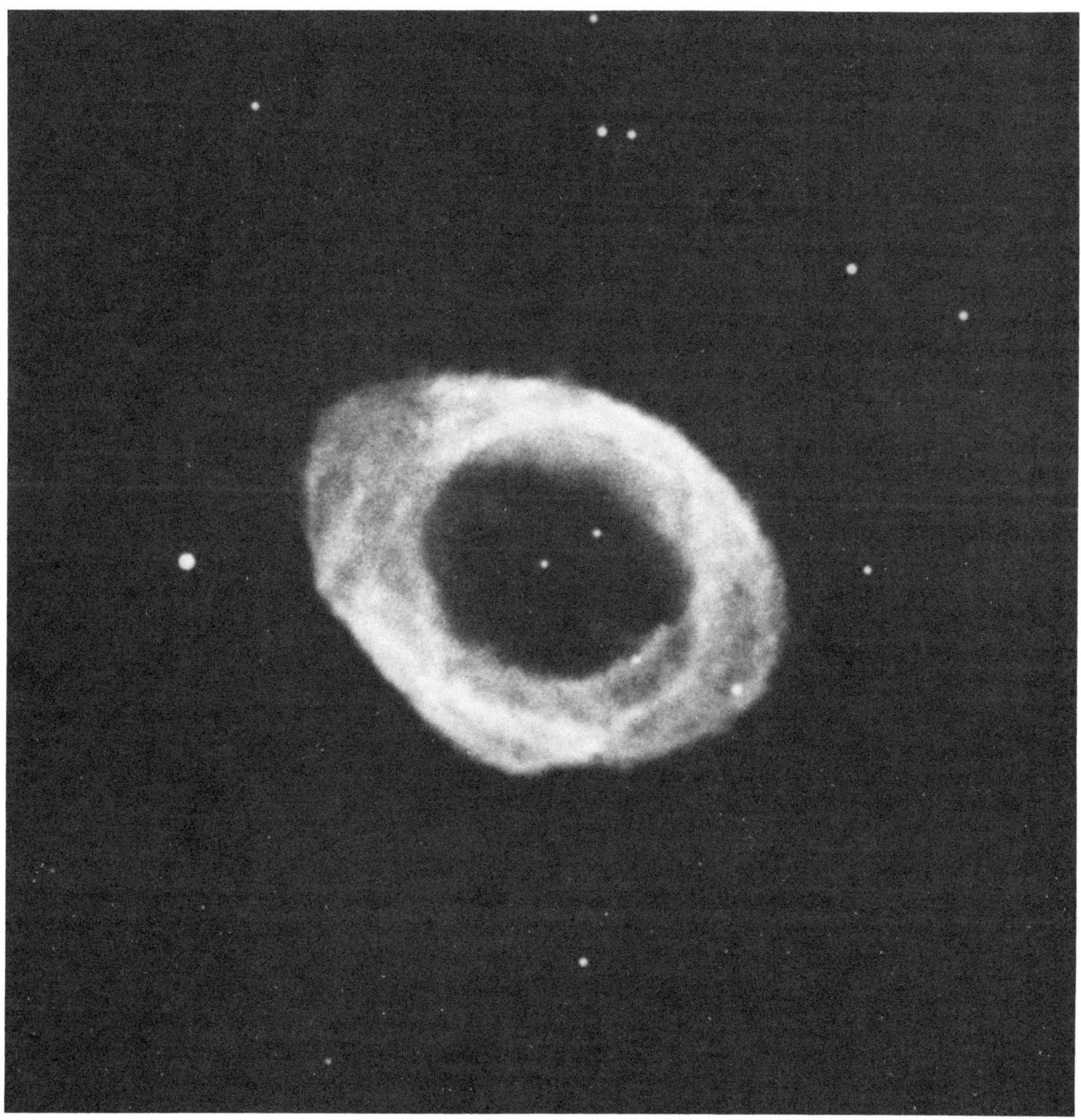

FIGURE 9.2
NGC 6720, the Ring Nebula in Lyra, a classical planetary nebula. Original photograph was taken with 3-m Shane reflector, red filter, and 103a-E plate, emphasizing Hα, [N II]$\lambda\lambda$6548, 6583. (*Lick Observatory photograph.*)

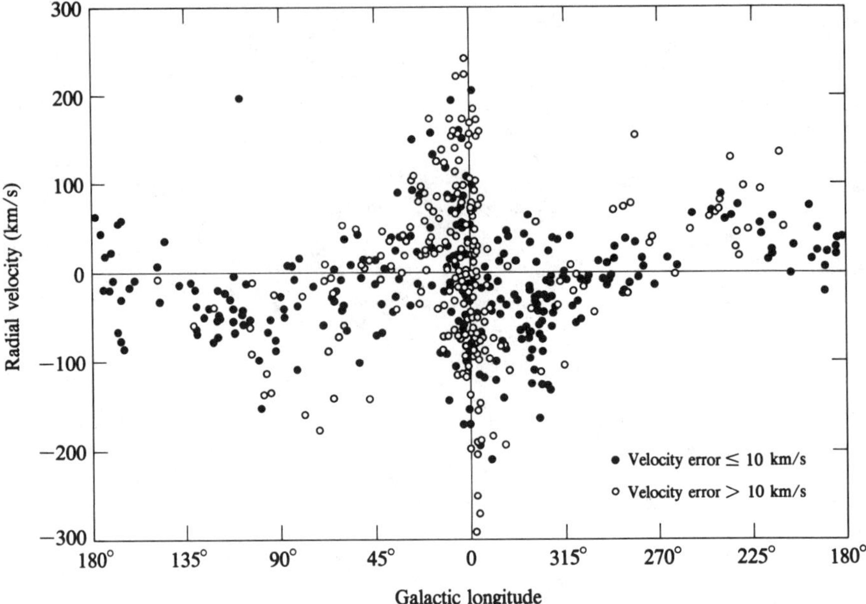

FIGURE 9.3
Observed radial velocities of planetary nebulae, relative to the Sun, plotted against galactic longitude.

Though interstellar extinction does not allow us to survey the entire Galaxy for planetary nebulae, the discovery statistics should be fairly complete up to a distance of about 1,000 pc in the Galactic plane. The observed total number of planetaries within a cylinder of radius 1,000 pc centered on the Sun, perpendicular to the Galactic plane, is 41, corresponding to a surface density of planetary nebulae (projected on the Galactic plane) near the Sun of 1.3×10^{-5} pc^{-2}. The statistics are increasingly incomplete at large distances because of interstellar extinction, though some planetaries are known at very great distances at high Galactic latitudes. We can find the total number of planetaries in the whole Galaxy only by fitting the local density to the model of their Galactic distribution based on stars of approximately the same kinematical properties. The number of planetary nebulae in the whole Galaxy found in this way is approximately 2.5×10^4. This number is, of course, not nearly so well determined as the local surface density which is more nearly directly observed. However, a radio-frequency survey for planetary nebulae near the center of our Galaxy, interpreted by similar methods, gives a total number 2.1×10^4. This figure, derived nearly completely independently and from observations of the region where the space density of planetaries is high, confirms very well the value based on measurements in the region near the Sun.

The height distribution of the planetary nebulae in the Galaxy may be derived from the known distances of planetaries. The average distance from the Galactic plane $< |z| >$ of the planetary nebulae within 1,000 pc projected distance from the sun in the Galactic plane is about 150 pc, while the root-mean-square distance $(< z^2 >)^{1/2} \approx 215$ pc. This is a fairly strong concentration to the Galactic plane, approximately the same as the Intermediate Population I of Oort. It should be noted again, however, that there is a tremendous range in properties of the planetaries; in spite of their rather strong concentration to the Galactic plane, one nebula, Haro's object Ha 4-1, is at a height $z \approx 10$ kpc from the plane.

9.3 The Origin of Planetary Nebulae and the Evolution of Their Central Stars

Observations of the distances of planetary nebulae give information on the properties of their central stars, for the distance of a nebula, together with the measured apparent magnitude of its central star, gives the absolute magnitude of the star. Furthermore, the effective temperature of the central star, or at least a lower limit to this temperature, can be found by the Zanstra method described in Section 5.8. The idea is that the measured flux in a recombination line such as Hβ is proportional to the number of recombinations in a nebula and hence to the whole number of ionizing photons absorbed in the nebula; if the nebula is optically thick to ionizing radiation, this number is, in turn, equal to the whole number of ionizing photons emitted by the central star. Comparison of the number of ionizing photons with the number of photons in an optically observed wavelength band gives T_*, the effective temperature of the central star, which in turn determines the bolometric correction, that is, the difference between the visual and bolometric magnitudes of the star. If the nebula is optically thin, so that all the ionizing photons are not absorbed, this method gives only a lower limit to the relative number of ionizing photons emitted by the central star and hence a lower limit to T_*.

If He II lines are observed in a nebula, the same method may be used for the He$^+$-ionizing photons with $h\nu \geq 4\ h\nu_0 = 54$ eV. Almost all nebulae with He II are optically thick to He$^+$-ionizing radiation; nebulae in which [O II] lines are observed are almost certainly optically thick to H^0-ionizing radiation. In this way, a luminosity-effective temperature diagram of the central stars of planetary nebulae can be constructed; the results of a recent study are as shown in Figure 9.4. The individual uncertainties are great, as can be seen from the error bars, because of the difficulties of making photometric measurements of these faint stars immersed in nebulosity, and also because

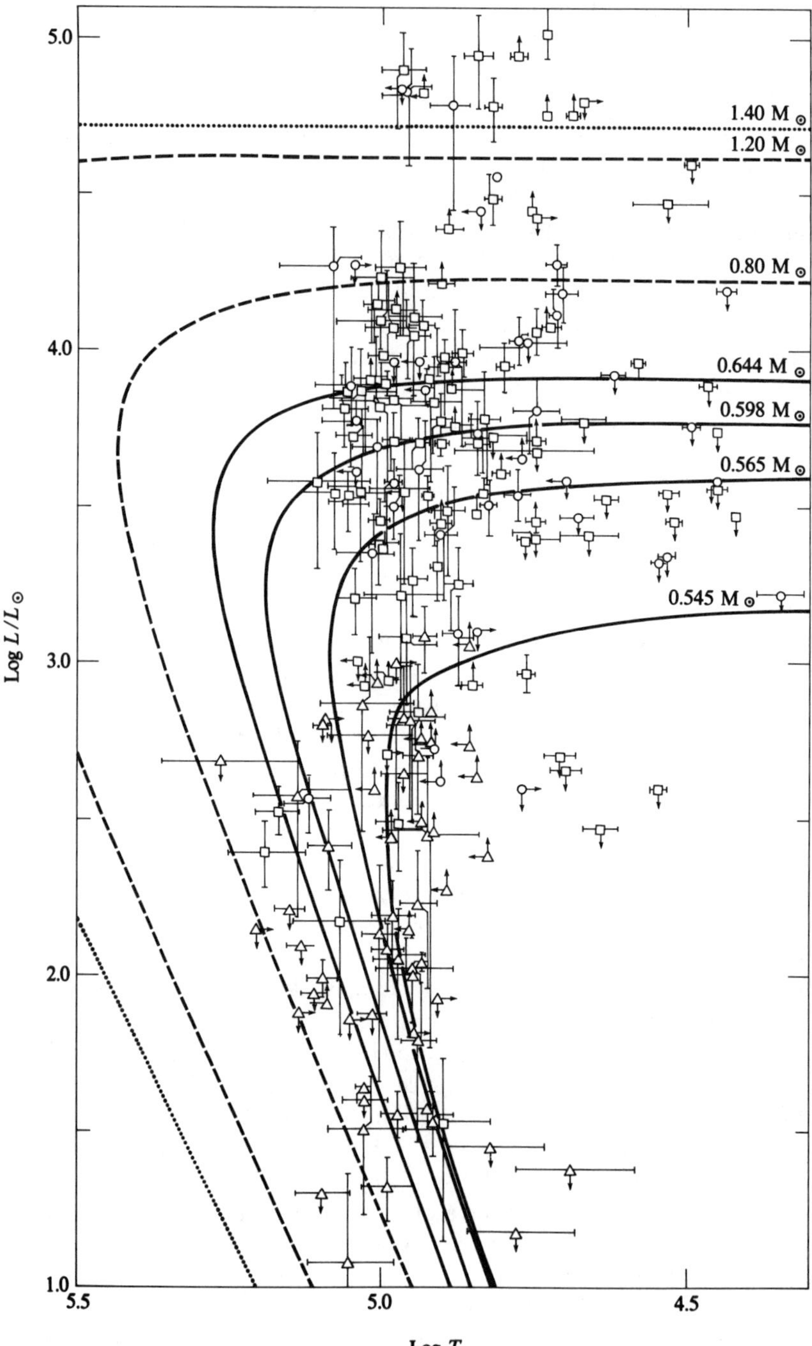

1.40 M ⊙
1.20 M ⊙
0.80 M ⊙
0.644 M ⊙
0.598 M ⊙
0.565 M ⊙
0.545 M ⊙
5.0
4.0
3.0
2.0
1.0
Log L/L ⊙
5.5
5.0
4.5
Log T*

of uncertainties as to the completeness of absorption of the ionizing photons. Furthermore, as we now understand, the distribution of the points in Figure 9.4 is strongly biased by selection effects. Nevertheless, it is clear that the effective temperatures range up to 2×10^5 °K; they are as high as, if not higher than, temperatures found for any other types of stars. The Stoy method, or energy-balance method of determining T_* shows generally good agreement with the Zanstra method based on He^+ ionizing photons, confirming that these are the best determined temperatures. (They are the values used, if available, in Figure 9.4.) The luminosities of the central stars are much higher than that of the sun, and at their brightest they are as luminous as many supergiant stars.

Furthermore, it is possible to attach a time since the planetary was "born" (that is, since the central star lost the shell that became the planetary nebula) to each point plotted on this diagram, since the radius of each planetary is known from its distance, and the radius, together with the mean expansion velocity, measures the time since the expansion began. When this is done it is seen that the youngest planetaries are those with central stars around $L \approx 10^4 \, L_\odot, T \approx 50{,}000°$ K, while the somewhat older planetaries have central stars around the same luminosity but hotter, and still older planetaries have successively less-luminous central stars. This must mean that the central stars of planetary nebulae evolve around the path shown in Figure 9.5, a schematic $L - T_*$ diagram, in the same time that the nebula expands from essentially zero radius at formation to a density so low that it disappears at $r_N \approx 0.7$ pc. This time, with a mean expansion velocity of 20 km sec^{-1}, is about 3.5×10^4 yr, much shorter than almost all other stellar-evolution times. It shows that the planetary-nebula phase is a relatively short-lived stage in the evolution of a star.

FIGURE 9.4 (facing page)
Luminosities L and effective temperatures T_* of planetary-nebula central stars, derived from the Zanstra method. Arrows indicate limits; bars indicate probable errors. Different symbols indicate different surveys from which the data were taken. Solid lines are computed evolutionary tracks of stars contracting and then cooling along the white-dwarf cooling lines.

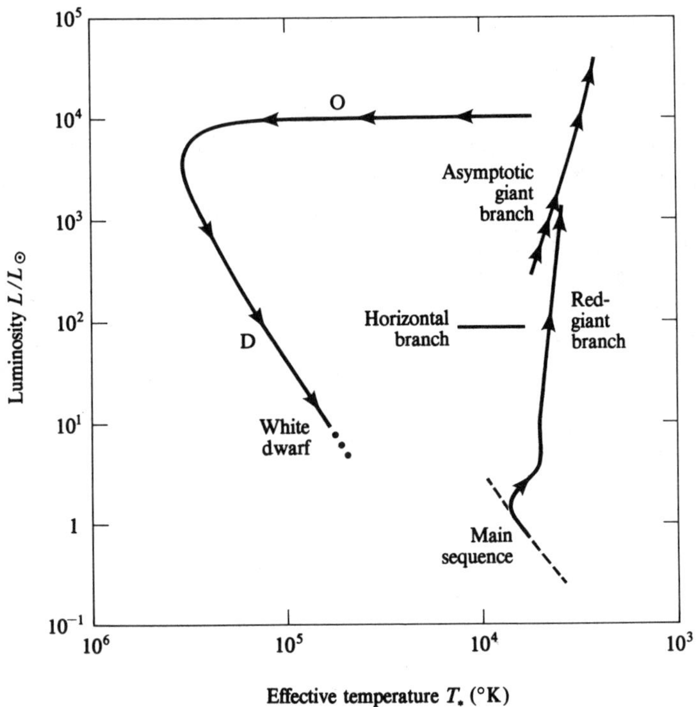

FIGURE 9.5
Schematic luminosity-effective temperature diagram showing the evolution of an approximately solar-mass star through the main-sequence, red-giant, asymptotic-giant-branch, and planetary-nebula nucleus stages to the final white dwarf cooling line.

At D the observed L and T_* correspond to a stellar radius $R \approx 0.03\ R_\odot$; so in the final stage of a planetary nebula, the nebular shell expands and merges with the interstellar gas, while the central star becomes a white dwarf. Separation between the shell and remnant star must occur at very nearly a composition discontinuity, because the nebula has approximately normal abundance of H, but a white-dwarf star can contain almost no H at all except at its very surface. As Figures 9.4 and 9.5 show, the oldest planetary-nebula central stars are in the luminous part of the white-dwarf region of the $L - T_*$ diagram, just above the region in which many white dwarfs lie. In this region a white dwarf's radius, which is fixed by its mass, remains constant, and the star simply radiates its internal thermal energy, becoming less luminous along a cooling line

$$L = 4\pi R^2 \sigma T_*^4. \tag{9.7}$$

Two of the most interesting problems in the study of planetary nebulae are the nature of their progenitors and the process by which the nebula is formed. The velocity of expansion of the nebular shell is approximately 20 km sec^{-1}. This is much lower than the velocity of escape of the present planetary-nebula central stars, of order 1,000 km sec^{-1}. Any impulsive process that could throw off an outer shell from these stars would not be likely to provide just a little more energy than the necessary energy of escape. On the other hand, the velocities of expansion are comparable with the velocities of escape from extreme red giants, which strongly suggests that the shell is ejected in the red-giant or supergiant stage.

Let us recall the evolution of a low- or intermediate-mass star with $M_\odot < M < 8\ M_\odot$ and its track in the Hertzsprung-Russell diagram as shown schematically in Figure 9.5. After contraction to the main sequence, hydrogen burning continues in the central core for most of the star's lifetime, until all the H in the core is exhausted. The nearly pure He core then begins to contract, and for low-mass stars ($M < 2\ M_\odot$) becomes degenerate. H burning begins in a shell source just outside the core; the star then evolves into the red-giant region, and develops a deep outer convection zone that increases in depth as the H-burning shell moves outward in mass, and as the core becomes denser and hotter. While in the red-giant stage, the star loses a significant amount of mass from its outer boundary, as a wind. When the central temperature $T_c \approx 1 \times 10^8$ °K is reached near the red-giant tip, He burning begins in the helium flash, and the star rapidly moves to lower L and higher T_* in the Hertzsprung-Russell diagram, to a position on the "horizontal branch." The planetary-nebula shell is not ejected at the time of the helium flash; both theory and observation show that the stars immediately after their first excursion to the red-giant tip are horizontal-branch stars, not incipient white dwarfs. A horizontal-branch star initially burns He in its central core and H in its outer shell source, and the direction of its evolution immediately after arriving on the horizontal branch (toward larger or smaller T_*) depends on the relative strengths of these two energy sources. Somewhat more massive stars ($2\ M_\odot < M < 8\ M_\odot$) do not form degenerate He cores and hence do not have a He flash, but nevertheless arrive at this same two-energy-source state. In either case, after the star burns out all the He at its center, it consists of a central C + O-rich core, an intermediate He-rich zone, and an outer H-rich region. There are He-burning and H-burning shell sources at the inner edges of the two latter regions, and the star evolves with increasing L and decreasing T_* along the "asymptotic-giant branch" toward the red-giant tip again. This evolution would terminate when a central temperature $T_c \approx 6 \times 10^8$ °K is reached and C detonation begins, but before this happens the ejection of the planetary nebula occurs.

A likely mechanism by which the shell is ejected from the supergiant star is dynamical instability against pulsations. When this instability occurs in very extended red giants, the energy stored in ionization of H and He may be large enough that the total energy of the outer envelope of the star is positive, and the pulsation amplitude can then increase without limit, lifting the entire envelope off the star and permitting it to escape completely. Calculations of this process show that the boundaries between stable and dynamically unstable stars in the mass range $M_\odot \leq M \leq 3\ M_\odot$ all occur around $L \approx 10^4 L_\odot$ for cool red giants near the region in the $L - T_*$ diagram occupied by observed long-period variables. Detailed calculations show that stars in this region have envelopes with positive total energies corresponding to velocities at infinity (if the material were expanded adiabatically) of about 30 km sec^{-1}. Furthermore, these calculations show that if the outer part of the envelope is removed, then the remaining model with the same luminosity and the same mass in its burned-out core, but with a smaller mass remaining in the envelope, is more unstable, so that if mass ejection starts by this process, it will continue until the entire envelope, down to the bottom of the H-rich zone, is ejected. The process is stopped at this level by the discontinuity in density due to the discontinuity in composition. Detailed evolutionary calculations have been made that actually follow the growth of these pulsations, and lead to the expulsion of an outer shell. Thus, according to these ideas, the mass left in the planetary-nebula central star is the mass in the burned-out core of the parent star when it first becomes unstable. The available calculations suggest that a star of 0.8 $M_\odot$ produces a remnant star of 0.6 $M_\odot$ and a nebular shell of 0.2 $M_\odot$; a 1.5 $M_\odot$ star produces a remnant star of 0.8 $M_\odot$ and a shell of 0.7 $M_\odot$; and a 3 $M_\odot$ star produces a remnant star of 1.2 $M_\odot$ and a shell of 1.8 $M_\odot$; while stars with $M \geq 4\ M_\odot$ begin burning C explosively, and presumably become supernovae before they become dynamically unstable. These masses are not the masses the original stars had on the main sequence; direct observations of red-giant stars as well as the deduced masses of horizontal-branch stars, particularly of RR Lyrae variables, show that significant mass loss occurs in the form of stellar winds during the red-giant and asymptotic-giant stages. Stars with $M \leq 0.6\ M_\odot$ cannot become dynamically unstable at all; in any event, stars with $M \geq 0.9\ M_\odot$ or so that have evolved to the planetary-nebula stage in the lifetime of the Galaxy evidently have not lost sufficient mass to get down to this limit. The presence of dust in planetary nebulae suggests that the material in the shell came from a cool stellar atmosphere, and thus strengthens the evidence for the evolution of planetary nebulae from red-giant stars.

As was discussed above, there is good observational evidence that red-giant stars are losing mass in stellar winds. This evidence comes partly from optical absorption lines, but chiefly from mm-wave emission lines. In addition, there is indirect evidence for stellar winds from the fact that the envelope masses must be reduced before a red giant can become a planetary nebula. In the

Sun there is a weak solar wind driven by the high temperature of the corona; winds much stronger (in mass flux) are observed in some highly evolved red giants, and even stronger "superwinds" are required at some stage to reduce the envelope masses sufficiently. Thus these very winds or superwinds, even though their physical cause is not well-understood, may also be the origin of some or even all planetary nebulae.

The observed winds in red-giant and asymptotic-giant branch stars have relatively low velocities, ~ 10 km sec^{-1} as measured by mm-wave radial velocities. This process thus produces a large, cool, neutral, slowly expanding shell around the star. If it continues to grow until the hot central core is exposed, after most of the envelope mass has been lost, a new much faster wind will then arise, as in other hot central stars. These fast winds are driven by radiation pressure in the resonance absorption lines. The interaction of this new, fast wind with the expanding envelope will speed up its expansion, as described in Section 6.6, at the same time that the high-energy photons from the central star ionize it. This interacting stellar-wind process is an alternative possible mechanism for the formation of planetary nebulae.

An object apparently in the very early stage of transition from a red giant to a planetary nebula is the object AFGL 618, with measured expansion velocity in Hα, [N II] and [S II] about 80 km sec^{-1}. Determined searches to very faint intensity levels, using CCD detectors and narrow-band emission-line filters have shown that as many as half the well resolved planetary nebulae have large, faint outer haloes, which perhaps represent the material lost in the stellar wind before planetary formation.

Many models have been calculated of the rapidly evolving central star after the planetary nebula has been ejected. The essential features of the evolution can be reproduced qualitatively by a star with a degenerate C + O core, or a degenerate core consisting of an inner C + O region and a small outer He region, together with possibly a very small-mass, H-rich envelope. The inner core hardly evolves at all as the envelope burns H rapidly in a shell source and contracts, while the He region, if it is small enough, contracts without igniting. Neutrino processes rapidly cool the interior, and the entire star becomes a white dwarf and then simply cools at constant radius. Some problems remain with the time scale and the details of the track, but the general features of the observed track are approximately reproduced.

9.4 Mass Return from Planetary Nebulae

The gas in a planetary-nebula shell, which, as we have seen, is enriched slightly in He, more so in N, and probably also in C, ultimately mixes with and thus returns to interstellar matter. The present rate of this mass return, calculated from the local density of planetaries derived from distance measurements in the way explained in Section 9.2, is 8.2×10^{-10} planetary shells pc^{-2} yr^{-1}

projected on the Galactic plane near the Sun, or $1.6 \times 10^{-10}\ M_\odot\ \mathrm{pc}^{-2}\ \mathrm{yr}^{-1}$. This rate is larger than the rate of mass return observed from any other type of object except for the estimated rate $6 \times 10^{-10}\ M_\odot\ \mathrm{pc}^{-2}\ \mathrm{yr}^{-1}$ from long-period variables, which is derived from infrared observations, together with very specific assumptions about the nature and optical properties of the dust particles in these stars. The rate of mass loss by planetaries is, however, considerably smaller than the estimated total rate of mass loss from all stars to interstellar matter, based on a model of the Galaxy, the luminosity function or distribution of star masses, and calculations of their lifetimes, which gives approximately $10 \times 10^{-10}\ M_\odot\ \mathrm{pc}^{-2}\ \mathrm{yr}^{-1}$.

Integrated over the entire Galaxy, the rate of planetary nebulae "deaths" is approximately 25 per year; in other words, the rate of mass return to interstellar space is approximately $5\ M_\odot\ \mathrm{yr}^{-1}$. This may be compared with the rate of mass return to interstellar matter from supernovae, using the rate of one supernova per 20 yr estimated for our Galaxy. Approximately half of the Galactic supernovae are Type I and half are Type II; the average mass in either type of shell is not at all well known, but is estimated from current theoretical ideas as roughly $5\ M_\odot$ for Type II supernovae and much less for Type I. According to these estimates, the rate of mass return to interstellar space from supernovae would be of order $0.1\ M_\odot\ \mathrm{yr}^{-1}$ integrated over the whole Galaxy, and this is less by a factor of approximately 100 than the rate of mass return from planetaries. However, the material in supernova shells has been processed much more thoroughly by nuclear reactions, and has abundances that are considerably more extreme than those within a typical planetary-nebula shell.

The death rate of planetary nebulae may also be compared with the birth rate of white-dwarf stars derived from observational data on the local density of white dwarfs in each color range. This quantity is only very poorly known because of the intrinsic faintness of the white dwarfs and the small region that may therefore be surveyed, but the best estimate is that, near the Sun, new white dwarfs form at a rate of about $2 \times 10^{-12}\ \mathrm{pc}^{-3}\ \mathrm{yr}^{-1}$. This estimate is very close to the birth rate of planetary nebulae, which is the same as their death rate, $2.4 \times 10^{-12}\ \mathrm{pc}^{-3}\ \mathrm{yr}^{-1}$, near the Galactic plane.

Overall, it seems that many stars with original masses in the range 1 to $8\ M_\odot$ apparently end their evolution by throwing off shells and becoming white dwarfs. Do all stars in the stated mass range go through this evolution? Probably not, because pre- and postnovae are hot blue stars, in the same general region of the Hertzsprung-Russell diagram as planetary-nebula central stars, apparently with masses of about $1\ M_\odot$, that do not occur anywhere in the evolutionary track leading to planetary nebulae outlined previously. Observational and theoretical evidence strongly suggests that all novae are relatively close binaries, so a reasonable interpretation is that all single stars (or members of wide binary pairs) evolve through the planetary-nebula stage, while all members of close binaries evolve through the nova stage.

In close binary-star evolution, when the more massive star becomes a red giant, its radius becomes so large that it "overflows its Roche lobe"; that is, its outer parts lose their gravitational binding and flow, in the rotating reference system, to the other star, thus transferring mass from one component to the other, rather than losing it in a wind. When the second star becomes a red giant, the direction of mass flow may reverse. Many observed features of old novae, and of stellar X-ray sources can be understood on this basis. It has been suggested that some planetary nebulae may also be formed in this way, particularly those with pronounced cylindrical symmetry. Some planetary-nebula central stars have been found to be spectroscopic binaries, adding credence to this idea. Planetary nebulae have a very wide range of observed structures, and probably several quite different evolutionary paths lead to the formation of this class. Obviously, much further research, both observational and theoretical, is still needed on this problem.

9.5 Planetary Nebulae with Extreme Abundances of the Elements

Most planetary nebulae have fairly "normal" abundances of the elements, as described briefly in Section 5.9. However, a small minority of observed planetaries have abundances that fall outside this range. They are worthy of further discussion here for the information they provide on stellar and planetary-nebula evolution.

Three planetary nebulae are members of Extreme Population II, the oldest population of stars known in our Galaxy. One is K 648 in M 15, the only planetary nebula known in a globular cluster (which is in fact an extreme "metal-poor" globular cluster). A second is Ha 4-1, Haro's planetary nebula in Coma, an object approximately 10 kpc from the Galactic plane, and a third is BB-1. In all three of these nebulae, as noted in Section 5.9, the He/H abundance ratios are in the range 0.10-0.11, essentially indistinguishable from the ratios in normal planetaries. On the other hand, the abundances of the heavy elements are much smaller than in typical planetary nebulae, as shown in Table 9.1. Low "metal" (heavy-element) abundance is one of the characteristics of the old Population II, along with large mean distance from the Galactic plane, and high velocity relative to the local standard of rest defined by young Population I stars. Thus these abundance determinations simultaneously suggest that the abundances of the elements are not greatly changed in the planetary shells from their original composition, and that the helium abundance in Extreme Population II is not very different from that in much younger populations. The minimum values in Table 9.1 are probably the best available estimates of the abundances of these elements (which are difficult to determine in stars) in Extreme Population II matter that has not yet been processed by nuclear reactions in the stars to which it currently belongs.

TABLE 9.1
Extreme abundances in planetary nebulae

	He/H	10^4 O/H	10^4 C/H	10^4 N/H
K 648	0.104	0.47	5.4	0.03
Ha 4-1	0.106	2.2	17.0	0.63
Typical PN	0.11	4.2	7.1	1.3
NGC 1976	0.10	5.6	3.3	0.63
Type I PN	0.14	6.3	–	10.0

Looking at "normal" planetary nebulae in more detail, we see what appear to be real abundance differences. The objects with the highest He/H and N/H ratios have been isolated and classified as "Type I planetary nebulae." Two objects of this type are NGC 2440 and NGC 6302; they, like other members of this group, exhibit pronounced filamentary structure. On the average they have low velocities and small distances from the Galactic plane; evidently they represent the remnants of the most massive, youngest stars that become planetary nebulae. Comparing the abundances in them with those in the Orion Nebula, we find that the planetary shells are slightly enriched in He/H and in N/O as a result of nuclear reactions within their own progenitor stars. The products of these reactions, such as He and N, are mixed into the material of the outer part of the star (which later becomes the planetary nebula shell) only when the outer convective envelope penetrates down into the burned or burning region. The first such "dredge-up" phase is calculated to occur when a star exhausts its central H and becomes a red giant. In this first dredge-up, material enriched in N and with its C abundance correspondingly reduced, but with the He/H ratio hardly changed, is brought to the surface layers. In the second dredge-up, following exhaustion of central He and the formation of a degenerate C + O core, large amounts of He and more N are mixed to the surface layers. A third dredge-up phase is calculated to occur during the stage in which He burning occurs in a convective shell, leading to higher C and lower N abundances, and to the thermal pulses that may eventually remove the planetary shell. These calculations are highly detailed, however, and depend upon a very simplified theory of turbulent convection; so it is not surprising that their results do not agree with all the observational data. Planetary nebulae are a laboratory in which some of these results of nuclear burning can be examined in more detail than in the atmospheres of red-giant stars, and no doubt further information that will help to refine our understanding of stellar evolution will come from them.

Two most interesting examples are Abell 30 and Abell 78, both of which are large low-surface-brightness planetaries. The former has several fairly bright knots within a few seconds of arc of its nucleus. These knots have typical forbidden emission-line spectra, with strong [O III] and [Ne III], plus weaker [N II] and [O II]. They also have strong He II $\lambda\lambda$4686, 5411 and weaker He I $\lambda\lambda$5876, 6678, but Hα is exceedingly weak and only barely detectable in one of the knots, and absent from the spectra of the others; Hβ and the fainter H I lines are not detected in any of them. Abell 78 has a small inner ring with a similar emission-line spectrum. Abundance analyses of these spectra confirm that the H abundance in these features is extremely low; in both these nebulae the inner structures have $N_{\rm H}/N_{\rm He} \leq 0.03$. Evidently they represent parts of the core of the progenitor stars, outside the He shell-burning layer but inside the H shell-burning layer. From their position near the centers of these nebulae, they clearly left the star at the end of the nebular formation process, or more probably in a second or even later expulsion. The abundances of N and O in these He-rich structures are typical of their abundances in other planetary nebulae, if they are expressed with respect to the sum of the masses of H and He, that is, as if all the H were converted into He. The temperatures derived from [O III] (λ4959+λ5007)/λ4363 in these features is $T \approx 1.2 - 1.6 \times 10^4$ °K; in Abell 30, a large part of the heating is due to the stopping of the stellar wind, as described in Section 6.6, in addition to photoionization heating of He^+. The spectra of the outer parts of both these nebulae are too faint for detailed analyses, but they show Hα and Hβ in typical strength with respect to He II λ4686, and thus have approximately normal composition.

9.6 Planetary Nebulae in Other Galaxies

Planetary nebulae are far less luminous than giant H II regions, and they are correspondingly less easily observed in other galaxies. However, the Large and Small Magellanic Clouds are close enough that numerous planetaries have been discovered in them by objective-prism surveys, and individual slit spectra have been obtained for many of these planetaries. Many more faint planetaries have been found in the clouds by comparison of direct images of fields taken in [O III] λ5007 and just off this line in the neighboring continuum. To date the total numbers of known planetary nebulae are 51 in the SMC, and 137 in the LMC.

These numbers can be converted to total numbers of planetaries by comparing the luminosity range covered by the surveys with the total expected luminosity range (based on planetary nebulae in our Galaxy), and also by scaling from the fields surveyed to faint luminosities to the total areas and numbers of stars in the Clouds. These methods agree well in giving totals of about 300 planetaries in the SMC, and 1,000 in the LMC. Expressed in terms of planetary nebulae per unit mass in stars, these totals lead to 2.6×10^{-7} planetary nebulae $M_\odot^{-1}$ in the SMC, and 1.8×10^{-7} $M_\odot^{-1}$ in the LMC. If one or the other of these same ratios applied in our Galaxy, the total number of planetaries in it would be 2.3×10^4 or 3.4×10^4, both values being quite similar to the directly determined figures mentioned in Section 9.2. In the areas in the Magellanic Clouds surveyed to faint magnitude limits, the luminosity function of planetary nebulae can be directly derived; it is shown in Figure 9.6. It agrees reasonably well with the simple Shklovsky model of an expanding sphere of constant mass of ionized gas, described by equation (9.1). As was stated in Section 9.2, with assumed constant velocity of expansion, this predicts a uniform distribution of the number of planetaries with each radius. Therefore, since $L_{\mathrm{H}\beta}$, the luminosity in a recombination line such as Hβ, follows

$$L_{\mathrm{H}\beta} \propto N_e N_p r_N^3 \propto N_e^2 r_N^3 \propto r_N^{-3}, \tag{9.8}$$

the predicted luminosity function is

$$N(L_{\mathrm{H}\beta}) \propto L_{\mathrm{H}\beta}^{-4/3}. \tag{9.9}$$

If the luminosity in Hβ of the most luminous planetary nebula observed in the Magellanic Clouds is designated L_0, this gives for the fraction of the total number brighter than any specified luminosity $L_{\mathrm{H}\beta}$

$$\int_{L_{\mathrm{H}\beta}}^{L_0} N(L_{\mathrm{H}\beta}) dL_{\mathrm{H}\beta} \propto \left(\frac{L_{\mathrm{H}\beta}}{L_0}\right)^{-1/3} - 1. \tag{9.10}$$

Actually the luminosity function available from the observational data is for the luminosity in [O III] $\lambda\lambda$4959, 5007, not Hβ, but on the simple assumption that these luminosities are proportional to one another, the comparison shown in Figure 9.6 is quite satisfactory.

The He abundances in the planetary nebulae in both Magellanic Clouds are closely the same and also nearly the same as in the planetaries in our Galaxy, slightly enriched with respect to H II regions. The abundances of the heavier elements are also the same in the LMC and SMC. In them, O is less abundant than in planetaries in our Galaxy by about a factor 2, and N is also underabundant, but by larger factors, ranging from about 3 to 8. The distance of the Clouds, about 6×10^4 pc, is large enough that most of the planetary nebulae in our Galaxy would not be resolved at this distance; indeed, only a very few of the planetaries in the Clouds have been resolved in direct photographs. They are objects similar to the very largest Galactic planetary nebulae, like NGC 7293, with diameter about 0.5 pc, which would just barely be resolved with apparent angular diameters of 1″ to 2″ if they were in the Magellanic Clouds.

In more distant galaxies, four planetary nebulae were identified in M 31 many years ago by Baade, who compared plates taken in [O III] $\lambda\lambda$4959, 5007 with plates taken in the nearby continuum. These nebulae are among the very brightest planetary nebulae in M 31. More recently, determined searches with very narrow [O III] λ5007 filters (which of course must be matched to the redshift of the galaxy) and comparison continuum filters have yielded many more planetaries in M 31. In addition, 16 planetaries have been identified in its round companion M 32, 19 in its elliptical companion NGC 205, and one in the Fornax dwarf spheroidal galaxy. Still more have been found in other Group members, such as M 33, and even further out, as in M 81.

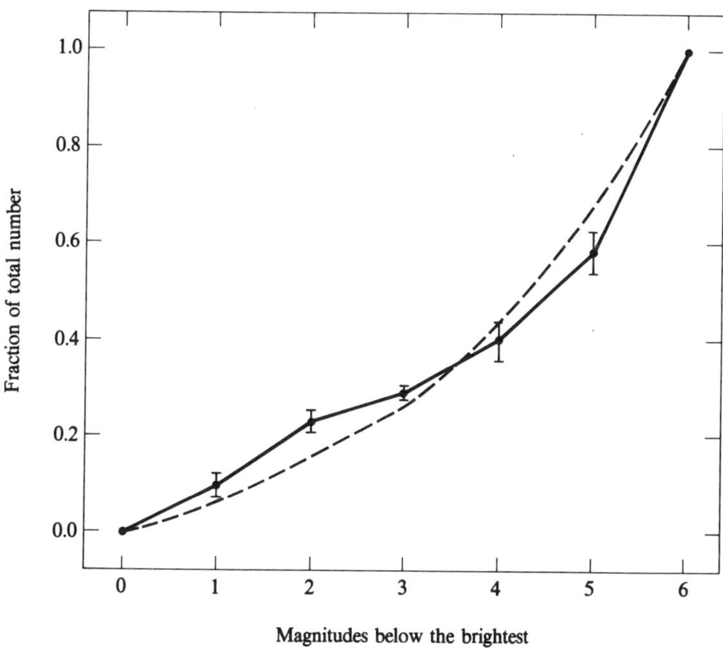

FIGURE 9.6
Observed [O III] luminosity function of planetary nebulae in the Large Magellanic Cloud (*solid line and circles with error bars*) and predicted luminosity function of equation (9.10) (*dashed line*).

Velocities of the planetaries in M 31 provide important information on its dynamical properties. Measurements of 30 planetary nebulae within 250 pc of the nucleus of M 31, where no H II regions are available, are particularly important. They give a radial-velocity dispersion of 155 ± 22 km sec^{-1}, which in turn leads to a good determination of the mass-to-light ratio for the inner nuclear bulge of this galaxy.

The luminosity functions of the planetary nebulae in the various Local Group galaxies are fairly similar; hence planetaries may be used as distance indicators. Extrapolating these luminosity functions to fainter magnitudes by using the Magellanic Cloud determinations gives the total number of planetaries in each Local Group member that has been surveyed. They range from roughly 10 in Fornax and 250 in M 32 to 2×10^4 in M 31. The average number of planetaries per unit stellar visual luminosity found in this way is $6 \pm 2 \times 10^{-7}$ planetary nebulae $L_\odot^{-1}$. Applied to our own Galaxy, with estimated visual absolute magnitude $M_V = -20.6$, this predicts $9 \pm 3 \times 10^3$ planetaries, which is about a factor of three less than the other figures estimated in this chapter for this quantity.

Abundance determinations have been carried out for many of the brighter planetary nebulae in other galaxies, particularly in M 31. The general results are that, as in our Galaxy and the Magellanic Clouds, the He abundances are slightly larger in planetary nebulae than in H II regions in the same galaxies, and N is overabundant by larger factors, ranging from 4 to 10^2.

9.7 Molecules in Planetary Nebulae

In the simplest approximation planetary nebulae consist entirely of ionized gas, and no molecules are expected to be present in them. However, to a better approximation optical and infrared observations indicate the presence of dust, and of neutral gas also, in dense globules or neutral condensations. These are especially evident in direct images of some planetaries taken with narrow-band filters in the [O I] $\lambda 6300$ emission line. The ideas of mass loss discussed above suggest that there may be fairly large neutral shells around planetary nebulae, and that these shells might well contain molecules as well as neutral atoms and dust particles.

Though planetary nebulae contain much less gas than H II regions, so that the molecular emission lines are correspondingly more difficult to detect, progress in infrared and mm-wave systems has led to the observation of CO and H_2 in NGC 7027, NGC 6853, and a few other planetary nebulae in recent years. The CO lines are pure rotational transitions $J = 1 \rightarrow J = 0$ $\lambda 2.60$ mm and $J = 2 \rightarrow J = 1$ $\lambda 1.30$ mm, which have very low excitation energies that can be collisionally excited even in cool gas ($T \geq 10$ °K). The H_2 lines are forbidden quadrupole transitions in the $v = 1 \rightarrow v = 0$ vibration-rotation band at $\lambda\lambda 2.12\ \mu$, $2.41\ \mu$, $2.42\ \mu$ that can be excited collisionally at somewhat higher temperatures ($T \geq 1,000°$ K), or by resonance fluorescence with ultraviolet radiation even in cold gas.

In addition, the H I hyperfine-structure line $\lambda 21.106$ has been observed in absorption in several planetary nebulae, and in emission in IC 418. No doubt in the near future progress will be rapid in measuring these molecular and neutral atomic lines with greater accuracy in more planetary nebulae, and will lead to better understanding of the neutral gas in and around these objects.

References

General references on planetary nebulae include

Kaler, J. B. 1985. *Ann. Rev. Astr. Ap.* **23,** 89.

Miller, J. S. 1974. *Ann. Rev. Astr. Ap.* **12,** 331.

Aller, L. H. 1984. *Physics of Thermal Gaseous Nebulae.* Dordrecht: Reidel.

Pottasch, S. 1984. *Planetary Nebulae.* Dordrecht: Reidel.

Flower, D. R. 1983. *Planetary Nebulae* (IAU Symposium No. 103). Dordrecht: Reidel.

Terzian, Y. 1978. *Planetary Nebulae* (IAU Symposium No. 76). Dordrecht: Reidel.

Perek, L., and Kohoutek, L. 1967. *Catalogue of Galactic Planetary Nebulae.* Prague: Czechoslovak Academy of Sciences.

This is an excellent source that contains practically all observational data on planetary nebulae published through mid-1966. Updates are given in the various IAU Symposium volumes on planetary nebulae listed above.

General references on surface distribution and kinematics, include

Minkowski, R. and Abell, G. O. 1963. *P.A.S.P.* **75,** 488.

(Figure 9.1 is based on this reference.)

Minkowski, R. 1965. *Galactic Structure.* A. Blaauw and M. Schmidt, eds. Chicago: University of Chicago Press, p. 321.

Schneider, S. E., Terzian, Y., Purgathofer, A., and Perinotto, M. 1983. *Ap. J. Suppl.* **52,** 399.

(Figure 9.3 is based on the second of these references.)

Some references on the distance scale include

Shklovsky, I. S. 1956. *Russian Astron. J.* **33,** 515.

Osterbrock, D. E. 1960. *Ap. J.* **131,** 541.

O'Dell, C. R. 1962. *Ap. J.* **135,** 371.

Cudworth, K. M. 1974. *A. J.* **79,** 1384.

Cahn, J. H., and Kaler, J. B. 1971. *Ap. J. Suppl.* **22,** 319.

Acker, A. 1978. *Astr. Ap. Suppl.* **33,** 367.

The first reference describes the idea of obtaining distances of planetaries from the hypothesis that they all contain the same amount of ionized gas. The second describes the observational check of the correlation between electron density and surface brightness predicted by that hypothesis. The last reference is the most complete existing catalogue of distances of planetary nebulae, and the numerical values quoted in the text are mostly based on it. The evolution of central stars of planetary nebulae is discussed by

Salpeter, E. E. 1971. *Ann. Rev. Astr. Ap.* **9,** 127.

Kwok, S. 1983. *Planetary Nebulae* (IAU Symposium No. 103). Dordrecht: Reidel, p. 293.

Schoenberner, D., and Weidemann, V. 1983. *Planetary Nebulae* (IAU Symposium No. 103). Dordrecht: Reidel, p. 359.
Iben, I., and Renzini, I. 1983. *Ann. Rev. Astr. Ap.* **21,** 271.
Shaw, R. A. 1985. *The Evolution of Planetary Nebula Nuclei.* University of Illinois Ph.D. thesis.

Figure 9.4 is based on the last of these references.

Important generalizations and applications of the Stoy or energy-balance method of determining the effective temperatures of the central stars of planetary nebulae are given in

Pottasch, S. R. 1981. *Astr. Ap.* **94,** L13.
Preite-Martinez, A., and Pottasch, S. R. 1983. *Astr. Ap.* **126,** 31.

The basic idea that the low velocities of expansion of planetary nebulae indicate that their progenitors are probably extended red giants was first stated by Shklovsky 1956, and was further developed by

Abell, G. O., and Goldreich, P. 1966. *P.A.S.P.* **78,** 232.

Later, more detailed treatments include

Renzini, A. 1983. *Planetary Nebulae* (IAU Symposium No. 103). Dordrecht: Reidel, p. 267.
Tuchman, Y. 1983. *Planetary Nebulae* (IAU Symposium No. 103). Dordrecht: Reidel, p. 281.
Kwok, S., Purton, C. R., and Fitzgerald, P. M. 1978. *Ap. J.* **219,** L125.
Kwok, S. 1982. *Ap. J.* **258,** 280.
Kwok, S. 1983. *Planetary Nebulae* (IAU Symposium No. 103). Dordrecht: Reidel, p. 293.

The first of these treats the general problem, the second is concerned with the dynamical-instability picture, and the last three with the interacting stellar-wind mechanism. Surveys to detect faint outer envelopes or haloes of planetary nebulae, and the resulting statistics of their incidence, are described by

Jewitt, D. C., Danielson, G. E., and Kuperman, P. N. 1986. *Ap. J.* **302,** 727.
Chu, Y.-H., Jacoby, G. H., and Arendt, R. 1987. *Ap. J. Suppl.* **64,** 529.

The return of matter from planetary nebulae to interstellar matter, formation of white dwarfs, and the resulting changes in the abundances of the elements in the interstellar matter from which stars are formed, are discussed by

Serrano, A. 1983. *Planetary Nebulae* (IAU Symposium No. 103). Dordrecht: Reidel, p. 463.
Audouze, J., and Tinsley, B. M. 1976. *Ann. Rev. Astr. Ap.* **14,** 43.
Tinsley, B. M. 1979. *Ap. J.* **229,** 1046.

The abundance variations in different types of planetary nebulae are discussed by

Peimbert, M., and Torres-Peimbert, S. 1983. *Planetary Nebulae* (IAU Symposium No. 103). Dordrecht: Reidel, p. 233.

Jacoby, G. H., and Ford, H. C. 1983. *Ap. J.* **266,** 298. (Abell 30 and Abell 78).

Adams, S., Seaton, M. J., Howarth, I. D., Auriere, M., and Walsh, J. R. 1984. *M.N.R.A.S.* **207,** 471. (K 648).

Barker, T., and Cudworth, K. M. 1984. *Ap. J.* **278,** 610. (Population II planetaries.)

The significance of Abell 30 and Abell 78 as the strongest observational evidence for planetary-nebula central stars undergoing a post-asymptotic giant-branch helium flash and repeating their evolution, all the way back to the asymptotic-giant branch, is discussed by

Iben, I., Kaler, J. B., Truran, J. W., and Renzini, A. 1983. *Ap. J.* **264,** 605.

Excellent summaries of our knowledge of planetary nebulae in other galaxies are

Jacoby, G. H. 1980. *Ap. J. Suppl.* **42,** 1.

Jacoby, G. H. 1983. *Planetary Nebulae* (IAU Symposium No. 103). Dordrecht: Reidel, p. 427.

Ford, H. C. 1983. *Planetary Nebulae* (IAU Symposium No. 103). Dordrecht: Reidel, p. 443.

Figure 9.6 is based on the first of these references.

Papers on observations of molecular emission lines in planetary nebulae include

Mufson, S. L., Lyon, J., and Marionni, P. A. 1975. *Ap. J.* **201,** L85.

Treffers, R. R., Fink, U., Larson, H. P., and Gautier, T. N. 1976. *Ap. J.* **209,** 793.

Thronson, H. A., and Mozurkewich, D. 1983. *Ap. J.* **271,** 611.

Huggins, P. J., and Healy, A. P. 1986. *Ap. J.* **305,** L29.

The observations of the H I λ21.106 cm hyperfine-structure line in emission in IC 418 is described in

Taylor, A. R., and Pottasch, S. R. 1987. *Astr. Ap.* **176,** L5.

10

Nova and Supernova Remnants

10.1 Introduction

Novae and supernovae are observed objects in which shells of gas are cast off from an evolving star and returned to interstellar space. They are much more violent events than planetary nebulae; the velocities of expansion of nova shells are typically of order 10^3 km sec^{-1}, and of supernova shells of order 10^3 to 10^4 km sec^{-1}. The masses of nova shells are much smaller than planetary-nebula shells, typically perhaps 10^{-4} $M_\odot$ or less, but the masses of supernova shells are considerably larger, perhaps 1 $M_\odot$ or more. All these objects are different in physical nature from the typical gaseous nebulae (H II regions and planetary nebulae) we have studied up to this point. The main physical energy-input mechanisms are different in each. Nevertheless, the observed line spectra have general similarities because heated, ionized gas tends to radiate more or less the same emission-line photons, however its high temperature and ionization were produced. We will discuss each of these types of shells in turn.

10.2 Nova Shells

Novae are episodes in which an evolving star suddenly (within a few days) becomes much brighter, reaching peak luminosities of order $\gtrsim 10^4$ $L_\odot$. Their spectra show that material leaves the star with velocities of order 10^3 km sec^{-1}. Their emission-line profiles have considerable structure, indicating multiple "shells," that is, complex spatial density and velocity structure. A typical nova may liberate an energy $\geq 10^{45}$ erg over a time interval of a year, and gradually returns to its pre-outburst state over a time $\sim 10^2$ yr.

The physical mechanism of novae is fairly well understood. They are close binary stars, in which one component is a white dwarf, the other a red dwarf or subgiant, which is overflowing its Roche lobe and losing mass into the other lobe. The inflowing mass spirals into the white dwarf as an accretion disk, which radiates much of the light seen in the pre- and post-outburst phases. The build-up of hydrogen-rich material on the surface of the white dwarf raises the temperature at its inner edge, until it eventually becomes high enough to start an explosive thermonuclear runaway that energizes the nova outburst.

Soon after the peak luminosity, the absorption-line spectrum of the nova begins to show emission lines, indicating the appearance of large regions of hot, optically thin gas. As the continuum weakens, the emission lines strengthen with respect to it, and as the development continues, typical nebular lines such as [N II], [O III] and [Ne III] appear and become stronger relative to the fading continuum. Initially [N II] $\lambda5755$ and [O III] $\lambda4363$ are relatively strong, but they soon weaken with respect to [N II] $\lambda\lambda6548$, 6583 and [O III] $\lambda\lambda4959$, 5007, respectively. The spectrum gradually changes to a nearly entirely nebular-type spectrum, with broadened emission lines resulting from the high expansion velocity, plus the faint blue continuum of the remaining or reformed accretion disk.

Within a few years after the outburst, a small, faint nebulous shell can often be seen on direct images, surrounding the postnova star. This shell increases in size at an approximately uniform rate, indicating constant-velocity expansion. An example is Nova DQ Her 1934, whose shell, photographed in 1963 and reproduced in Figure 10.1, appeared as a faint ellipse approximately $9'' \times 14''$ in size. These nova shells are often fairly symmetric, with structures that can be interpreted as "equatorial" rings and "polar" condensations. The nova shells gradually become fainter as they expand, eventually disappearing below the threshold of sensitivity of detection. Physically, in time they merge into and become part of the interstellar matter of the Galaxy.

The distance of a nova shell can be determined by comparing the measured radial velocity of expansion, determined in the nebular stage, with the proper motion (or angular velocity) of expansion. For instance, for Nova Her 1934, the radial expansion velocity, as measured from the separation of emission lines on spectra taken in 1949 when the shell was small but relatively bright, is 320 ± 20 km sec^{-1}. This, together with the dimensions on a direct photograph taken in 1977, when the dimensions of the shell were $11'' \times 17''$, leads to a distance 420 ± 100 pc. The range of uncertainty corresponds to the difference between assuming that the radial velocity of expansion corresponds to the major or minor axis of the elliptical shell, as seen on the plane of the sky.

The decrease in surface brightness of the shell with time, though not measured quantitatively, corresponds qualitatively to that expected for a constant-mass shell, as derived in equation (9.2) for planetary nebulae. The mass of ionized gas, M_N as estimated from the total flux, distance, and angular size using equation (9.6) is typically of order 10^{-4} or $10^{-5} M_\odot$, much smaller than

FIGURE 10.1
Nova HQ Herculis 1934 shell, photographed in 1963 with the Shane 3-m reflector and an ultraviolet filter transmitting chiefly [O II] $\lambda 3727$. Dimensions of shell are approximately $9'' \times 14''$, corresponding to 0.02×0.03 pc. (*Lick Observatory photographs.*)

the mass of a planetary nebula.

Nova shells are very faint. Thus in the days of photographic spectroscopy, it was impossible to record more than a very few of the brightest emission lines in any of them, even with the largest telescopes. However, since the development of more sensitive solid-state imaging detectors, such as CCDs and Reticons, deeper quantitative spectral data have become available. In Table 10.1 approximate measurements of relative emission-line strengths are listed, corrected for interstellar extinction, in the spectra of Nova Her 1934 and of Nova Pup 1942. These spectra date from 1977 - 1984. One ultraviolet spectrum from the IUE satellite is included. It can be seen that the nova shells have quite unusual nebular spectra, with many permitted lines of various stages of ionization of N and C, in addition to the normal H I, He I and He II lines. Only a very few forbidden lines are detected, [N II] $\lambda\lambda$6548, 6583 and [O II] λ3727, which, in the Nova Puppis spectrum, may arise largely in a surrounding H II region, rather than in the shell itself.

TABLE 10.1
Line strengths in nova shells

Ion	Transition	λ Å	Nova Pup	Nova Her	Model
C II	$2p\ ^2P - 2p^2\ ^2D$	1335		270	82
C III]	$2s^2\ ^1S - 2s2p\ ^3P$	1909		13	21
[O II]	$2p^3\ ^4S - 2p^3\ ^2D$	3727	21	100	24
He I	$2s\ ^3S - 3p\ ^3P$	3889	20		
Hζ	$2 - 8$				
Hϵ	$2 - 7$	3969	13		
Hδ	$2 - 6$	4101	28		
N III	$3s\ ^2S - 3p\ ^2P$				
C II	$3d\ ^2D - 4f\ ^2F$	4267		29	16
Hγ	$2 - 5$	4341	26	65	40
N III	$4f\ ^2F - 5g\ ^2G$	4379	19		
N IV	$5g\ ^3G - 6h\ ^3H$	4606	52		
N III	$3p\ ^2P - 3d\ ^2D$	4640			
He II	$3 - 4$	4686	92	18	22
Hβ	$2 - 4$	4861	100	100	100
N II	$3p\ ^3D - 3d\ ^3F$	5005	44	29	24
He II	$4 - 7$	5411	7		
N II	$3s\ ^3P - 3p\ ^3D$	5678	9		
He I	$2p\ ^3P - 3d\ ^3D$	5876	19	26	45
[N II]	$2p^2\ ^3P_1 - 2p^2\ ^1D_2$	6548		10	3
Hα	$2 - 3$	6563	300:	146	250
[N II]	$2p^2\ ^3P_2 - 2p^2\ ^1D_2$	6583	200:	29	9

There are only two ways to understand the nearly complete absence of forbidden lines from these optical spectra. One is to assume that the electron density is so high that they are all suppressed by collisional deexcitation. The other is to suppose that the temperature is so low that they all are suppressed because their thresholds for excitation χ are much larger than thermal energies, so that $\exp(-\chi/kT) \ll 1$ for all of them. The first possibility is eliminated by the observation that forbidden lines were strong in the early nebular stages of many of these shells, and that as they expanded the forbidden lines weakened, not strengthened, as the density decreased. Also, the electron densities derived from the known distances and masses of the shell, using the equivalent of equation (9.5), are of order $\sim 10^2$ cm^{-3}, far too low for collisional deexcitation to be important.

Therefore, the near absence of optical forbidden lines from the nova shell spectra must indicate that their mean temperatures are low. For $\exp(-\chi/kT) < 10^{-3}$ for a collisionally excited line near $\lambda6000$, for instance, requires $T < 3,500°$ K. At such low temperatures, only recombination lines are expected to be visible in the spectra of nebulae. For H and He, as described in Chapter 4, the strongest expected lines in the optical spectrum are Hα, Hβ, He I $\lambda5876$, and He II $\lambda4686$, as observed in nova shells. As explained in Chapter 4, all the recombination cross sections vary approximately as v^{-2} (in terms of electron velocity v), and the recombination coefficients therefore vary approximately as $T^{-1/2}$. The relative strengths of different lines of the same atom or ion, which depend mainly on ratios of cross sections and transition probabilities, are therefore nearly independent of T. Detailed calculations including collisional effects are available down to $T = 1,000°$ K for H I, and to $T = 3,000°$ K for He II, and confirm this statement. Calculated results for H I at $T = 500°$ K in the low-density limit (omitting collisions), which is as accurate as needed for this application, are listed in Table 10.2.

For He I, because of its more complicated energy-level diagram, calculations are available only down to $T = 5,000°$ K, but they can also be extrapolated reasonably safely to lower temperatures. For the various stages of ionization of the heavier elements, such as C, N, and O, no detailed calculations of the recombination spectra, comparable to those for H, are available in published form. However, it is easy to see in qualitative terms their main expected features, and even to make semi-quantitative calculations on this basis, following the physical ideas expressed in Sections 2.2, 2.4, 2.7 and 4.2. The upper levels ($n > 2$) of the ions of C, N, O, etc., are close to H-like. Most of the captures occur to these levels, and their cross sections can be well-approximated by the cross sections for one-electron ions. These captures occur to a wide range of levels n, and preferentially to the terms with large L and S. Ions thus produced in these levels cascade down to lower levels by permitted radiative transitions, following the selection rules $\Delta l = \pm 1$, $\Delta L = 0, \pm 1$, $\Delta S = 0$. The transition probabilities tend to be largest for $\Delta n = 0$ or -1, and as all the transitions go downward in n, they converge toward the

TABLE 10.2
H I recombination lines (Case B)[a]

Balmer-line intensities relative to Hβ		Paschen and Brackett line intensities	
$j_{H\alpha}/j_{H\beta}$	4.22	$j_{P\alpha}/j_{H\beta}$	0.839
$j_{H\gamma}/j_{H\beta}$	0.417	$j_{P\beta}/j_{H\gamma}$	0.607
$j_{H\delta}/j_{H\beta}$	0.221	$j_{P\gamma}/j_{H\delta}$	0.529
$j_{H\epsilon}/j_{H\beta}$	0.134	j_{P8}/j_{H8}	0.476
$j_{H8}/j_{H\beta}$	0.0885	j_{P10}/j_{H10}	0.452
$j_{H9}/j_{H\beta}$	0.0619	$j_{Br\alpha}/j_{H\gamma}$	0.650
$j_{H10}/j_{H\beta}$	0.0453	$j_{Br\beta}/j_{H\delta}$	0.450
$j_{H15}/j_{H\beta}$	0.0142	$j_{Br\gamma}/j_{H\epsilon}$	0.373
$j_{H20}/j_{H\beta}$	0.0063	j_{Br8}/j_{H8}	0.336

[a]$T = 500°$ K; $N_e \rightarrow 0$;
$4\pi j_{H\beta}/N_p N_e = 1.06 \times 10^{-24}$ erg cm^3 sec^{-1};
$\alpha_{H\beta}^{eff} = 2.60 \times 10^{-13}$ cm^3 sec^{-1}.

levels with $L = n - 1$, the highest possible value. Thus the strongest emission lines expected to arise in the optical spectral region from recombination are those with small n, large l and L, large S, and $\Delta n = 0$ or -1, exactly as observed in the nova shells.

A very strong confirmation of the idea that the temperatures in nova shells are low is provided by a broadened emission feature that peaks near λ3640 in the spectrum of the shell of Nova Pup 1942. Its profile is relatively sharp on the long wavelength side, and tails off more gradually on the short wavelength side. Its width is perhaps 50 Å. This can only be the Balmer discontinuity at λ3646 and the continuum at wavelengths just below it. At typical nebular temperatures $T \approx 10^4$ ° K, the Balmer continuum strength decreases slowly to higher frequencies, as shown in Figure 4.1. The main frequency dependence is through the factor $\exp(-h\nu/kT)$ in equation (4.22). The temperature indicated by the narrow Balmer continuum peak in the shell of Nova Pup 1942 is thus only $T \approx 800°$ K.

The strengths of [N II] λλ6548, 6583 appear to present a problem for this low-temperature interpretation of the shell spectrum. However, they can be understood qualitatively as resulting from recaptures also. All those electrons recaptured by N^{++} which form N^+ in *singlet* terms must ultimately lead, through downward allowed radiative transitions of N II, to population of the lowest singlet level $2p^2\ ^1D$. Decay of this level can then occur only by the forbidden transitions $2p^2\ ^3P - {}^1D$ λλ6548, 6583, whose strengths should thus be comparable with the strongest permitted triplet recombination line of N II, $3p\ ^3D - 3d\ ^3F$ λ5005, as observed.

The relative abundances of the elements in the nova shells can be approximately determined from the relative strengths of their emission lines, as described in Section 5.9. Calculated effective recombination coefficients for specific lines (corresponding to $\alpha^{eff}_{H\beta}$ for H I) are not available in published form for heavy ions like N^+ and N^{++}, but can be estimated with fair accuracy. At the low temperatures of nova shells, dielectronic recombination is not a major effect except for a few specific ions. The most important is C II $2p\ ^2P - 2p^2\ ^2D\ \lambda1335$, which is strong because of dielectronic recombination of C^{++} through the C^+ $2s2p3d\ \ ^2D$ term, which is only 0.22 eV above the ground level of C^{++}. These observational data show that nova shells are somewhat enriched in He, and greatly enriched in C, N, O, relative to unevolved stars. This is illustrated in Table 10.3, which gives the derived abundances in the shell of Nova Her 1934. Evidently the material in the shell has been strongly processed by prior nuclear reactions in the star, and has been rather thoroughly mixed before being returned to interstellar space. Though the mass in a single nova shell is small, the number generated per year in the Galaxy (about 25) is large enough that they are probably significant contributors to the present heavy-element abundances (especially of N) in interstellar matter.

TABLE 10.3
Relative abundances in Nova Her 1934 shell

Element	N/N_H	$(N/N_H)/(N/N_H)_\odot$
He	2.5×10^{-1}	3.7
C	1.5×10^{-2}	35
N	3.0×10^{-2}	350
O	4.0×10^{-2}	60
Ne	6.0×10^{-4}	6

The large abundances of C, N, O, and Ne listed in Table 10.3 immediately explain the unusually low temperature, $T \approx 500°$ K, derived from the observed Balmer continuum and from the near absence of forbidden lines in the observed spectrum. Even at low temperatures, the rate of collisionally excited far-infrared line radiation is very large, and the resulting equilibrium temperature is therefore quite small. This can be seen qualitatively by imagining all the radiative cooling rates in Figure 3.2 increased by a factor of roughly 10^2. If the effective heating rate is at all similar to that in stars, it is clear that the equilibrium temperature must be low.

Since nova shells are expanding rapidly, we might imagine that their ionization is "frozen-in," that is, that recombination time is long compared with the expansion time, and that no further photoionizations are taking place. This is not correct however, because, for instance, in Nova Her 1934 He II $\lambda4686$ has been observed for the entire observed history of the shell. This observation requires the presence of He^{++}. Its recombination coefficient at $T = 500°$ K is $\alpha_B = 1.21 \times 10^{-11}$ cm^{-3} sec^{-1}, corresponding to a recombination time of 26 yr at $N_e = 10^2$ cm^{-3}, while the expansion time scale is 50 yr. Thus significant continued ionization has occurred. (Because of the small amount of material and the large internal velocity spread in nova shells, Case A is often a better approximation than Case B, even for H I, He I, and He II. However, we use Case B values throughout this chapter, for consistency with the rest of the book.) An even stronger indication that the ionization is not frozen-in is provided by the cooling time, obtained from the ratio of the present thermal energy to the total rate of radiation in the observed lines. It is only about 3 yr, indicating that continued heating, which can occur only by photoionization, is going on.

Thus it is apparent that a photoionization model of a nova shell can be calculated, along lines similar to those of the model H II regions and planetary nebulae discussed in previous chapters. In a nova, the source of the ionizing radiation is not the white-dwarf star, but the accretion disk around it. The theory of such disks is not advanced enough for us calculate their ultraviolet continuous spectra, as can be done for O stars and planetary-nebula central stars. However, the near-ultraviolet continuum of Nova Her 1934 has been observed with the IUE satellite, and it has approximately a power-law form, $L_\nu \propto \nu^{-2}$. Extrapolating this power law into the unobserved ionizing ultraviolet spectral region, $h\nu > 13.6$ eV, with a turnover to a steeper dependence above about 50 eV, gives a model that fits the observed ionization distribution, line spectrum, and Balmer continuum satisfactorily. This is the model listed in the last column of Table 10.1. It is based on the abundances listed in Table 10.3, and an assumed density $N_e = 10^2$ cm^{-3}. The resulting calculated mean temperature in the shell is $T = 613°$ K. Though this model does not agree in every detail with the observed spectrum of Nova Her 1934, it reproduces qualitatively all its main features, which are quite different from those of the spectra of ordinary H II regions and planetary nebulae.

All of the high-energy photon processes, to be described in some detail in Chapter 11, were taken into account in calculating this Nova Her 1934 shell model. Most of these are not especially important in this nova shell, but it is interesting to note that photoionization of O^0 by photons with $h\nu > 16.9$ eV, leaving O^+ in the excited $2p^3\ ^2D$ term, is the main source of excitation of [O II] $\lambda3727$ according to this model, rather than collisional excitation as in typical H II regions and planetary nebulae. The difference is due to the relatively high flux of photons at high energies in the assumed spectrum of the accretion disk $L_\nu \propto \nu^{-2}$, compared with stellar atmospheres, which always have an approximately exponential cutoff ($\exp[-h\nu/kT]$) at high energies. The largest discrepancy between the observed spectrum of Nova Her 1934 and the calculated model of Table 10.1 is in the strengths of [N II] $\lambda\lambda 6548$, 6583. They are predicted to be considerably weaker than observed. According to the model, they arise from recombination to all the singlet levels, as described above. Perhaps additional excitations occur by some process (possibly analogous to the situation in [O II]) whose cross section has not yet been correctly calculated.

According to the adopted model, the main cooling in nova shells occurs by collisional excitation and line radiation from the fine-structure levels of N^+, N^{++}, and O^{++}. The predicted strengths of the far infrared lines [O III] $\lambda\lambda 52\ \mu$, $88\ \mu$, [N II] $\lambda\lambda 122\ \mu$, $204\ \mu$, and [N III] $\lambda 57\ \mu$ are all relatively large. The nova shells, however, are quite faint, and to date it has been impossible to observe these far-infrared emission lines. At some time in the future it should be possible to either measure them or, if not, to set firm upper limits to their strengths, thus either confirming or disproving this currently highly attractive physical picture of nova shells.

10.3 The Crab Nebula

NGC 1952, the Crab Nebula (Figure 10.2) is the remnant of a supernova that was seen as a bright star in the daytime sky by the Chinese in A.D. 1054. It is in the same position in the sky as the "guest star" to within the accuracy of the Chinese records. The spectrum of NGC 1952 shows nebular emission lines together with a relatively strong continuum that extends far into the blue and violet spectral regions. Although NGC 1952 was generally included in planetary nebula catalogues until the 1940s, it was long recognized as peculiar, its true nature a puzzle for many years.

Ultimately, however, photographic spectra to faint light levels showed that the emission lines are concentrated in the filaments seen in Figure 10.2*a*, while the continuum arises in the amorphous gas that fills the nebula, shown in Figure 10.2*b*. Direct CCD images taken in the very finest seeing show that the filaments consist of, or contain, strings of small, dense bright knots, arranged like beads on a string, the filament. The line spectra show a velocity pattern, with a maximum of Doppler splitting, at the center of the nebula, falling to zero at the edges, that indicates expansion, geometrically very similar to the situation in planetary nebulae. However, in the Crab Nebula the radial velocity splitting observed at the center is approximately 2,900 km sec^{-1}, indicating an expansion velocity of 1,450 km sec^{-1}, much larger than in planetaries. The expansion velocity is so large that it has proven possible to measure the proper motions of the filaments as well. They indicate the same pattern of expansion about the center, with maximum values approximately $\pm\ 0\overset{''}{.}222\ \text{yr}^{-1}$ at the ends of the major axis. Comparing the angular and linear expansion rates gives the distance to the nebula; since it appears elliptical in the sky an assumption is involved about the true three-dimensional form of the nebula.

A

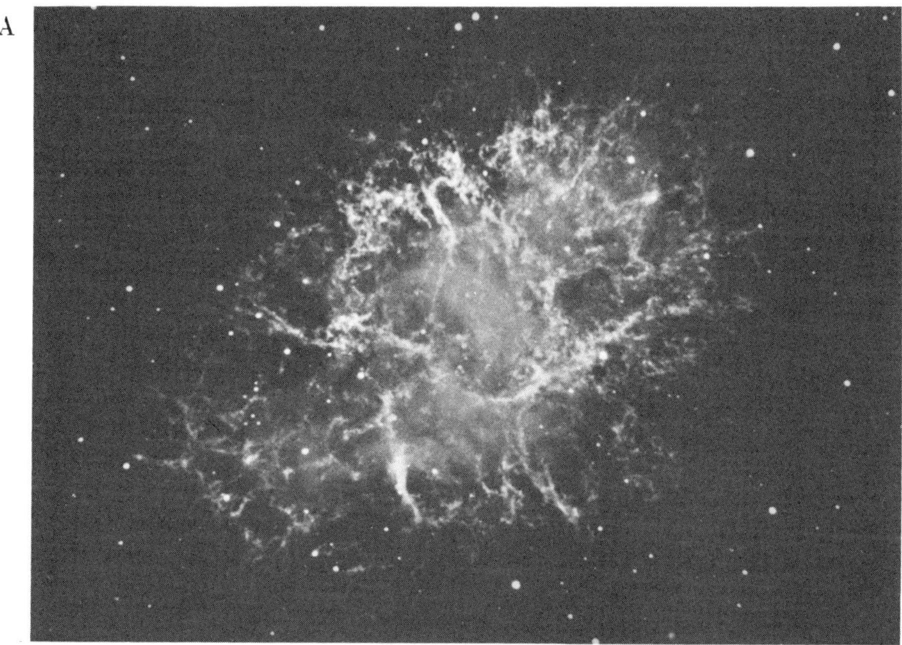

B

FIGURE 10.2
NGC 1952 (Crab Nebula). Both photos were taken with the Shane 3-m reflector. The upper photo (*a*) was taken with a red filter-plate combination transmitting chiefly Hα, [N II]. The lower photo (*b*) was taken with a yellow filter-plate combination transmitting chiefly continuum radiation. (*Lick Observatory photograph.*)

The range of assumptions from oblate to prolate spheroid corresponds to distances from 1,500 pc to 2,200 pc; 1,850 pc is a good representative value to adopt.

Supernovae are distinguished from novae by the fact that they reach much higher luminosities, of order $M_{bol} \approx -18$ or $L \approx 10^9 L_\odot$. Type II supernovae are understood as end stages in the evolution of massive stars ($M \gtrsim 8 M_\odot$), in which after H burning, He burning, and further thermonuclear burning stages that form heavy nuclei, the central core collapses to a neutron star or black hole, and a shell is expelled with high velocity. Type I supernovae, on the other hand, result from the thermonuclear destruction of white dwarf stars. They appear to be accreting white dwarfs, in binary systems, that grow to the critical mass, and ignite a C (or possibly He) detonation. The entire star is disrupted, and no neutron star or black hole is left.

There is a pulsar very near the center of the Crab Nebula; clearly it is the neutron-star remnant of the original pre-supernova star. The Crab Nebula is also a very strong radio source; it is so bright in the radio-frequency spectral region that it was one of the first sources to be discovered, and then later to be identified. The emission mechanism that produces its radio-frequency continuum is well-understood as synchrotron emission, resulting from relativistic electrons spiraling in a magnetic field.

The observed emission-line spectrum of the Crab Nebula contains typical nebular lines of H I, He I, He II, [O II], [O III], [N II], etc., but with a wide range of ionization, including relatively strong [O I] and [S II], and [Ne V]. There are variations in relative intensities of the lines from point to point. In the sky NGC 1952 is in the constellation Taurus, close to the plane of the Milky Way, and the interstellar extinction at its distance is significant. From the [S II] $I(^2D - ^2P)/I(^4S - ^2P)$ ratio method described in Section 7.2, as well as from the strength of the λ2175 extinction "bump," the reddening can be estimated as $E_{B-V} = 0.47 \pm 0.04$, corresponding to extinction $A_V = 1.46 \pm 0.12$. An average (over the summed) emission-line spectrum of the whole filamentary system is listed in Table 10.4. It is based on the work of many observers, and extends from the satellite ultraviolet, through the ground-based optical and infrared spectral regions. The [O II]$\lambda\lambda$3726, 3729 and [S II] $\lambda\lambda$6716, 6731 line ratios can be measured in many of the filaments; they indicate typical electron densities $N_e \approx 10^3$ cm^{-3}. The [O III] (λ4959 + λ5007)/λ4363 ratio gives a mean temperature $T = 15{,}000°$ K, while the corresponding ratio for [N II] indicates $T = 7{,}400°$ K. Temperatures and densities of the same order also match the other observed [O II] and [S II] line ratios well. The derived abundances are $(N_{\mathrm{He}^+} + N_{\mathrm{He}^{++}})/N_{\mathrm{H}^+} = 0.47$, $(N_{\mathrm{O}^+} + N_{\mathrm{O}^{++}})/N_{\mathrm{H}^+} = 10^{-3.5}$, and $N_{\mathrm{N}^+}/N_{\mathrm{H}^+} = 10^{-4.0}$. Thus it is clear that the material in the filaments is He-rich, no doubt as a result of nuclear processing, but that the abundances of N and O, with respect to H are approximately normal, but are *low* if expressed by mass fraction with respect to H + He. These are average abundances over the entire nebula; a detailed study based on spectra taken over a fine grid of points shows that the He abundance is higher in the filaments in the inner part of the nebula, but more nearly normal in the filaments in the outer part.

TABLE 10.4
Relative emission-line intensities in NGC 1952

Ion	Transition	Wavelength (Å)	Relative intensity[a]
C IV	$2\ ^2S - 2\ ^2P$	1549	140:
He II	$2 - 3$	1640	120:
C III]	$2\ ^1S - 2\ ^3P$	1909	150:
Mg II	$3\ ^2S - 3\ ^2P$	2798	$<$ 30:
[Ne V]	$2p^2\ ^3P_1 - 2p^2\ ^1D_2$	3346	5:
[Ne V]	$2p^2\ ^3P_2 - 2p^2\ ^1D_2$	3426	15:
[O II]	$2p^3\ ^4S - 2p^3\ ^2D$	3727	330
[Ne III]	$2p^4\ ^3P_2 - 2p^4\ ^1D_2$	3869	50
[Ne III]	$2p^4\ ^3P_1 - 2p^4\ ^1D_2$	3967	15
[S II]	$2p^3\ ^4S - 2p^3\ ^2P$	4072	10
[Fe V]	$3d^4\ ^5D_4 - 3d^4\ ^3H_4$	4227	5:
[O III]	$2p^2\ ^1D_2 - 2p^2\ ^1S_0$	4363	6:
He I	$2\ ^3P - 4\ ^3D$	4471	6:
[Fe III]	$3d^6\ ^5D_4 - 3d^6\ ^3F_4$	4658	4:
He II	$3 - 4$	4686	17
Hβ	$2 - 4$	4861	32
[O III]	$2p^2\ ^3P_1 - 2p^2\ ^1D_2$	4959	90
[O III]	$2p^3\ ^3P_2 - 2p^2\ ^1D_2$	5007	270
He I	$2\ ^3P - 3\ ^3D$	5876	15
[O I]	$2p^4\ ^3P_2 - 2p^4\ ^1D_2$	6300	24
[O I]	$2p^4\ ^3P_1 - 2p^4\ ^1D_2$	6364	8
[N II]	$2p^2\ ^3P_1 - 2p^2\ ^1D_2$	6548	50
Hα	$2 - 3$	6563	105
[N II]	$2p^2\ ^3P_2 - 2p^2\ ^1D_2$	6583	150
[S II]	$2p^3\ ^4S - 2p^3\ ^2D$	6725	160
[Ar III]	$3p^4\ ^3P_2 - 3p^4\ ^1D_2$	7136	12:
[O II]	$2p^3\ ^2D - 2p^3\ ^2P$	7325	14:
[Ni II]	$a\ ^2D_{5/2} - a\ ^2F_{7/2}$	7378	14:
[Fe II]	$a\ ^4F_{9/2} - a\ ^4P_{5/2}$	8617	3:
[S III]	$3p^2\ ^3P_1 - 3p^2\ ^1D_2$	9069	30:
[S III]	$3p^2\ ^3P_2 - 3p^2\ ^1D_2$	9531	60:
[C I]	$3p^2\ ^3P_1 - 3p^2\ ^2D_2$	9824	8:
[C I]	$3p^2\ ^3P_2 - 3p^2\ ^1D_2$	9850	16:

[a]Corrected for interstellar extinction.

The derived temperatures together with the abundances indicate clearly that photoionization is the main energy-input mechanism to the ionized gas in the filaments. The pulsar is much too faint to be the source of ionizing photons. However, the blue continuum of the amorphous region of NGC 1952 itself, shown in Figure 10.2*b*, is quite bright through the optical region into the near ultraviolet. Correcting it for the measured interstellar extinction, this continuum can be extrapolated to fit, very approximately, with the X-ray flux from the Crab Nebula, which has been measured down to energies of about 1 keV. The continuum between 10 and 10^3 eV found in this way can be crudely fitted by a power law

$$L_\nu = C\nu^{-n}, \tag{10.1}$$

with $n \approx 1.2$. Measurements of the optical continuum shows that it is fairly strongly polarized, and that the direction of polarization varies with position in a way that depends strongly upon the structure of the amorphous nebula. Its luminosity and polarization fit smoothly to those measured in the radio-frequency region. The optical continuum of the amorphous region of NGC 1952 is thus due to synchrotron emission by relativistic electrons spiraling in the magnetic field of the nebula, just as the radio continuum is, and as the ultraviolet and X-ray continua are.

Precise radio measurements of the pulsar's period show that it is slowing down; that is, the neutron star's period of rotation is increasing. The calculated rate of rotational energy loss is about four times the energy radiated in the observed synchrotron spectrum, integrated over all frequencies. The interpretation is that the magnetic field of the rotating neutron star delivers energy to the gas near it, part of which goes into expanding the nebula, and the rest into accelerating electrons to the relativistic energies that produce the synchrotron radiation. From the nebular point of view, then, the filaments of the Crab Nebula may be regarded as high-density regions photoionized by a diffuse source of continuum radiation, rather than by a central star as in an H II region or a planetary nebula. As the direct photographs show, the synchrotron source is extended; the detailed kinematic study shows that the filaments are in a thick shell that surrounds this extended source. The earliest models were calculated with an assumed input power-law spectrum of the form given by equation (10.1). A better fit to the observed line spectra of the filaments can be obtained with more-complicated input radiation fields. A recent very good model calculation uses an assumed composite input radiation field

$$L_\nu = \begin{cases} L_1 \left(\dfrac{\nu}{\nu_0}\right)^{-0.5} e^{-\nu/20\,\nu_0} & \text{for } \nu \leq 40\,\nu_0, \\[2ex] L_1 \left(\dfrac{\nu}{\nu_0}\right)^{-1.1} & \text{for } \nu > 40\,\nu_0, \end{cases} \tag{10.2}$$

in terms of $h\nu_0 = 13.60$ eV, the ionization potential of H^0. Its limiting values have the form of equation (10.1) with $n = 0.5$ for $\nu \ll 20\ \nu_0$, and $n = 1.1$ for $\nu > 40\ \nu_0$ respectively, which match the observed optical-near ultraviolet and X-ray measured spectra much better than any single power law.

As was mentioned above, detailed spectroscopic measurements show apparent differences in the He abundance (relative to H) in different filaments. Model calculations were therefore made for two different well-observed filaments, one with relatively high He abundance, the other relatively low. Comparison between the observed (corrected for interstellar extinction) and calculated line spectra are given in Table 10.5. The abundances assumed in these models are listed in Table 10.6. The agreement between the calculated and

TABLE 10.5
Observed and calculated relative line intensities in Crab Nebula filaments

Ion	Line	He-poor filament		He-rich filament	
		Observed[a]	Model	Observed[a]	Model
[O II]	3727	922	291	1490	828
[Ne III]	3869	104	103	141	193
[S II]	4072	51	18	34	31
[O III]	4363	<26	13	30	21
He I	4471	6	6	17	15
He II	4686	30	33	29	41
Hβ	4861	100	100	100	100
[O III]	4959	184	200	312	313
[O III]	5007	640	586	982	918
[N I]	5199	<11	50	<9	49
He I	5876	17	16	44	41
[O I]	6300	75	95	157	183
[S III]	6312	–	8	<13	10
[O I]	6364	28	30	51	58
[N II]	6548	233	73	94	93
Hα	6563	249	363	347	380
[N II]	6583	695	211	286	269
[S II]	6716	218	167	253	251
[S II]	6731	210	157	305	342

[a]Corrected for interstellar extinction.

TABLE 10.6
Assumed relative abundances in Crab Nebula photoionization models

Element	log N/N_H	
	He-poor filament	He-rich filament
H	0.00	0.00
He	− 0.82	0.00
C	− 3.18	− 3.18
N	− 4.34	− 4.34
O	− 3.78	− 3.48
Ne	− 4.38	− 4.08
S	− 4.80	− 4.80

the observed spectra is qualitatively, and in some cases quantitatively, satisfactory. In particular these models reproduce the observed great strengths of [O I] λ6300, 6364 and [S II] $\lambda\lambda$6716, 6731, not observed in any planetary nebula with [O III] $\lambda\lambda$4959, 5007 as much stronger than Hβ as in NGC 1952. This is a result of the relatively large fraction of high-energy photons (with $h\nu \gg h\nu_0$) in the photoionizing spectrum (10.2) or (10.1), while in any stellar spectrum, there is always an $\exp(-h\nu/kT)$ cutoff at high energies. The high-energy photons have a small cross section for absorption (from equation [2.4] or its approximate ν^{-3} dependence), and thus create a "transition region" – in which the fraction of neutral H, $\xi = N_{H^0}/(N_{H^0} + N_p)$, increases from nearly 0 to nearly 1 – that is significantly larger than for the H II regions of Figure 2.3. It is in this transition region that H^0 and therefore O^0, H^+ and therefore e^-, and S^+ can all coexist, and hence in which collisionally excited [O I] and [S II] lines are emitted.

The models of Table 10.5 show the calculated Hα/Hβ ratio to be significantly larger than the recombination value 2.85. This results from collisional excitation of H^0 from its ground level in this same transition zone. This process is analyzed in more detail in Chapter 11, as are the effects of Auger absorptions of high-energy photons by heavy ions, which are also significant in these Crab Nebula filament models.

Table 10.6 shows that He is significantly enriched in some filaments of the Crab Nebula, but that the heavy elements are not. If anything, they may be less abundant by mass relative to H plus He. The total mass of the gas in the its luminosity and mean density. The result is approximately 1.5 $M_{\odot}$; it may be larger, however, if there are significant amounts of nearly neutral gas in the "middle" of the filaments, or in the knots that form them, protected from the surrounding ionizing radiation.

Within the past few years, the object N 157 B = SNR 0540−693, near the 30 Doradus nebula in the Large Magellanic Cloud, has been identified as an object very similar to the Crab Nebula. It has similar radio, optical, and X-ray spectra, and no doubt is the remnant of a fairly recent supernova, photoionized by its own high-energy synchrotron radiation, just as is the case in NGC 1952.

10.4 The Cygnus Loop

The Cygnus Loop is a large (3° in diameter) emission nebula with pronounced filamentary structure (Fig. 10.3). Its brightest regions are catalogued as NGC 6990, 6992 and 6995. Proper motions of the filaments show that the Loop is expanding with a proper motion of approximately $0\rlap{.}''03$ yr^{-1} (at the edge with respect to the center). Radial velocities confirm the expansion, and fix its amount as approximately 115 km sec^{-1} (also at the edge with respect to the center). Comparison of these two measures of the expansion lead to a distance of approximately 770 pc, and diameter 40 pc. The inside is "hollow" (nearly devoid of optically observable gas), and to a first approximation the Cygnus Loop may be regarded as composed of filaments that are located in a spherical shell about 20 pc in outer radius and 10 pc thick.

No central star that might be responsible for photoionizing the nebula has ever been detected. The present velocity and size indicate an approximate age of 2.5×10^4 yr since the Cygnus Loop was a point source. The optical line spectra of the filaments are unusual in showing a wide range of ionization. There are great variations in the spectrum from one filament to another, but two fairly characteristic ones are listed in Table 10.7. They represent a high-ionization and a low-ionization filament, but note that [O II] λ3727 is relatively strong relative to [O III] $\lambda\lambda$4959, 5007 even in the high-ionization filament, and that [O I] $\lambda\lambda$6300, 6364 and [S II] $\lambda\lambda$6716, 6731 are stronger in both of them than in typical planetary nebulae. (The observed relative line intensities have been corrected for interstellar extinction $E_{B-V} = 0.08$, a value derived from measured color excesses of stars out to the distance 770 pc.)

A striking feature of the spectra of all the filaments is their relatively strong [O III] λ4363, or correspondingly, the small values of the [O III] (λ4959 + λ5007)/λ4363 intensity ratios. The electron densities indicated by [O II]

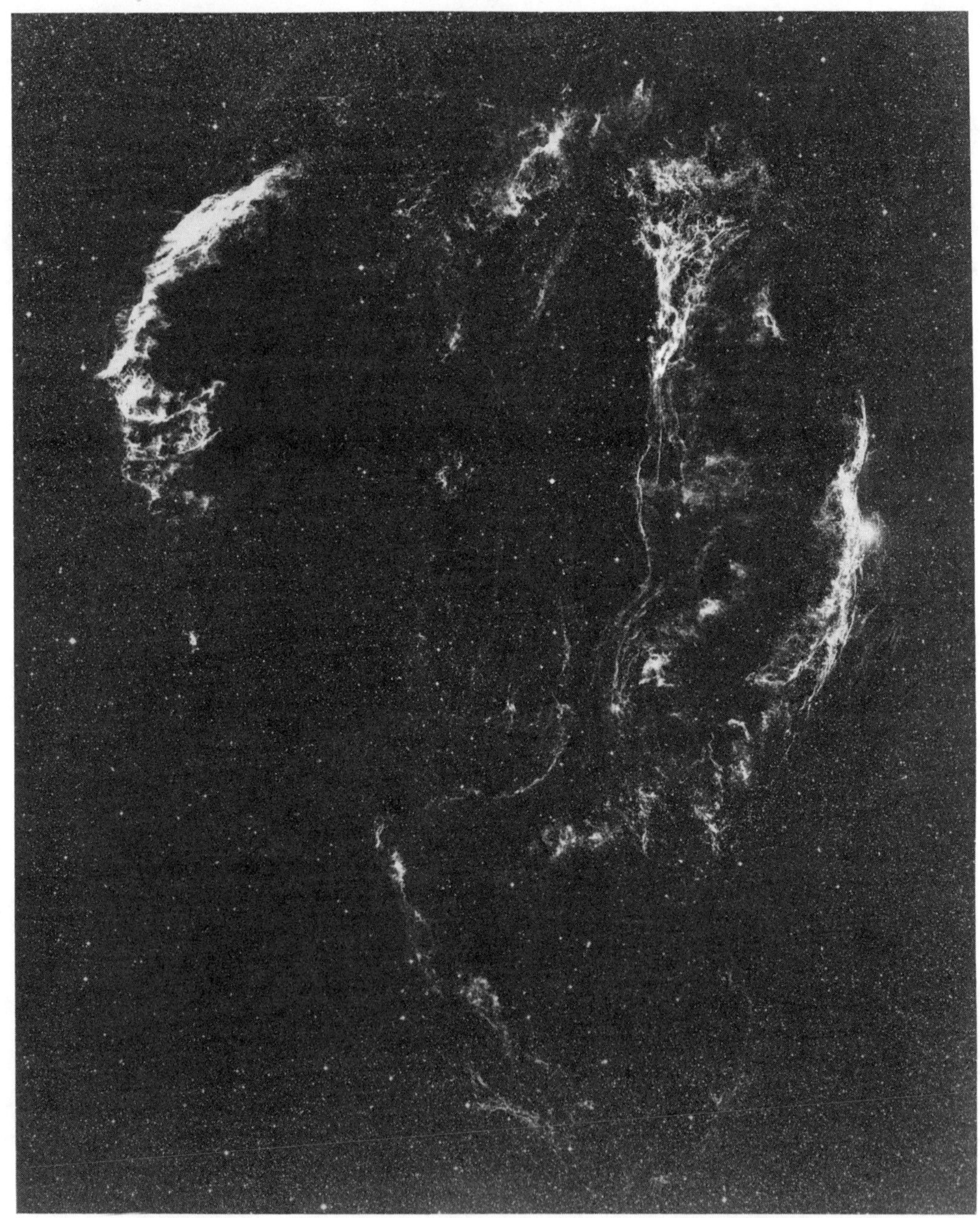

FIGURE 10.3
NGC 6960-6992-6995 (Cygnus Loop). Taken with Palomar 48-inch Schmidt telescope in red light, chiefly Hα, [N II], [S II]. (*Palomar Observatory photograph*).

TABLE 10.7
Observed relative line intensities in Cygnus Loop[a]

Ion	Wavelength	High-ionization filament	Low-ionization filament
[O II]	3727	2200	919
[Ne III]	3869	174	61
[S II]	4072	<37	–
Hγ	4340	30	49
[O III]	4363	106	14
[Fe III]	4658	18	–
He II	4686	10	–
Hβ	4861	100	100
[O III]	4959	604	73
[O III]	5007	1812	221
[N I]	5199	<7	10
[N II]	5755	10	6
He I	5876	6	10
[O I]	6300	34	56
[O I]	6364	9	19
[N II]	6548	97	78
Hα	6563	361	288
[N II]	6583	267	226
[S II]	6716	246	164
[S II]	6731	168	111
[Ar III]	7136	8	9
[O II]	7325	38	15

[a] Corrected for interstellar extinction.

$\lambda 3729/\lambda 3726$ and [S II] $\lambda 6716/\lambda 6731$ are low, typically $N_e \lesssim 300$ cm^{-3}. The only possible interpretation of the [O III] line ratio therefore is that the temperature in the O^{++} zone is relatively high. The derived temperatures from [O III] $(\lambda 4959 + \lambda 5007)/\lambda 4363$ are $T = 29,000°$ K and 31,000° K in the high- and low-ionization filaments of Table 10.7, respectively. These temperatures are too high to be understood as resulting from heating by photoionization. The only other plausible mechanism of energy input to the ionized gas is shock-wave heating, that is, the conversion of kinetic energy to heat. The observed expansion velocity of the Cygnus Loop agrees with this physical picture. The energies involved are so large that it must be the remnant of a supernova.

Here we can give only the very simplest sketch of the evolution of a supernova remnant. Let us imagine a supernova going off in an initially homogeneous, low-density interstellar medium with number density $N_0 = N_H + N_{He}$. We can idealize the explosion as a large amount of energy released instantaneously at a point, in the form of kinetic energy plus radiation. This causes a strong shock wave which expands radially, compressing and heating the medium, and setting it into outward motion. The hydrodynamic equations and the jump conditions across the shock front of Chapter 6 apply, except no ionizing radiation is involved.

The shock wave runs out into undisturbed gas ahead of it as a strong blast wave, expanding adiabatically. The Sedov-Taylor similarity solution, developed originally for the physically very similar situation of an "atomic" (fission) bomb releasing instantaneously a large amount of energy in the atmosphere, has the radius R_s of the shock front, the shock velocity v_s, and the temperature T_s and density $N_s = N_H + N_{He}$ just behind the shock front given by the equations

$$R_s = 12.8\left(\frac{t}{10^4}\right)^{2/5}\left(\frac{E_{51}}{N_0}\right)^{1/5} \text{ pc}, \tag{10.3}$$

$$v_s = 500\left(\frac{t}{10^4}\right)^{-3/5}\left(\frac{E_{51}}{N_0}\right)^{1/5} \text{ km sec}^{-1}, \tag{10.4}$$

$$T_s = 3.4 \times 10^6\left(\frac{t}{10^4}\right)^{-6/5}\left(\frac{E_{51}}{N_0}\right)^{2/5} \text{ °K} \tag{10.5}$$

$$N_s = 4\, N_0\ , \tag{10.6}$$

where t is the time since the outburst in years, and E_{51} is the energy released by the supernova outburst, in units of 10^{51} erg. As the gas passes through the front it is compressed, heated, and ionized; generally it is a reasonably good first approximation to suppose that collisions are frequent enough in the front so that the newly shocked gas rapidly approaches its equilibrium thermal collisional ionization. This is similar to the situation in the solar corona, and the equilibrium is sometimes called "coronal." There are no ionizing photons, and collisional ionization is balanced by recaptures:

$$N(X^{+i})N_e q_{\text{ion}}(X^{+i},T) = N(X^{+i+1})N_e \alpha_G(X^{+i},T). \tag{10.7}$$

Here $q_{\text{ion}}(X^{+i},T)$ is the collisional-rate ionization coefficient

$$q_{\text{ion}}(X^{+i},T) = \int_{\frac{1}{2}mv^2=\chi}^{\infty} v\sigma_{ion}(X^{+i},v)f(v)\, dv, \tag{10.8}$$

with $\sigma_{ion}(X^{+i}, v)$ the cross section for collisional ionization of ion X^{+i} by electrons of velocity v, and χ is its ionization potential. The resulting degree of ionization is independent of electron density, from equation (10.7), and depends only on temperature. The collisional-ionization cross sections can be calculated approximately, and from them the ionization coefficients and the equilibrium ionization as a function of post-shock temperature. As an example, the degree of collisional ionization of O is plotted in Figure 10.4. It can be seen that the maximum ionization to the stage O^+ occurs at $T \approx$ 30,000° K, and to O^{++}, at $T \approx 90,000°$ K.

Behind the front the shocked gas cools by radiation, and the temperature, density, and velocity fields may be followed as functions of time by methods similar in principle to those described for expanding H^+ regions in Section 6.3. However, because of the low density, and the fact that the recombination cross sections are small compared with collisional-excitation cross sections, the degree of ionization does not remain in equilibrium with the temperature, but lags behind it. The time evolution of each stage of ionization can be followed by numerical integration of the equations

$$\frac{D}{Dt}[N(X^{+i})] = \frac{\partial}{\partial t}[N(X^{+i})] + \mathbf{v} \cdot \nabla N(X^{+i}) \tag{10.9}$$

where

$$\begin{aligned}\frac{\partial}{\partial t}[N(X^{+i})] &= N_e N(X^{+i+1})\alpha_G(X^{+i}, T) - N_e N(X^{+i})\alpha_G(X^{+i-1}, T) \\ &- N_e N(X^{+i})q_{\rm ion}(X^{+i}, T) + N_e N(X^{+i-1})q_{\rm ion}(X^{+i-1}, T).\end{aligned} \tag{10.10}$$

Generally the last two (collisional ionization) terms are small compared with the first two (recombination) terms, because the degree of ionization is higher than the coronal equilibrium value at the present temperature.

Although this is a good first approximation, the actual situation is more complex. Integrating equations (10.9) and (10.10) from the front shows that the peak ionization occurs not at the front, but close behind it. In addition, in the hot region ($T \approx 10^5$ °K) just behind the front, collisional excitation of heavy ions produce ionizing photons with $h\nu > 13.6$ eV which preionize the gas ahead of the front and, more importantly, maintain the ionization and temperature further behind it. In addition, charge exchange couples the ionization of different ions behind the front, as explained in Section 2.8. All these effects are taken into account in the actual computed models.

Models calculated on this basis agree reasonably well with the observed Cygnus Loop spectra. In particular, the predicted effective temperature in the [O III] emitting region is generally in the $T \approx 30,000°$ K region as observed, while in the [N II] region it is more like $T \approx 10,000°$ K, also as observed. From

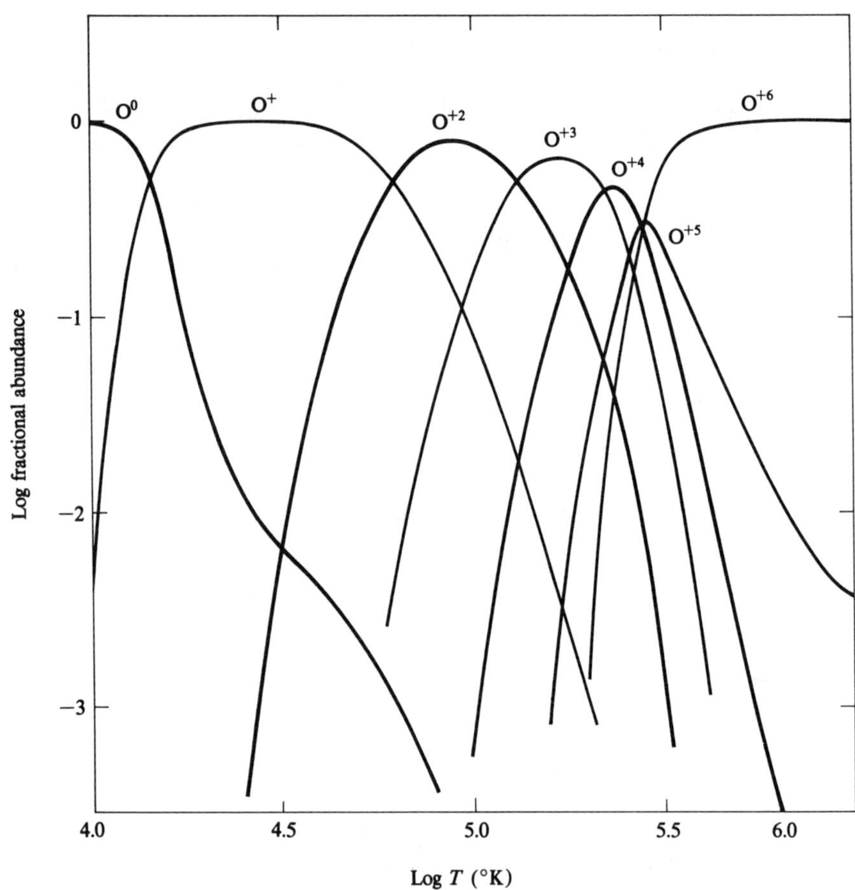

FIGURE 10.4

Calculated fractional ionization of various ions of O, as functions of temperature, for thermal collisional ionization.

these fits a shock velocity $v_s \approx 100$ km sec^{-1}, an average ambient density $N_0 \approx 7$ cm^{-3}, an initial energy $E_0 = 0.4\ E_{51} \approx 4 \times 10^{50}$ erg, and an age $t \approx 4 \times 10^4$ yr are derived for the Cygnus Loop. There are almost certainly slower shocks in isolated regions of higher ambient density, while regions of lower density have higher shock velocities. At the lowest densities v_s and T_s are so high that the shocked gas radiates primarily X-rays, which are observed. The X-ray continuum and line measurements indicate temperatures in the range $T \approx 1 - 3 \times 10^6$ °K. Correspondingly, the optical coronal emission lines [Fe X] λ6375 and [Fe XIV] λ5303 have been detected weakly as diffuse features outside the bright optical filaments. Clearly they arise in lower-density gas (smaller N_0) where the shock travels faster and reaches higher temperatures, according to equations (10.3) to (10.5).

A better physical model for the Cygnus Loop than a supernova outburst in a homogeneous medium is an outburst in a low-density medium, with higher-density condensations ("clouds") in it. The best value for the ambient intercloud density is 0.2 cm^{-3}; in it the shock travels fastest, and the bulk of the X-ray emission arises. In the clouds the ambient densities are higher ($\sim$ 10 cm^{-3}), the shock moves more slowly, and the gas cools more quickly and emits optical line radiation more strongly. The apparent "filaments" are actually regions in which the relatively thin shock front is seen approximately edge-on.

The total mass of interstellar gas swept past by the shock and now included within the Cygnus Loop is of order 10^2 or $10^3\ M_\odot$, no doubt considerably larger than the mass of the supernova shell, which presumably remains somewhere near its center. Correspondingly, the abundances observed in the filaments appear approximately normal.

10.5 Younger Shock-Wave Heated Supernova Remnants

Several other supernova remnants have been observed, including those of the supernovae observed at maximum light by Tycho Brahe in 1572, and by Johannes Kepler in 1604. Cas A is apparently the youngest supernova remnant observed optically as a nebula to date. It was originally discovered as a bright radio source; all known supernova remnants from it, through Tycho's, and Kepler's and the Crab Nebula, to the Cygnus Loop, are radio sources. Their continuous spectra are nonthermal; they are power laws of the form of equation (10.1), and this together with their polarization show that their observed radio-frequency emission occurs by the synchrotron process. Thus relativistic electrons and magnetic fields are present in all these remnants.

Optically Cas A appears as a cluster of "bits," "streaks," or "knots" of emission, covering an area about 4′ in diameter. Some of these "fast-moving knots" now show large proper motions (up to 0.″4 yr^{-1}), and/or Doppler shifts (up to 5,600 km s^{-1}). The radial velocities are highest at the center of the object, the proper motions at the edge, showing a typical expansion pattern. The indicated space velocity of expansion is about 6,000 km s^{-1}. The indicated distance of Cas A, from comparison of the radial velocities and proper motions of expansion, is about 2.8 kpc. It is near the Galactic plane and the interstellar reddening is strong; the extinction indicated by the [S II] lines is $A_V \approx 4.3$.

The spectra of these fast-moving knots are unique in that they do not show any emission lines of H I, He I or He II. The strongest emission lines observed in them are [O III] $\lambda\lambda$4959, 5007, with [O I], [O II], [S II], and [Ar III] also visible. Evidently the fast-moving knots are fragments of processed material from deep within the supernova, ejected with high velocities during the outburst. Approximate abundance analyses show that the material in the fast-moving knots is chiefly oxygen, with smaller amounts of other heavy elements, perhaps S/O $\approx$ 0.1, Ne/O $\approx$ 0.003, and Ar/O $\approx$ 0.01 (by mass) on the average. The upper limits to the relative strengths of the H I, He I and He II lines that are not detected correspond to upper limits to the relative abundances H/O $\leq$ 0.02 and He/O $\leq$ 0.4 (again by mass). The abundances are not the same in all the fast-moving knots; some show only [O I], [O II], and [O III] in their spectra, and evidently consist of nearly pure O. From the observed level of ionization and the deduced temperatures (up to $\sim 10^5$ °K), the indicated shock velocities are of order 150 km s^{-1}. Evidently the fast-moving knots strike much lower-density ambient material, and as a result a shock propagates back into the knot with this velocity.

In Cas A, in addition to the fast-moving knots, there are other bits or streaks of nebulosity with much smaller space velocities, of order 150 km s^{-1}. They are usually referred to as "quasi-stationary flocculi." Their spectra, in contrast to those of the fast-moving knots, show H I, He I, and He II emission lines as well as typical forbidden lines. They can be interpreted as resulting from shock heating of material with approximately normal abundances, but with N/H about ten times greater, and with He/H perhaps similarly more abundant, than in typical interstellar matter. Evidently the quasi-stationary flocculi are composed of material that was originally in the outer part of the precursor star. It was partly modified by nuclear reactions before leaving the star in a slow mass-loss process before the outburst, and is now being overtaken, struck, and heated by subsequently ejected material.

The proper motions of the observed fast-moving knots, projected back as constant in time, would date to the Cas A supernova outburst as occurring in the year 1658 $\pm$ 3. There may have been a supernova observed at its position by John Flamsteed in 1680; if so, this would indicate that some deceleration has occurred.

Studies of the composition, mass, and history of remnants such as Cas A, the Cygnus Loop and the Crab Nebula are important in giving direct evidence of the results of the nuclear processes that have gone on in supernovae and their progenitors. The abundances cannot be directly observed during the outburst itself, except in the outermost layers of the star. As the shell expands, deeper and deeper layers are revealed, and the heavy elements are seen, both in broad unresolved clumps of emission lines, and in the time scales set by radioactive decay. Different supernovae have different properties, depending on the mass and prior evolution of the progenitor star. The classification of supernovae by their light curves and spectra near outburst (into Types I and II, or Ia, Ib, IIp, and IIl) is difficult indeed from medieval or Chinese records, and highly theory-dependent from the spectrum of the remnant alone. Thus, though the remnants studied today are extremely interesting as unusual but understandable examples of nebular astrophysics, relating them to detailed supernova models is still far from straightforward.

Supernova 1987a was discovered in the Large Magellanic Cloud at the very time this chapter was being written. It subsequently brightened to apparent magnitude 3, and then started becoming fainter. It is the nearest supernova by far to have been observed in modern times, and undoubtedly the study of its remnant, as it develops, will yield much important new information on these fascinating objects.

10.6 Other Supernova Remnants

Many other supernova remnants have been identified, in addition to the few described above, in our Galaxy. They can readily be found in radio surveys as Galactic objects (concentrated to the plane of the Milky Way) with nonthermal, power-law spectra, indicating synchrotron emission. This radio emission is often strongly concentrated near the edge of the object, as in the Cygnus Loop. Most of the supernova remnants discovered in this way are not optically observable, because of the strong interstellar extinction near the Galactic plane. However, some of the nearest or least-absorbed have been photographed, particularly on deep exposures with narrow-band filters centered on characteristic emission lines. In these photographs they show highly filamentary structures, arranged more or less tangentially, and generally similar to the Cygnus Loop.

These supernova remnants have emission-line spectra. The strongest lines are generally [O III] $\lambda\lambda4959, 5007$ (if the object is not too highly reddened for this spectral region), Hα, [N II] $\lambda\lambda6548, 6583$ and [S II] $\lambda\lambda6716, 6731$. The latter two lines are much stronger relative to Hα than in typical H II regions. This can be understood to result from the difference between shock wave ionization and photoionization by a hot star. In the shock-wave case, gas is nearly instantaneously heated and ionized as the front strikes it, and then cools by radiation. The gas is not in temperature or ionization equilibrium. Recombination lags behind cooling, as explained above. Thus a long, partly ionized, still-heated region is created (and maintained by further photoionization by line photons emitted close to the front), in which H^+ is recombining to

H^0, O^+ to O^0, S^{++} to S^+, and S^+ to S^0. In this region there are appreciable amounts of O^0 and S^+, along with appreciable numbers of electrons from the H^+. Thus strong collisionally excited [O I] and [S II] lines are emitted in this partly ionized zone, in contrast to the situation in H II regions. Within them O^0, with ionization potential 13.62 eV, is completely ionized to O^+ or higher stages, and S^+, with ionization potential 23.33 eV (1.26 eV less than He^0), is mostly ionized to S^{++}; outside their boundaries O is entirely O^0, and S is entirely S^+, but there are almost no free electrons and the temperature is too low for appreciable excitation of [O I] $\lambda\lambda$6300, 6364, and [S II] $\lambda\lambda$6716, 6731.

Another very good diagnostic of shock-wave heating is [O III] (λ4959 + λ5007)/λ4363, which is relatively small in supernova remnants, as explained in the discussion of the Cygnus Loop. The ionization is thermal and the [O III] emission occurs mostly at temperatures in the vicinity of 30,000° K, rather than around 10,000° K as in H II regions. This behavior of the [O III] line ratio has been verified in the Cygnus Loop, and in a few other optically observed supernova remnants in our Galaxy, but most of them, particularly the heavily reddened objects, are too faint for λ4363 to be measured, even though it is presumably strengthened by the predicted amount.

In other galaxies supernova remnants have been identified by the same combination of radio and optical methods. Many candidates have also been located by their strong X-ray emission, known to be a good diagnostic from the Cygnus Loop and other supernova remnants in our Galaxy. The X-ray radiation arises in the very hot gas left behind the shock, particularly in the low-density regions which cool only very slowly. The most complete study of supernova remnants in other galaxies is in the nearby Large and Small Magellanic Clouds. One of these remnants, N 49 in the LMC, shows in its optical emission-line spectrum [Fe XIV] λ5303, the green coronal line. Many of the Magellanic Cloud supernova remnants were optically identified on direct images taken through narrow-band interference filters centered on the mean wavelength of [S II] $\lambda\lambda$6716, 6731. A total of 25 supernova remnants are known in the LMC and six in the SMC. From their sizes and an assumed mean expansion velocity, a mean lifetime can be estimated for the observed remnants, much as was done for observed planetary nebulae in Chapter 9. The rates of supernova occurrences may then be estimated from the present numbers of remnants and their lifetimes. The results are approximately one supernova per 275 yr in the LMC and one per 800 yr in the SMC, in good agreement with the numbers expected from their respective luminosities and the observed rates in more luminous Sc galaxies.

References

A good review of the general properties of novae is

Gallagher, J. S., and Starrfield, S. 1978. *Ann. Rev. Astr. Ap.* **16,** 171.

Papers specifically on nova shells include

Williams, R. E., Woolf, N. J., Hege, E. K., Moore, R. L., and Kopriva, D. A. 1978. *Ap. J.* **224,** 171.

Gallagher, J. S., Hege, E. K., Kopriva, D. A., Williams, R. E., and Butcher, H. R. 1980. *Ap. J.* **237,** 55.

Ferland, G. W., and Truran, J. W. 1981. *Ap. J.* **244,** 1022.

Williams, R. E. 1982. *Ap. J.* **261,** 170.

Ferland, G. J., Williams, R. E., Lambert, D. L., Shields, G. A., Slovak, M., Gondhalekar, P. M., and Truran, J. W. 1984. *Ap. J.* **281,** 194.

The observational data of Table 10.1 are from the first, fourth and fifth of these references, and the model is also from the fifth. The abundances of Table 10.3 are referred to in this paper, but are given in a still unpublished paper by P. G. Martin.

Some of the atomic data needed to analyze these very low-temperature gaseous nebulae are given in

Ferland, G. J. 1980. *P.A.S.P.* **92,** 596.

Nussbaumer, H., and Storey, P. J. 1984. *Astr. Ap.* **56,** 293.

Martin, P. G. 1988. *Ap. J. Suppl.* **66,** 125.

The first gives recombination coefficients for H^+ and He^{++} to very low T. The second gives effective recombination coefficients, including dielectronic effects, for various ions of C, N and O. Table 10.2 is taken from the last.

Two good recent review articles on supernovae and supernova remnants, respectively, are

Woosley, S. E., and Weaver, T. A. 1986. *Ann. Rev. Astr. Ap.* **24,** 205.

Raymond, J. C. 1984. *Ann. Rev. Astr. Ap.* **22,** 75.

Specific papers on the Crab Nebula are

Woltjer, L. 1958. *Bull. Astr. Inst. Netherlands* **14,** 39.

Mayall, N. U. 1962. *Science* **137,** 91.

Miller, J. S. 1978. *Ap. J.* **220,** 490.

Fesen, R. A., and Kirshner, R. P. 1982. *Ap. J.* **258,** 1.

Henry, R. B. C., and MacAlpine, G. M. 1982. *Ap. J.* **258,** 11.

Davidson, K., and Fesen, R. A. 1985. *Ann. Rev. Astr. Ap.* **23,** 119.

Clark, D. H., Murdin, P., Wood, R., Gilmozzi, R., Danziger, J., and Furr, A. W. 1983. *M.N.R.A.S.* **204,** 415.

Kafatos, M. C., and Henry, R. B. C., eds. 1985. *The Crab Nebula and Related Supernova Remnants.* Cambridge: Cambridge University Press.

van den Bergh. S., and Pritchet, C. J. 1986. *Nature* **321,** 46.

The first of these is the earliest discussion of the Crab Nebula in terms of photoionization of the filaments by high-energy synchrotron radiation of the surrounding amorphous nebulosity. The second is an excellent historical summary of the identification of NGC 1952 as the remnant of the Chinese supernova, and the early work on direct images and long-slit spectroscopy of it. The next two papers give spectrophotometric measurements of the filaments; the observational data of Tables 10.5 and 10.6 is from the second of these. The fifth paper in the list contains the photoionization models described in the text and listed in Tables 10.5 and 10.6. Many other photoionization models have been calculated; references to many of them, as well as a good recent review of other recent research on the Crab Nebula, are given in the sixth reference. The observational data of Table 10.4 are taken from it. The seventh paper is an excellent observational study of the three-dimensional structure of the nebula, which shows in particular the difference in He abundance between the inner and the outer filaments. The second to the last reference is a good overall review of recent research, and includes most of the available data on the Crab-Nebula-like object in the Large Magellanic Cloud. It was first discussed in

Danziger, I. J., Goss, W. M., Murden, P., Clark, D. H., and Boksenberg, A. 1981. *M.N.R.A.S.* **195,** 33P.

The last reference on the Crab Nebula describes the structure of the filaments as strings of knots, from direct images taken in very fine seeing.

The Cygnus Loop, because of its proximity and low interstellar extinction, is by far the most-studied shock-heated supernova remnant. Some key references on it are

Parker, R. A. R. 1964. *Ap. J.* **139,** 493.
Parker, R. A. R. 1967. *Ap. J.* **149,** 363.
Cox, D. P. 1972. *Ap. J.* **178,** 143, 159, 169.
Miller, J. S. 1974. *Ap. J.* **189,** 239.
McKee, C. F., and Cowie, L. L. 1975. *Ap. J.* **195,** 715.
Raymond, J. C. 1979. *Ap. J. Suppl.* **39,** 1.
Contini, M., Kozlovsky, R. Z., and Shaviv, G. 1980. *Astr. Ap.* **92,** 1980.
Fesen, R. A., Blair, W. P., and Kirshner, R. P. 1982. *Ap. J.* **262,** 171.
Hester, J. J., Parker, R. A. R., and Dufour, R. J. 1983. *Ap. J.* **273,** 219.
Hester, J. J., and Cox, D. P. 1986. *Ap. J.* **300,** 675.
Hester, J. J. 1987. *Ap. J.* **314,** 187.

The data of Table 10.7 were taken from the paper by Fesen, Blair, and Kirshner.

Collisional ionization rates and equilibrium are discussed in

Shull, J. M., and Van Steenburg, M. 1982. *Ap. J. Suppl.* **48,** 95.

Figure 10.4 is based on this reference.

Young shock-wave-heated supernova remnants, including Tycho, Kepler, and especially Cas A, are discussed in

Baade, W., and Minkowski, R. 1954. *Ap. J.* **119,** 206.

van den Bergh, S. 1971. *Ap. J.* **165,** 457.
Kamper, K., and van den Bergh, S. 1976. *Ap. J. Suppl.* **32,** 351.
van den Bergh, S. 1977. *Ap. J.* **218,** 617.
Kirshner, R. P., and Chevalier, R. A. 1977. *Ap. J.* **218,** 142.
Chevalier, R. A., and Kirshner, R. P. 1978. *Ap. J.* **219,** 931.
Dennefeld, M., and Andrillat, Y. 1981. *Astr. Ap.* **103,** 44.

Two good papers containing many direct photographs of supernova remnants in our Galaxy are

van den Bergh, S., Marscher, A. P., and Terzian, Y. 1973. *Ap. J. Suppl.* **26,** 19.
Fesen, R. A., Gull, T. R., and Ketelson, D. A. 1983. *Ap. J. Suppl.* **51,** 337.

The supernova remnant identifications in the Magellanic Clouds, and the conclusions on their rates of occurrence, are from

Mathewson, D. S., and Clarke, J. N. 1973. *Ap. J.* **180,** 725.
Mathewson, D. A., Ford, V. L., Dopita, M. A., Tuohy, I. R., Long, K. S., and Helfand, D. J. 1983. *Ap. J. Suppl.* **51,** 345.

The identification of [Fe XIV] $\lambda5303$ emission in N 49 in the Large Magellanic Cloud was made in

Danziger, I. J., and Dennefeld, M. 1976. *P.A.S.P.* **88,** 44.

Three excellent papers on shock-wave-model calculations, and the interpretation of supernova-remnant spectra in terms of them, are

Dopita, M. 1977. *Ap. J. Suppl.* **33,** 437.
Daltabuit, E., MacAlpine, G. M., and Cox, D. P. 1978. *Ap. J.* **219,** 372.
Binette, L., Dopita, M. A., and Tuohy, I. R. 1985. *Ap. J.* **297,** 476.

11

Active Galactic Nuclei: Diagnostics and Physics

11.1 Introduction

H II regions, planetary nebulae, novae and supernova shells occur in our Galaxy, in which they were all first identified and studied, and in other galaxies as well. H II regions are the largest and most luminous nebulae and the easiest to observe at large distances, and hence they are the gaseous nebulae that have been most thoroughly studied in other galaxies. As mentioned in Chapter 8, H II regions occur primarily in spiral and irregular galaxies, in which there are large amounts of interstellar gas, and also young O and B stars which can ionize it. The gas and the young stars that form from it are strongly concentrated to the plane of a spiral galaxy, and in our Galaxy, M 31, and most other spirals are found in the outer parts of the system, far from the nucleus. The population of the central regions of these galaxies consists almost entirely of old stars, and little ionized gas is apparent. Some spiral nuclei contain large numbers of OB stars and with them, ionized gas. These are the starburst galaxies, also discussed in Chapter 8.

But in addition to the "extragalactic H II regions" or "H II-region galaxies," there are a few galaxies with ionized gas in their nuclei that is not associated with O and B stars. Examples are Seyfert galaxies, radio galaxies, quasars, and quasistellar objects, collectively called active galactic nuclei. They are rare in space, hence on the average distant from us, and consequently faint. Only in recent decades have they been recognized and studied intensively. Understanding their nature is one of the most interesting and important subjects in astrophysics today.

Observations and measurements at all wavelengths have contributed to what we now know about active galactic nuclei. Every spectral band, from the far infrared through the optical and ultraviolet, and to the X-ray region, has provided information that has helped in understanding these objects. Since the ionized gas within them emits a prominent emission-line spectrum, the methods of nebular astrophysics have been particularly useful in studying them. Much of the framework we have developed for analyzing H II regions, planetary nebulae, and nova and supernova remnants may be carried over to active galactic nuclei. However, extensions are necessary to high-energy photons especially, and also to high densities, large volumes and hence large optical depths, and high internal velocities.

We will apply the methods of nebular astrophysics to analyzing the nature of active galactic nuclei in the last two chapters of this book. This chapter deals with the primary observational data on their optical spectra, the preliminary diagnostic conclusions that can be drawn from them, and the additional physical processes that occur with the high-energy photons that are present in active galactic nuclei, but absent from H II regions and planetary nebulae. The final chapter applies these methods and concepts to analyzing the spectra of active galactic nuclei.

11.2 Historical Sketch

The observational study of active galactic nuclei (or AGNs) began with the work of Edward A. Fath at Lick Observatory in 1908. Using a small, photographic spectrograph on the Crossley (36-inch) reflector, he was studying the spectra of the nuclei of the brightest "spiral nebulae," now known to us to be galaxies. Most of them showed absorption-line spectra, which Fath realized could be understood as resulting from the integrated light from large numbers of stars (similar to star clusters, he called them) too distant and therefore too faint to be seen individually. But in the spectrum of the nucleus of one galaxy, NGC 1068, he recognized six emission lines, one of them Hβ, the other five well-known from their wavelengths to him and to other astronomers of his time as characteristic of gaseous nebulae. Today they are instantly recognizable as [O II] $\lambda3727$, [Ne III] $\lambda3869$, and [O III] $\lambda\lambda4363$, 4959, 5007. V. M. Slipher obtained much better spectra of this same nucleus in 1917, and in 1926 Edwin Hubble, in his monumental study of "extragalactic nebulae," particularly noted the planetary-nebula-type emission-line spectra of three AGNs (as we now know them), NGC 1068, 4051, and 4151. Nearly two decades later Carl K. Seyfert published his important paper, in which he clearly stated that a very small fraction of galaxies, including these three, have nuclei whose spectra show many high-ionization emission lines. These nuclei are invariably especially luminous, he noted, and their emission lines are wider than the absorption lines in normal galaxies. These properties, broad emission lines arising in a bright, small ("semi-stellar" in appearance) nucleus and covering a wide range of ionization, define the class of objects which we now call Seyfert galaxies. They are the most common type of active galactic nuclei, but are very rare compared with typical (inactive) galactic nuclei, as shown by the relative numbers in Table 11.1.

Very rapid advances in radio astronomy in the decade immediately after World War II led to the first optical identifications of the strong radio sources. Among them was Cygnus A, identified by Walter Baade and Rudolph Minkowski with a faint galaxy with redshift $z = \lambda/\lambda_0 - 1 = 0.057$, where λ is the observed wavelength and λ_0 the laboratory or rest wavelength. The rich emission-line spectrum of Cyg A proved to be very similar to the spectra of Seyfert galaxies. Other identifications of similar objects quickly followed. The spectra of the small, highly luminous nuclei of many radio galaxies show emission lines that cover a wide range of ionization, and that are wider than the lines in the spectra of normal galaxies. They are also active galactic nuclei, but they are considerably rarer in space than the nuclei of Seyfert galaxies, which are typically radio-quiet (but not radio-silent) and are identified by their optical spectra. Table 11.1 lists very approximate space densities of both these types of AGNs "here and now" in the universe. Such numbers depend, of course, strongly on the lower limits of absolute magnitude adopted for the various groups. They are also no doubt incomplete to different degrees that depend upon the discovery methods. Nevertheless, they are useful for general orientation.

TABLE 11.1
Approximate space densities here and now

Type	Number Mpc^{-3}
Field galaxies	10^{-1}
Luminous spirals	10^{-2}
Seyfert galaxies	10^{-4}
Radio galaxies	10^{-6}
QSOs	10^{-7}
Quasars	10^{-9}

A certain fraction of the early optically identified radio sources were apparently stellar objects, with no sign of a galaxy or nebula in their images. Their spectra were continuous, with no absorption lines, and with broad emission lines at wavelengths that resisted identification. Efforts to understand these "stellar radio sources" were based on the idea that they were stars, perhaps white dwarfs with unusual abundances of some normally rare heavy elements.

However, in 1963 Maarten Schmidt broke the puzzle by identifying several well-known nebular emission lines in the spectrum of the stellar-appearing, thirteenth-magnitude radio source 3C 273 with the then unusually large redshift $z = 0.158$. Jesse L. Greenstein and Thomas A. Matthews soon afterward identified similar lines in 3C 48, with redshift $z = 0.367$. This was a larger redshift than that of even the faintest galaxy known at the time. It was immediately clear that these objects are highly luminous and can be observed to very great distances. They are not stars, but quasistellar radio sources, commonly called *quasars* for short. In fact, we now understand most if not all of them as AGNs, so luminous and so distant that the galaxy in which they are cannot (or could not) be detected on photographic plates. With CCDs, which reach faint light levels and have linear outputs, the stellar-appearing nucleus can be subtracted with good precision, and for many of the nearby quasars the galaxy has been revealed.

Corresponding radio-quiet high-luminosity stellar-appearing objects were found soon afterward. Initially they were called "quasistellar objects" by most research workers in the field, but they are now commonly referred to as quasars, even though they are radio-quiet. In this book we will usually preserve the distinction, and call them QSOs.

From the bright apparent magnitudes of quasars like 3C 273 and 3C 48, it was immediately apparent that they are beacons that can be observed out to the distant reaches of the universe. Today, as a result of systematic discovery programs, we know many quasars and QSOs with redshifts up to $z = 3.5$. There appears to be a fairly sharp cutoff at larger redshifts so only a very few are known with $z > 4$. Thus they are the most distant objects we know in the universe, and there seems to be some kind of a limit to their distance, or light-travel time, from us. However, observations soon showed that quasars and QSOs do not all have the same absolute magnitude; like stars they are spread over an enormous range in luminosity. Therefore, to study the overall structure of the universe, we can measure the distances to individual quasars only by understanding their nature, and thus from their spectra recognizing their absolute magnitudes.

In Table 11.1 the space densities of quasars here and now are listed along with those of Seyfert and radio galaxies. The table shows that the relative numbers of quasars and radio galaxies are in the same ratio as the relative numbers of QSOs and Seyfert galaxies. This is but one of the many observational indications that the radio-loud and radio-quiet objects each form physically continuous sequences, covering a wide range of luminosities, much as stars do. The quasars and QSOs represent the rarest but most luminous forms of AGNs; the radio and Seyfert galaxies are more common in space, but less luminous. That they form a sequence does not mean that they are physically identical objects differing only in scale, any more than O supergiants and K or M dwarfs are physically identical stars, differing only in scale.

11.3 Observational Classification of AGNs

The first-known Seyfert galaxies were discovered, or recognized, on slit spectra of individual galaxies taken mostly in radial-velocity programs. Since only a few percent of normal spirals are Seyferts, the total number found in this way in the days of photographic spectroscopy was very small. More were discovered when spectra were obtained of galaxies with unusually bright nuclei, so-called "compact galaxies," since many of them turned out to fit the spectroscopic criteria for Seyfert galaxies. Objective-prism surveys with a Schmidt camera by B. E. Markarian and his collaborators at Byurakan Observatory turned up many additional Seyfert galaxies. In their survey they catalogued galaxies with strong ultraviolet continuous spectra; roughly 10 percent of these turned out to be Seyfert galaxies when individual slit spectra were obtained. Most of the rest proved to be starburst galaxies, of the type mentioned in Section 11.1. More recently spectral surveys of individual galaxies, with modern, fast detectors, such as CCDs and enhanced Reticons, carried out primarily to measure radial velocities, have identified even more Seyferts. Other Schmidt-camera objective-prism surveys, some using filters to isolate galaxies with ultraviolet excesses, others using objective prisms to look for strong emission lines, especially Hβ, [O III] $\lambda\lambda$4959, 5007, and Hα, have found still more candidates. Slit spectra of these candidates have then led to the identification of more Seyfert galaxies.

The emission-line spectra of Seyferts can be classified into two types, following a scheme first proposed by E. Khachikian and D. W. Weedman. Seyfert 1 galaxies are those with very broad H I, He I, and He II emission lines, with full widths at half maximum (FWHM) of order 1 to 5 $\times$ 10^3 km sec^{-1}, while the forbidden lines, like [O III] $\lambda\lambda$4959, 5007, [N II] $\lambda\lambda$6548, 6583, and [S II] $\lambda\lambda$6716, 6731, typically have FWHMs of order 5 $\times$ 10^2 km sec^{-1}. Thus the forbidden lines, though narrower than the very broad permitted emission lines, nevertheless are broader than the emission lines in most starburst galaxies. An example of a Seyfert 1 galaxy is NGC 3227, whose spectrum is shown in Figure 11.1. Seyfert 2 galaxies, on the other hand, have permitted and forbidden lines with approximately the same FWHMs, typically 500 km sec^{-1}, similar to the FWHMs of the forbidden lines in Seyfert 1s. An example of a Seyfert 2 galaxy is Mrk 1157, whose spectrum is shown in Figure 11.2.

This classification into two types, Seyfert 1 and 2, may be further subdivided. Some Seyfert galaxies have H I emission-line profiles that can only be described as composite, consisting of a broad component (as in a Seyfert 1) on which a narrower component (as in a Seyfert 2) is superimposed. An example is Mrk 926, whose spectrum is shown in Figure 11.3. A wide range of relative strengths of broad and narrow H I emission-line components exists in nature, from those in which the broad component is very strong (Seyfert 1s like NGC 3227), through intermediate objects like Mrk 926, to those with quite strong narrow components (Seyfert 2s like Mrk 1157). The Seyfert

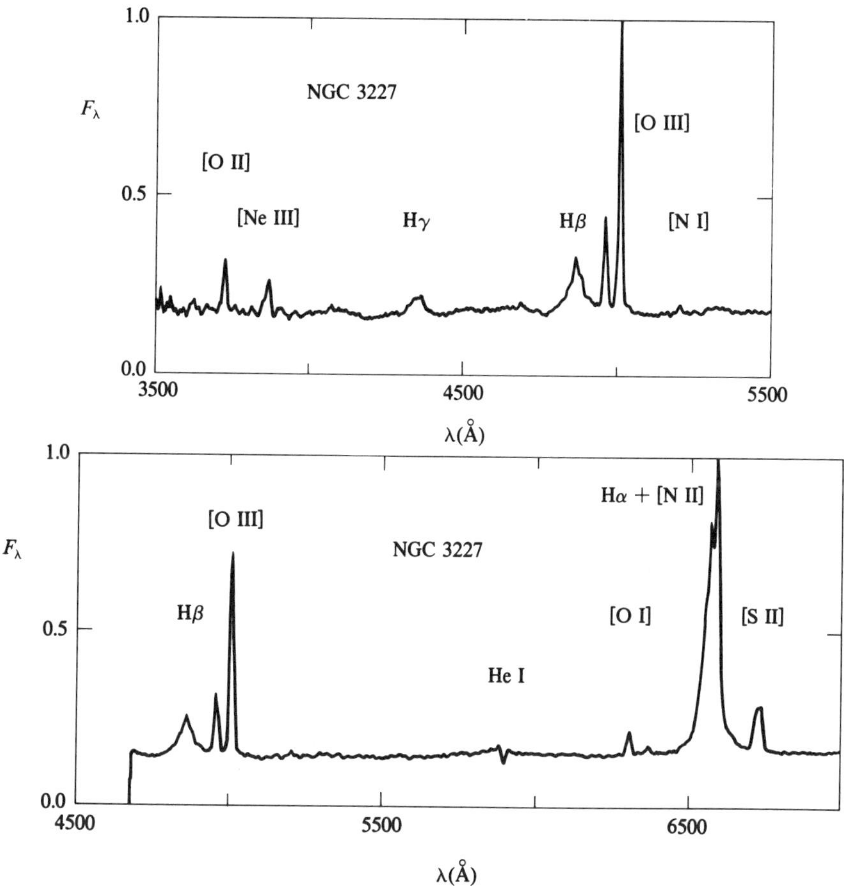

FIGURE 11.1
Spectral scan of NGC 3227, a Seyfert 1 galaxy. Relative flux per unit wavelength interval is plotted against wavelength in the rest system of the object; so areas under the curve correspond to energy radiated. Note that the two parts of the scan overlap in the Hβ, [O III] spectral region.

galaxies with intermediate-type H I profiles in which both components can easily be recognized are now generally classified as Seyfert 1.5 galaxies. Those with strong narrow components, and very weak but still visible broad components of Hα and Hβ, are often called Seyfert 1.8; and those in which a weak broad component may be seen at Hα, but none at Hβ, Seyfert 1.9. Certainly we can imagine that many if not all Seyfert 2s have even weaker broad H I components, too faint to detect against the continuous spectrum of the nucleus.

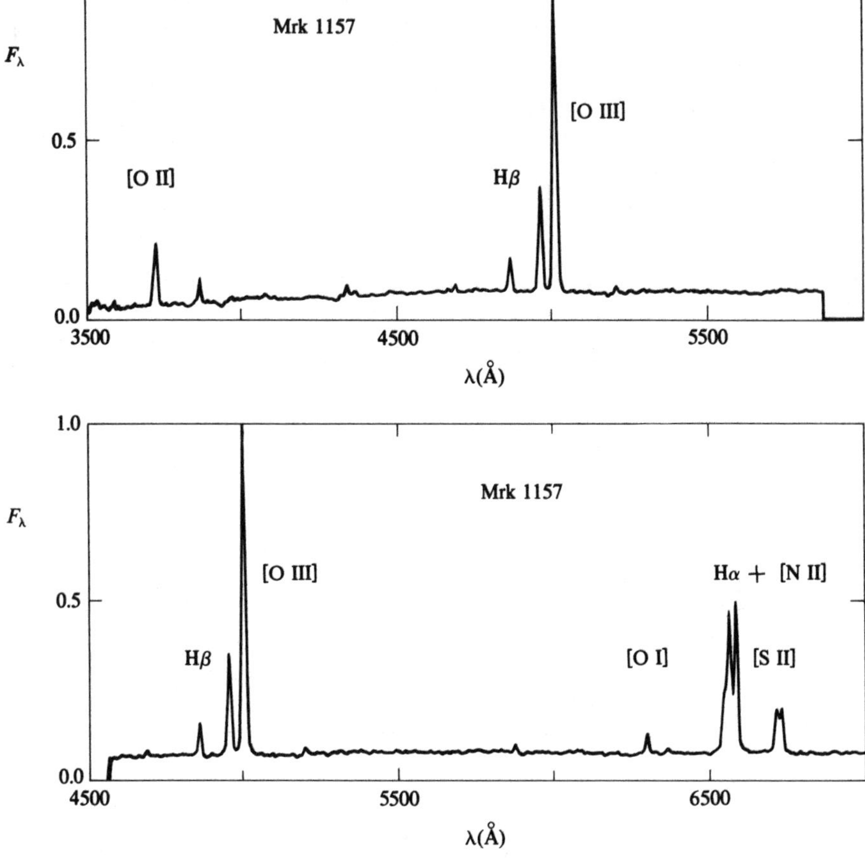

FIGURE 11.2
Spectral scan of Mrk 1157, a Seyfert 2 galaxy. Axes as in Figure 11.1.

In radio galaxies the synchrotron radio-frequency emission typically comes from two large, diametrically opposite lobes, far outside the limits of the galaxy as recorded on optical images. Sometimes very faint optical emission has been detected in the radio lobes. Frequently there is a weak, compact (flat radio-frequency spectrum) radio source in the nucleus of the galaxy. Strong optical-emission lines together with the featureless continuum spectrum (discussed later in this Chapter) come from this active galactic nucleus.

The optical spectra of radio-galaxy AGNs can be classified into two types, analogous to the classification of Seyfert AGNs. One type is the radio-loud equivalent of a Seyfert 1, with broad H I, He I, and He II emission lines, but

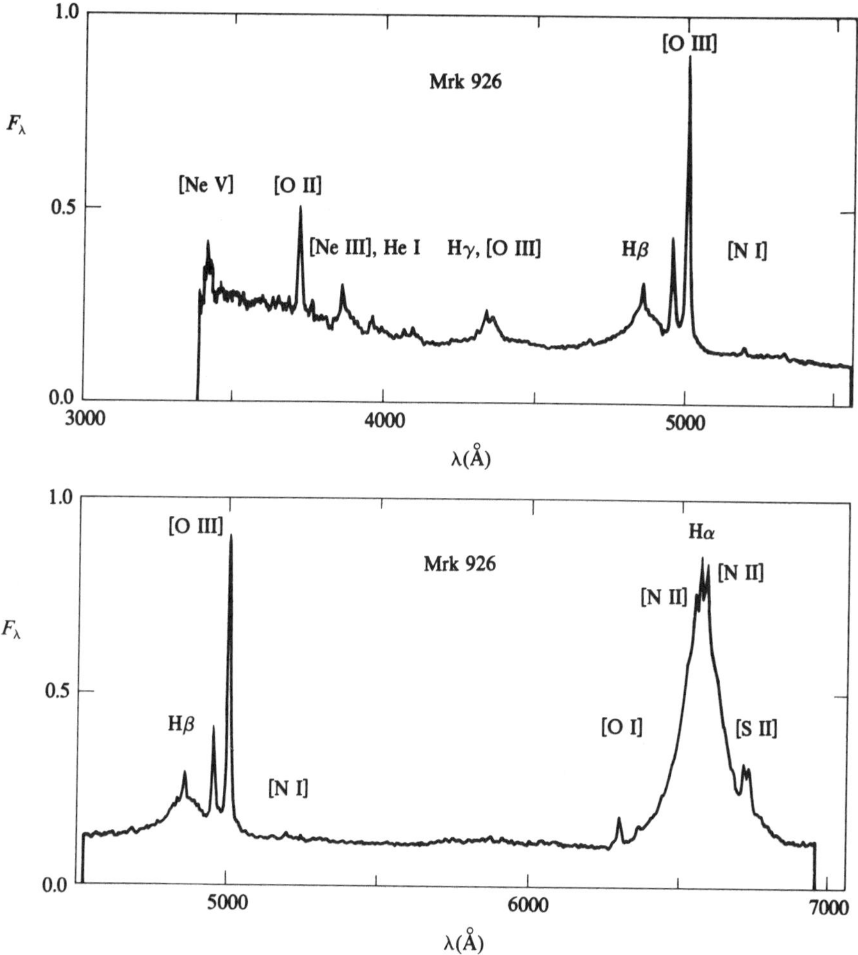

FIGURE 11.3
Spectral scan of Mrk 926, a Seyfert 1.5 galaxy. Axes as in Figure 11.1.

narrower forbidden lines. An example is 3C 390.3, whose spectrum is shown in Figure 11.4. These objects are called broad-line radio galaxies (BLRGs). The other type, similar in optical emission-line spectrum to Seyfert 2 galaxies, has "narrow" permitted and forbidden lines (but broader than in typical starburst galaxies). These are narrow-line radio galaxies (NLRGs). The best known example is Cyg A = 3C 405, the first identified radio galaxy that was described in the previous section. Its spectrum is shown in Figure 11.5.

Comparison of these figures shows that there are some differences between

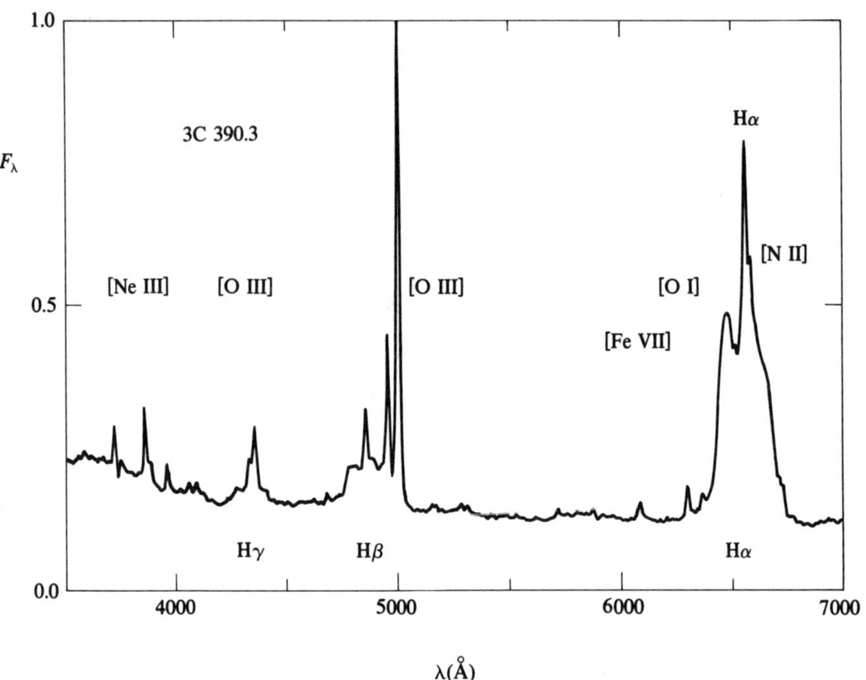

FIGURE 11.4
Spectral scan of 3C 390.3, a broad-line radio galaxy. Here the entire optical spectrum has been combined; otherwise axes as in Figure 11.1.

the spectra of radio and Seyfert galaxies, despite their general similarity. Most strikingly, almost all the observed BLRGs have composite H I profiles that are actually more similar to Seyfert 1.5 spectra than to those of Seyfert 1s. Furthermore, the broad components are typically broader than in "normal" Seyfert 1s, and often are more nearly square-shaped, or flat-topped, as well as more irregular, or structured. Besides their broad H I, He I, and He II emission lines, many Seyfert 1 AGNs also show broad permitted Fe II emission lines in their spectra. These Fe II lines, discussed in more detail in Chapter 12, come from several strong multiplets of Fe II. They overlap in two broad "bands" or "features" near $\lambda 4570$ and $\lambda 5250$. These features are relatively weak in the spectrum of NGC 3227 (Figure 11.1), but are quite strong in the spectrum of Mrk 376, another Seyfert 1 galaxy (Figure 12.4). Nearly all Seyfert 1 AGN spectra show these Fe II emission features with strengths between these two extremes, but they are typically much weaker in the spectra of BLRGs.

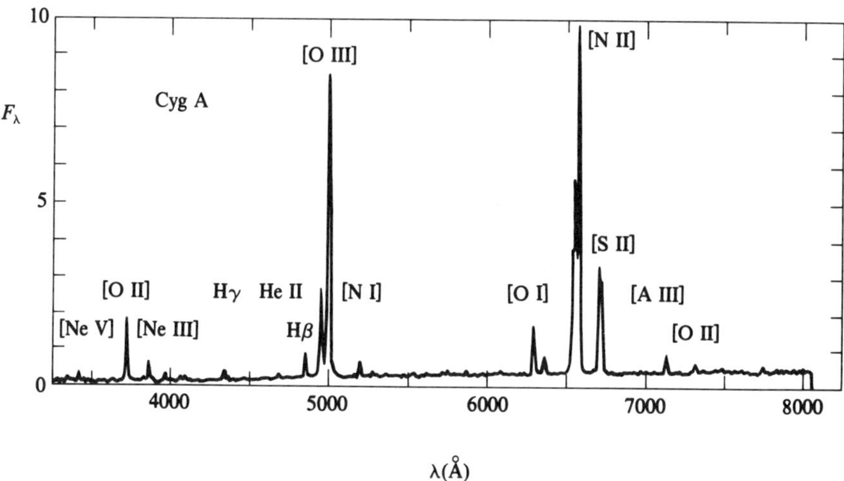

FIGURE 11.5
Spectral scan of 3C 405 = Cyg A, a narrow-line radio galaxy. Axes as in Figure 11.4.

Another, less obvious difference is that the ratio of the strengths of the broad Hα and Hβ emission components in BLRGs is on the average larger than in Seyfert 1 galaxies. There is a good deal of overlap, but mean values are perhaps $I(\text{H}\alpha \text{ broad})/I(\text{H}\beta \text{ broad}) \approx 3.5$ in Seyfert 1s and ≈ 6 in BLRGs. These observed differences are not yet fully understood, but they indicate that, on the average, radio-loud and radio-quiet broad-line AGNs differ in their optical properties as well. Any differences between the optical spectra of NLRGs and Seyfert 2s are much smaller, if they exist at all.

However, there are significant differences between the Seyfert and radio galaxies in which these two types of AGNs are located. Practically all Seyfert galaxies that are close enough to us in space to be resolved on direct images and classified as to morphological type are spirals. Many of them are distorted, but they are still basically spirals. Most are closer to Sb type than to either Sa or Sc. Many are barred spirals, especially of type SBb. Many have "companion" galaxies, or galaxies close enough to be interacting gravitationally with them. In contrast, almost none of the strong radio galaxies are spirals. Most of the NLRGs that are close enough to be classified are "giant ellipticals" of types cD, D or E. Practically all of the BLRGs are morphologically classified as type N, systems with brilliant "starlike" nuclei containing most of the luminosity of the system, but with faint, barely visible "fuzzy" or "nebulous envelopes" associated with them. Thus N galaxies are nearly quasars.

It is well-known that spiral galaxies contain more interstellar matter than giant ellipticals, and that they are more condensed to their principal planes. Very probably a difference between a Seyfert and a radio galaxy may be more in the near-nuclear environment – the former flattened, rotating, and rich in interstellar matter, the latter more nearly spherical and poor in interstellar matter – than in the structure of the nucleus itself.

All AGNs have a featureless continuous spectrum in the optical region, in addition to their emission lines. Naturally, they also have the typical integrated stellar continuous plus absorption-line spectrum of a normal galaxy. The featureless continuum comes from a tiny, unresolved object within the nucleus. It is evidently the seat of energy "generation" (release) distinctive to AGNs. The featureless continuum is very strong in typical Seyfert 1 galaxies, often so much stronger than the integrated stellar absorption-line spectrum that the latter is nearly invisible. In typical Seyfert 2s the featureless continuum is much fainter, and in many it can be detected only by careful analysis of the observed continuum. The broad emission lines are closely associated with the featureless continuum. This is a strong observational result that must be closely connected with the nature of AGNs.

As a result of their strong featureless continua, the AGNs of Seyfert 1 galaxies are generally more luminous than the AGNs of Seyfert 2s. The additional light of the nuclei makes a typical Seyfert 1 galaxy, *as a whole,* significantly more luminous than a typical Seyfert 2. The best available luminosity function of Seyfert 1 galaxies has its (weak) maximum near $M_B = -21$, while for Seyfert 2 galaxies it is near $M_B = -20$.

Although quasars and QSOs were regarded as new types of objects when first discovered, observational data since obtained have proved them to be simply the rarest and most luminous AGNs. The luminosity function of Seyfert 1 galaxies fits smoothly onto the luminosity function of QSOs (or "optically selected quasars") around absolute magnitudes $M_B = -21$ or -22. As a practical working definition, QSOs are objects with $M_B < -23$. All galaxies more luminous than about $M_B \approx -22$ are Seyferts. There are essentially no known QSOs analogous to Seyfert 2 galaxies, with narrow permitted and forbidden lines. Practically all known quasars and QSOs are of the BLRG or Seyfert 1 type. This is consistent with the observational data (discussed above), that if the featureless continuum is so bright that the light of the AGN completely dominates the total light of the galaxy, *broad* permitted emission lines are almost certain to be present. Similarly, though the numbers are much smaller and the statistics correspondingly less certain, radio-loud quasars seem to be the extension of the BLRGs to high optical luminosity.

11.4 Densities and Temperatures in the Ionized Gas

The "narrow" emission lines observed in Seyfert 2 and narrow-line radio galaxies are much the same as the emission lines observed in H II regions and planetary nebulae, except that in the AGNs the range of ionization is considerably greater. Not only are [O II] and [O III], [N II] and [Ne III] observed, but also [O I], [N I], [Ne V], [Fe VII] and frequently [Fe X]. Furthermore, [S II], which is a relatively low stage of ionization (the ionization potential of S^0 is only 10.4 eV), is generally much stronger in the AGNs than in nebulae. In addition to the forbidden lines, permitted lines of H I, He I, and He II are moderately strong. Thus these narrow lines are emitted by a highly ionized gas, with roughly "normal" abundances of the elements. The standard nebular diagnostic methods we have discussed in earlier chapters may be used to analyze it further. They depend on spectrophotometric measurements of the relative strengths of the emission lines.

A particularly well-studied example is the NLRG Cyg A. Its measured line intensities are listed in Table 11.2. As can be seen there, the measured H I Balmer-line relative strengths do not fit the recombination predictions of Table 4.4, which are almost independent of T and N_e over a wide range of physical conditions. The observed Balmer decrement is steeper than the calculated recombination decrement, just as it is in most observed planetary nebulae and H II regions. There the discrepancy is well-understood as resulting from the effects of extinction by interstellar dust; it is natural to assume that the same is true for the AGNs in general and Cyg A in particular. We know no other way to explain the observed emission-line spectrum listed in Table 11.2 quantitatively; and dust is present in all ionized-gas nebulae in our own or other nearby galaxies.

Therefore the amount of extinction has been calculated from the Balmer decrement, to give the best overall fit with the recombination decrement for $T = 10^4$ °K, $N_e = 10^4$ cm^{-3}, an apparently reasonable value which, as will be seen, agrees with other line ratios. The result for Cyg A is $E_{B-V} = 0.69 \pm 0.04$. As a first approximation we shall assume that this same amount of reddening applies to all the ionized gas in Cyg A, and correct the observed line ratios for this amount of extinction using the standard Whitford law of Chapter 7. The resulting corrected relative emission-line strengths, found in this way, are also listed in Table 11.2. Note that in deriving the amount of extinction, more weight has been given to the $H\gamma/H\beta$ and $H\delta/H\beta$ ratios than to $H\alpha/H\beta$; the corrected value of the latter is 3.08, slightly larger than the calculated recombination value 2.85. The slight increase is real; it results from an additional contribution due to collisional excitation of H^0, as will be discussed in Section 11.7.

Note that Cyg A has a relatively large E_{B-V}; in the sky it is near the Galactic equator, and some of its reddening undoubtedly is due to dust within our Galaxy. From observations of elliptical galaxies near Cyg A in the sky, about half its extinction arises within our Galaxy, and the other half in Cyg A itself.

TABLE 11.2
Observed and calculated relative line fluxes in Cyg A

Ion	λ Å	Relative Fluxes Measured	Corrected	Crab Nebula	Photoionization Model
[Ne V]	3346	0.14	0.38	–	0.12
[Ne V]	3426	0.36	0.95	0.46	0.34
[O II]	3727	2.44	5.00	10.3	0.24
[Ne III]	3869	0.66	1.23	1.56	0.53
[Ne III]	3967	0.22	0.40	0.47	0.16
[S II]	4072	0.14	0.23	0.31	–
Hδ	4101	0.17	0.28	0.31	0.26
Hγ	4340	0.32	0.46	0.61	0.47
[O III]	4363	0.16	0.21	0.19	0.19
He I	4471	$\leq$0.07	$\leq$0.09	0.28	0.02
He II	4686	0.25	0.28	0.53	0.18
Hβ	4861	1.00	1.00	1.00	1.00
[O III]	4959	4.08	3.88	2.81	6.3
[O III]	5007	13.11	12.30	8.43	18.1
[N I]	5199	0.40	0.32	–	–
[Fe XIV]	5303	$\leq$0.10	$\leq$0.08	–	0.01
[Fe VII]	5721	$\leq$0.10	$\leq$0.06	–	0.03
[N II]	5755	0.14	0.09	0.11	–
He I	5876	0.13	0.08	0.79	0.06
[Fe VII]	6087	$\leq$0.07	$\leq$0.04	–	0.04
[O I]	6300	2.10	1.10	1.20	1.24
[O I]	6364	0.69	0.35	0.33	0.41
[Fe X]	6375	0.10	0.05	–	0.07
[N II]	6548	3.94	1.90	1.56	0.29
Hα	6563	6.61	3.08	3.28	2.85
[N II]	6583	13.07	6.15	4.69	0.86
[S II]	6716	3.65	1.66	5.00 (6716 + 6731)	–
[S II]	6731	3.29	1.51		–
[Ar III]	7136	0.64	0.25	0.38	–
[O II]	7325	0.35	0.13	0.50	–
[Ar III]	7751	0.13	0.043	–	–

There is no doubt that dust is present in galaxies, and that it causes extinction. The same Balmer-line method is used to correct for its effects on the measured line intensities in the spectra of H II regions in other galaxies, in the nuclei of starburst galaxies, and in AGNs. Quantitatively, the results may be less certain than they seem, for the extinction law used assumes that the optical properties of the dust in these objects are similar to the mean properties derived from observations of the extinction by dust in our Galaxy within approximately 1 kpc of the Sun. This is an extreme simplification, since the properties of the dust no doubt depend upon the physical conditions and past history of the region in which it is situated. Furthermore, the reddening law was derived from measurements of stars, for which extinction occurs both by absorption and by scattering all along the line of sight, but in the AGN measurements the situation is different. Dust within the AGN absorbs and scatters light, but scattering not only removes photons from the beam toward the observer, but also adds photons to it that were originally going in other directions. Thus, at least for simple, spherically symmetric systems, scattering within the AGN has no effect. However, dust outside the AGN both absorbs and scatters photons from the beam. Finally, in nearly all observed nebulae, the gas and dust have clumpy, irregular distributions. To use one mean extinction for them is an extreme oversimplification. No doubt the same is true for AGNs. Our only justification for using it is that we do not yet know how to make more sophisticated corrections for the effects of dust extinction on the observed strengths of emission lines from AGNs. The method used is at least automatically very nearly correct at Hα and Hβ (or at Hα and a mean wavelength somewhere between Hβ, Hγ and Hδ), and, since the extinction probably varies fairly smoothly with wavelength, cannot be too far off anywhere in the limited optical range. In time, comparisons of measurements over a much wider wavelength range, from the satellite ultraviolet to the infrared, will undoubtedly give better information on the effects of extinction by dust on the spectra of AGNs.

The corrected intensities in Table 11.2 may be used to derive diagnostic information on the physical conditions in the ionized gas in Cyg A. The [O III] intensity ratio $(\lambda4959 + \lambda5007)/\lambda4363 = 77$ gives a mean temperature $T = (1.5 \pm 0.1) \times 10^4$ °K in the [O III] emitting region in the low-density limit $N_e < 10^4$ cm^{-3}, or lower temperatures at higher electron density. The [N II] ratio $(\lambda6548 + \lambda6583)/\lambda5755 = 89$ corresponds to a mean temperature $T = 1.0 \times 10^4$ °K, also in the low-density limit.

The [O II] $\lambda3729/\lambda3726$ ratio, which is a good electron-density diagnostic in H II regions and planetary nebulae, cannot be applied in Cyg A or other AGNs because the line widths are comparable to or larger than the separation of the two lines, 2.8 Å, which corresponds to about 300 km sec^{-1}. The [S II] ratio $\lambda6716/\lambda6731 = 1.10$ corresponds to $N_e = 3 \times 10^2$ cm^{-3} at $T = 1.0 \times 10^4$ °K, or to $N_e = 4 \times 10^2$ cm^{-3} at $T = 1.5 \times 10^4$ °K. Probably much of the [S II] emission arises in a less highly ionized region outside the [O III] emitting zone, so that the mean electron density derived from this ratio is not representative of the entire ionized volume, but it seems unlikely that $N_e > 10^4$ cm^{-3} through much of it.

The relative abundances of the ions responsible for observed lines may next be estimated by the methods of Section 5.9. To make some allowance for the evidently different temperatures in different zones, an early model adopted $T = 8{,}500°$ K for the [S II], [O I] and [N I] emitting regions, 12,000° K for H I, He I and He II, 15,000° K for [O III] and [Ne III], and 20,000° K for [Ne V] and [Fe VII]. The derived ionic abundances are shown in Table 11.3. With a rough allowance for unobserved stages of ionization, these ionic abundances may be combined to give the approximate elemental abundances listed in Table 11.4. Though the probable errors are necessarily large and uncertain because the physical picture is so schematic, Table 11.4 certainly shows that Cyg A has approximately the same composition as in our Galaxy and other observed galaxies with H II regions or starburst nuclei. H is the most abundant element; He is about ten times less abundant; O, Ne, N, and presumably C are the most abundant heavy elements; etc. These abundances are useful as a starting point for model calculations based on a more specific physical model.

TABLE 11.3
Relative ionic abundances in Cyg A emission-line region

Ion	Abundance	Ion	Abundance
H^+	10^4	O^0	1.9
He^+	5.7×10^2	O^+	1.7
He^{++}	2.4×10^2	O^{++}	1.5
N^0	0.37	Ne^{++}	0.45
N^+	0.88	Ne^{+4}	0.16
		Fe^{+6}	≤ 0.008

Similar observational data is available for many other NLRGs and Seyfert 2 galaxies. Emission-line strengths have been measured in many of them, and the same diagnostic ratios may be used to derive information on mean temperatures and electron densities in their ionized gas. Table 11.5 gives a short list of mean values of T and N_e determined from the best overall fits to [O III], [N II], [O II], [S II], and [O I] line ratios in other Seyfert 2 and NLRGs. Values of the extinction, derived from the Balmer-line ratios and used to correct the observed line ratios, are also given.

TABLE 11.4
Approximate elemental abundances in Cyg A emission-line region

Element	Abundance	Element	Abundance
H	10^4	Ne	1
He	10^3	S	0.3
N	1	Fe	≤ 0.1
O	4		

TABLE 11.5
Mean temperatures, electron densities and extinctions in Seyfert 2 and narrow-line radio galaxies

Galaxy	log T (° K)	log N_e (cm^{-3})	E_{B-V}
Mrk 3	4.1	3.5	0.50
Mrk 34	4.1	3.2	0.30
Mrk 78	4.0	3.2	0.72
Mrk 198	4.1	2.6	0.24
Mrk 348	4.2	3.3	0.41
3C 33	4.1	2.9	0.38
3C 98	4.3	3.6	0.68
3C 327	4.3	3.2	0.43
3C 433	4.2	2.4	0.58

11.5 Photoionization

The temperatures in the ionized gas in Cyg A and other NLRGs and Seyfert 2 galaxies are of order $1 - 2 \times 10^4$ ° K. This is strong observational evidence that the main source of energy input is by photoionization. The only other energy-input mechanism known, shock-wave heating or collisional heating, which occurs in supernova remnants, results in collisional ionization and a direct relationship between temperature and degree of ionization. Under pure shock-wave heating the [O III] lines in particular would be radiated mostly at $T > 5 \times 10^4$ ° K, and would be expected to indicate a much higher representative temperature than the 1.5×10^4 ° K observed in Cyg A. In actual shock-wave heating, this effect is somewhat moderated by the additional transfer of energy due to photoionization by collisionally excited ultraviolet line radiation from the very hot gas close to the front, but both the observations and the models discussed in Chapter 10 show that even so the [O III] emitting zone always has $T \geq 3 \times 10^4$ ° K.

On the other hand, under photoionization conditions there is no direct relationship between gas temperature and ionization, but the thermostatic effect of radiative cooling by collisionally excited line radiation, which increases rapidly with increasing temperature, tends to keep $T \approx 1 - 2 \times 10^4$ ° K over a wide range of input ionization-radiation spectra.

In some Seyfert 2s and NLRGs the [O III] ($\lambda 4959 + \lambda 5007)/\lambda 4363$ ratio is smaller, indicating higher temperature (up to $T \approx 5 \times 10^4$ ° K in some objects) if $N_e < 10^4$ cm^{-3}, or alternatively, indicating higher densities (up to $N_e \approx 10^7$ cm^{-3} in the same objects) if in fact $T \approx 1 - 2 \times 10^4$ °K. It seems more plausible that one energy-input mechanism is operative in all (or most) AGNs, and since it cannot be collisional heating in most of them (such as Cyg A and the galaxies listed in Table 11.5), the mechanism must be photoionization.

Although this simple analysis strongly implies that photoionization is the main energy-input mechanism, it is clear that the main source of the radiation cannot be hot stars, as in H II regions and planetary nebulae. Radiation from such stars will not produce the wide range of ionization observed in NLRG and Seyfert 2 AGNs, with emission lines of low stages, such as [O I] and [S II], and high stages such as [Ne V] and [Fe VII], fairly strong in comparison with [O III], [N II], and [Ne III]. What is required is a source with a much "harder" spectrum, one that extends even further into the ultraviolet than the spectra of central stars of planetary nebulae, some of which have effective temperatures up to $T = 2 \times 10^5$ °K. The plentiful high-energy photons ($h\nu > 100$ eV) of such a hard spectrum will produce high ionization (up to Ne^{+4}, Fe^{+6} and even Fe^{+9}) near the source, as well as a long, partially-ionized "transition zone" in which H^0 and H^+ (and hence e^-), O^0, and S^+ all coexist, and strong [O I] and [S II] lines can be collisionally excited. Physically, the width of this transition zone is roughly one mean free path of an ionizing photon,

$$l = \frac{1}{N_{\rm H}^0 a_\nu({\rm H}^0)}, \tag{11.1}$$

as the discussion of Section 2.3 makes clear. Since both $N_{\rm H^0}$ and the mean frequency of the remaining photons vary rapidly with distance, the mean free path also varies rapidly, which is why the ionization and radiative-transfer equations have to be numerically integrated simultaneously to derive quantitative results. But it is clear that the higher the energy of the photons present, the longer their mean free path is (from equation 2.4), and the larger the transition zone is.

As was stated above, a featureless continuum is observed in essentially every AGN. The form of this spectrum in the observed optical region approximately fits a power law

$$L_\nu = C\nu^{-n}, \tag{11.2}$$

typically with $n \approx 1 - 2$. In Cyg A the observed featureless continuum has $n = 3.8$, but if it is corrected for the same amount of extinction as derived for the gas from the observed Balmer decrement, this becomes $n = 1.6$. If this spectrum continues with the same power-law form to high energies, it can explain qualitatively the observed Cyg A line spectrum. Unlike the radiation of OB stars or hot planetary-nebula stars, a power law can both fit the observed continuum and produce reasonably strong [O I], [S II], [Ne V], and [Fe VII] emission lines.

More quantitatively, as we will see in Chapter 12, a photoionization model calculated with "normal" abundances of the elements and an input power-law spectrum of the form of equation (11.2), with $n = 1.2$, does approximately fit the observed Cyg A emission-line spectrum. A more observational comparison can also be made. It is valuable, because the real physical situation in an AGN is no doubt much more complicated than can be represented in any simplified model. All nebulae we know have complicated density structure, often with large-scale gradients, and always with small-scale knots, filaments, and density condensations. It would be surprising indeed if the AGN in Cyg A did not have similar fine structure. Instead of trying to model these complications, we can instead compare the observed emission-line spectrum of Cyg A to that of the Crab Nebula. The latter, as we have seen in Chapter 10, is known to be photoionized by an ultraviolet synchrotron-radiation continuum with $n = 1.2$. Therefore a direct comparision of the two spectra is independent of any model, and at least roughly accounts for the fine-scale structure that we know the Crab Nebula has, and that we presume Cyg A also has to some degree.

This comparison is given in Table 11.2, where the column headed "Crab Nebula" is the best available estimate of the emission-line radiation integrated over the entire nebula. It has been corrected for interstellar extinction, and should be compared with the corrected column for Cyg A. It can be seen that the overall agreement is quite good, not in detailed numerical values but in which lines are strongest, which are weakest, etc. The main discrepancies are that the He I and He II lines are stronger in the Crab Nebula, but this is expected because of the large He abundance in this supernova remnant.

Another check on the photoionization idea is that enough ionizing photons must be emitted by the central source to balance the total number of recombinations in the ionized gas. These are, of course, related directly to the total number of Hβ photons emitted in the gas; this is the basis of the Zanstra method described in Section 5.8. The basic equation (5.23), which is quite straightforward physically, may be written

$$L_{\mathrm{H}\beta} = h\nu_{\mathrm{H}\beta}\frac{\alpha^{eff}_{\mathrm{H}\beta}(\mathrm{H}^0,T)}{\alpha_B(\mathrm{H}^0,T)}\int_{\nu_0}^{\infty}\frac{L_\nu}{h\nu}d\nu. \tag{11.3}$$

Observationally it is convenient to express $L_{\mathrm{H}\beta}$ in terms of its equivalent width with respect to the neighboring featureless continuum. The equivalent width W_0 is measured in wavelength units,

$$L_{\mathrm{H}\beta} = L_\lambda(\lambda4861)\,W_0(\mathrm{H}\beta) = L_\nu(\lambda4861)\frac{d\nu}{d\lambda}W_0(\mathrm{H}\beta), \tag{11.4}$$

where L_λ is the featureless continuum luminosity per unit wavelength interval. Thus when we substitute the power-law form of equation (11.2), this relation becomes

$$\begin{aligned} W_0 &= \frac{\lambda_{\mathrm{H}\beta}}{n}\frac{\alpha^{eff}_{\mathrm{H}\beta}(\mathrm{H}^0,T)}{\alpha_B(\mathrm{H}^0,T)}\left(\frac{\nu_0}{\nu_{\mathrm{H}\beta}}\right)^{-n} \\ &= \frac{568}{n}(5.33)^{-n}, \end{aligned} \tag{11.5}$$

where the numerical value is given in Å and is calculated for $T = 10^4$ ° K, though the ratio of recombination coefficients is almost independent of temperature. Equation (11.5) is satisfied if all the available photons are absorbed in ionization processes in the gas; if some escape (because the ionized region is density bounded in some or all directions) or are absorbed by dust, the right-hand side is an upper limit to W_0.

For Cyg A the observed equivalent width of Hβ with a slit 2.″7 × 4″ (projected on the sky) is 39 Å in the rest frame of the AGN. The observed continuum, however, is diluted by the integrated stellar absorption-line spectrum of the galaxy Cyg A. This is always a problem with NLRGs and Seyfert 2 galaxies, as well as for many Seyfert 1s. In Cyg A the featureless continuum is relatively strong for a narrow-line AGN; from an analysis of its spectrum the fraction of the observed continuum near λ4861 that is featureless continuum is approximately $f_{FC} = 0.6$, while the remainder, $f_G = 0.4$, is galaxy spectrum. Thus the corrected equivalent width of Hβ, expressed in terms of the featureless continuum, is $W_0 = 39/0.6 = 65$ Å. The Zanstra condition (11.5) is satisfied for $n = 1.2$, in agreement with the photoionization picture.

Similar approximate agreement exists between the observed spectra of essentially all NLRGs and Seyfert 2 galaxies that have been analyzed in detail and photoionization predictions. There is little doubt that a hard-photon spectrum, extending to X-ray energies, is the primary energy-input mechanism to the observed gas in Seyfert 2 and NLRGs.

The narrow emission-line spectra of BLRGs, Seyfert 1, and Seyfert 1.5 galaxies are quite similar to the spectra of NLRGs and Seyfert 2s. It therefore seems quite likely that the narrow-line regions (NLRs) in which the narrow lines arise in all these types of AGNs are photoionized by high-energy photons. The only known systematic difference is that on the average the ionization goes up to a high level of ionization (strong [Fe VII] and [Fe X]) in more of the NLRs in Seyfert 1 galaxies than in Seyfert 2s. This may indicate a difference in the shape of the ionizing spectrum at high energies or, more probably, in the fluxes of ionizing photons incident on the NLR. This can be expressed in terms of the ionization parameter

$$\Gamma = \frac{1}{4\pi r^2 c N_e} \int_{\nu_0}^{\infty} \frac{L_\nu}{h\nu} d\nu, \tag{11.6}$$

where L_ν is the luminosity of the source per unit frequency interval, and r is the distance from the source. Physically, Γ represents the ratio of the ionizing photon density to the electron density.

Finally, we may estimate the mass and size of the NLRs, using the same method as described for planetary nebulae in Section 9.2. The luminosity emitted in a recombination line, most conveniently Hβ, can be written

$$L(\mathrm{H}\beta) = N_e N_p \alpha_{\mathrm{H}\beta}^{eff} h\nu_{\mathrm{H}\beta} V \epsilon, \tag{11.7}$$

where V is the total volume of the NLR and ϵ is the filling factor defined in Section 5.11. The mass of ionized gas is essentially

$$M = (N_p m_p + N_{\mathrm{He}} m_{\mathrm{He}}) V \epsilon, \tag{11.8}$$

and to a sufficiently good approximation $N_{\mathrm{He}} = 0.1\ N_p$, and $N_e = (N_p + 1.5\ N_{\mathrm{He}})$. To fix our ideas we may visualize a spherical NLR with

$$V = \frac{4\pi}{3} R^3, \tag{11.9}$$

and the radius R is then a specific numerical value of the dimension. The most luminous Seyfert 2s or NLRs of Seyfert 1 have $L(\mathrm{H}\beta) \approx 2 \times 10^8\ L_\odot$, which gives $M \approx 7 \times 10^5\ (10^4/N_e) M_\odot$, and $R \approx 20\ \epsilon^{-1/3}\ (10^4/N_e)^{2/3}$ pc. Thus for such an object with $N_e = 10^4\ \mathrm{cm}^{-3}$, $M \approx\ 10^6\ M_\odot$, and for an assumed filling factor $\epsilon \approx 10^{-2}$, $R \approx 90$ pc. This agrees with the fact that a few of the nearest Seyfert 2 NLRs have been resolved on direct images, and apparently have diameters of order 10^2 - 10^3 pc.

11.6 Broad-Line Region

The characteristic spectral feature of Seyfert 1 and 1.5 galaxies, and BLRGs, is their broad permitted H I emission lines, as shown in Figures 11.1, 11.3, and 11.4. Weaker broad He I lines can also be seen, particularly $\lambda5876$, and usually broad He II $\lambda4686$ as well. In addition, most Seyfert 1s and 1.5s have broad Fe II $\lambda\lambda4570, 5250$ features as well, with a considerable range in strength from one galaxy to another. These Fe II features are usually considerably fainter in BLRGs, but weakly present. This can be deduced, even if they are too weak to see, from the fact that the much stronger Fe II features in the ultraviolet are always detected if this spectral region is observed. These same broad H I, He I, He II and Fe II emission lines are also seen in quasars and QSOs, including the blue bump. In fact, the "blue bump," or "little blue bump," an apparent feature in the continuum in the $\lambda\lambda2000$ - 4000 region, is actually composed of many more unresolved Fe II lines, plus the H I Balmer continuum.

All the broad emission lines observed in AGNs are permitted lines. None of the forbidden lines have similar broad profiles. The only interpretation known is that the broad lines arise in a region in which the density is so high that all the levels of abundant ions which might otherwise give rise to forbidden-line emission are collisionally deexcited. A more accurate way of expressing this conclusion is that the broad lines are emitted in a region in which the electron density is considerably higher than the critical densities of all these levels, so that lines which these levels emit are weakened, in the ratio N_c/N_e, from the strengths they would have with respect to the permitted lines, such as Hβ, at the same temperature and ionization but in the low-density limit. A quantitative estimate is rather difficult to make, but in well-observed broad-line objects any possible broad component of [O III] $\lambda5007$, for instance, is perhaps at most 1 percent as strong with respect to Hβ as this same line is in narrow-line objects. Since the critical density N_c (O III $^1D_2$) $\approx$ 10^6 cm^{-3}, a lower limit to the mean electron density in a broad-line region (BLR) is roughly $N_e > 10^8$ cm^{-3}.

There are no broad lines in the optical spectral region which can be used to set an upper limit to the density, but in the ultraviolet C III] $\lambda1909$ has been observed with a broad profile, similar to the H I profiles, in several Seyfert 1 and 1.5 galaxies, and BLRGs. Broad C III] $\lambda1909$ emission is also observed in the redshifted spectra of many QSOs and quasars. Thus in these objects the mean electron density $N_e < N_c$ (C III] $^3P_1$) $\approx 10^{10}$ cm^{-3}. An intermediate value, $N_e \approx 10^9$ cm^{-3}, may therefore be adopted as roughly representative of the mean electron density in observed BLRs. Needless to say, there may be regions of even higher density included in the BLR, but their contribution to the C III] $\lambda1909$ emission must be small.

There is practically no direct information on the temperature in the BLR. There are no straightforward diagnostics to determine T from the H I, He I and He II lines. The observed Fe II emission indicates that $T <$ 35,000° K, for at higher temperatures it would be nearly completely collisionally ionized to Fe III, even if there were no ionizing photons present. For approximate estimates, $T \approx 10^4$ ° K is a good figure to adopt. Although the observed Balmer decrements in BLRs show that other processes in addition to recombination must contribute to the emission in the H I lines, probably the simple recombination calculation of equation (11.7) gives a rough idea of the amount of ionized gas in the BLR. The most luminous AGNs of Seyfert 1 and BLRGs have $L\ (\mathrm{H}\beta) \approx 10^9\ L_\odot$, which gives $M \approx 36\ M_\odot\ (10^9/N_e)$ and $R = 0.015\ \epsilon^{-1/3}\ (10^9/N_e)^{2/3}$ pc.

The masses and dimensions of BLRs are thus extremely small. For a representative density $N_e \approx 10^9\ \mathrm{cm}^{-3}$, $M \approx 40\ M_\odot$, and for an assumed $\epsilon \approx 10^{-2}$, $R \approx 0.07$ pc ≈ 0.2 light yr. This is far too small to hope to resolve even for the nearest BLR, and in fact none has been resolved to date. Naturally, the attempt should be repeated every time there is a significant instrumental advance in resolving power, for this standard picture of AGNs though convincing, is far from certain. However, the broad Hβ and Hα emission line profiles, and the fluxes in them, have been observed to vary on time scales as short as a month or two (~ 0.1 yr) in a significant fraction of BLRs. This agrees with the light-travel times calculated on this simple model. On the other hand, there are no well-established examples of variations in the narrow-line spectra of AGNs, which is consistent with the much longer light-travel times estimated for these objects in the previous section.

The most luminous quasars and QSOs have observed luminosities up to $\mathrm{L(H}\beta) \approx 5 \times 10^{10}\ L_\odot$, which is 50 times larger than the representative value adopted above. Thus these most luminous objects have $M \approx 2 \times 10^3\ M_\odot$, and $R \approx 0.25$ pc ≈ 0.8 light yr, and in fact it does seem observationally that their broad lines do not vary as rapidly as such lines do in some of the typical Seyfert 1 AGNs.

The nature of the energy input to the small, dense BLR is not as obvious as it is for the NLR. Most probably, however, it is also photoionization by the high-energy extension of the observed featureless continuum. The most convincing evidence is provided by Figure 11.6, which is a plot of observed $L_{\mathrm{H}\alpha}$ (proportional to $L_{\mathrm{H}\beta}$) for AGNs versus their observed $L_{FC}(\lambda 4800)$, the featureless continuum near $\lambda 4800$. Different symbols show broad-line and narrow-line objects, extending over a range of 10^5 in luminosity, all clustered close to a line with slope 1. In other words, all these different AGNs have essentially the same Hα (or Hβ) emission equivalent width, expressed in terms of their featureless continuum. This is true not only for the Seyfert 2 and NLRGs, but also for the Seyfert 1.5s, BLRGs, Seyfert 1s, quasars, and QSOs, in which most of the Hβ and Hα emission is in the broad components, emitted in their BLRs. As is obvious physically, this is just the result expected if all the ionization, in the NLR and the BLR, is due to photoionization by essentially the same form of input spectrum. Equation (11.5) expresses this result quantitatively for the simple power-law form. Other evidence is that the variations in the continuum and variations in emission lines are well-correlated.

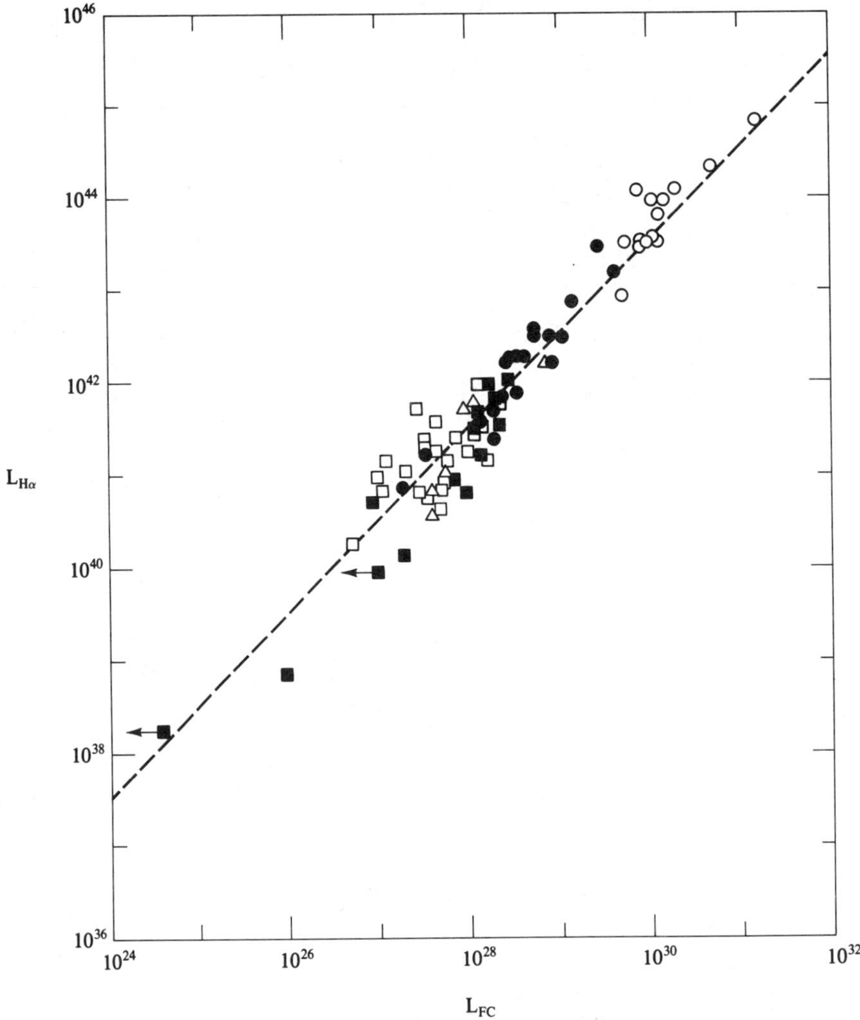

FIGURE 11.6
Luminosity in Hα emission line, $L_{H\alpha}$ (in erg sec^{-1}) *vs.* luminosity in featureless continuum at λ4800, L_{FC} (in erg sec^{-1} Hz^{-1}) for QSOs (open circles), Seyfert 1 galaxies (filled circles), Seyfert 2 galaxies (open squares), narrow-line radio galaxies (triangles), and additional Seyfert 2 and narrow-line radio galaxies (filled squares). The dashed line shows the predicted relationship for a power-law photoionizing continuum (equation 11.2) with exponent $n = 1.05$.

The observed results are thus consistent with photoionization being the energy-input mechanism to the BLR as well as the NLR. Of course they do not prove that photoionization *must* be the mechanism. However, the observational correlation shown in Figure 11.6 would show much more scatter, and would deviate systematically from the straight line, if only the luminosity in the narrow component of Hβ were plotted. Any interpretation other than photoionization would be considerably more complicated than this simple picture.

11.7 High-Energy Photons

The observational diagnostic information described in the first six sections of this chapter strongly suggests that photoionization by a spectrum that extends to very high energies is the main energy-input mechanism to the ionized gas. The aim of the astrophysical study of AGNs is to understand the nature of the source of this spectrum, the central "engine" that powers these objects. The optical emission-line spectrum can be analyzed by the methods of nebular astrophysics. Much of the methodology and many of the physical ideas can be carried over directly. However, some extensions are required, because certain physical processes that are not especially important in H II regions and planetary nebulae become increasingly relevant for the high-energy photons characteristic of AGNs.

One is that for very high-energy photons ($h\nu \gtrsim 3$ keV) the Coulomb binding of an electron to a proton in an H^0 becomes relatively unimportant. At these energies the standard bound-free absorption cross section of equation (2.4) approaches zero, but an additional Compton scattering term

$$\begin{aligned} a_{C\nu} &= \frac{8\pi}{3} r_0^2 \left(1 - \frac{3}{4}\frac{m_0 c^2 \nu_0}{h\nu^2}\right) \\ &= 0.66 \times 10^{-24} \left(1 - \frac{3}{4}\frac{m_0 c^2 \nu_0}{h\nu^2}\right) \text{ cm}^2, \end{aligned} \qquad (11.10)$$

where r_0 is the classical radius of the electron, and $\sigma_T = 8\pi r_0^2/3 = 0.66 \times 10^{-24}$ cm^2 is the Thomson electron-scattering cross section, must be added to the photoionization cross section.

More importantly, at high energies photoionization of heavy ions, especially of C, N, O, and Ne, dominates the opacity, rather than photoionization of H^0, He^0, and He$^+$ as in H II regions and planetary nebulae. Furthermore, photoionization by high-energy photons may remove electrons from inner shells while low-energy photons remove electrons only from the outermost shell. For instance, photons with energy $h\nu > 55$ eV can ionize O^{+2} by removing an outer $2p$ electron:

$$O^{+2}(2s^2 2s^2 2p^2\ {}^3P) + h\nu \rightarrow O^{+3}(1s^2 2s^2 2p\ {}^2P) + e. \qquad (11.11)$$

Photons with energy $h\nu > 64$ eV can also cause photoionization by removing a $2s$ electron:

$$O^{+2}(1s^22s^22p^2\ ^3P) + h\nu \rightarrow O^{+3}(1s^22s2p^2\ ^4P) + e, \qquad (11.12)$$

and photons with energy $h\nu > 570$ eV can also remove a $1s$ electron:

$$O^{+2}(1s^22s^22p^2\ ^3P) + h\nu \rightarrow O^{+3}(1s2s^22p^2\ ^2P) + e \qquad (11.13)$$

The calculated cross sections for these high-energy photoionization processes can, to a fair approximation, be averaged over all possible energy levels of the final ion, and calculated with an average threshold.

A simpler (yet sufficiently accurate) way than equation (2.31) to express the high-energy cross sections is by a single product

$$a_\nu = a_T \left(\frac{h\nu}{E_0}\right)^{-s} \qquad h\nu > E_0, \qquad (11.14)$$

where E_0 is either the threshold energy or 100 eV, whichever is smaller. As examples, values of the high-energy photoionization cross section parameters for ions of C and O are listed in Table 11.6. Cross sections are mostly of order 10^{-18} cm^2, which may be compared with $a_\nu(\mathrm{H}^0) = 1.9 \times 10^{-20}$ cm^2 at $h\nu = 10^2$ eV, and 1.1×10^{-23} cm^2 at 1 keV.

When a $1s$ electron (or in the terminology of X-ray spectroscopy, a K electron) is removed by a photoionization process such as equation (11.13), the resulting ion is left in a highly excited level, from which it decays rapidly by a radiationless Auger transition, releasing a second electron:

$$O^{+3}(1s2s^22p^2\ ^2P) \rightarrow O^{+4}(1s^22s^2\ ^1S) + e. \qquad (11.15)$$

Thus $1s$–shell photoionization of most light ions of the second row of the periodic table produces not one but two electrons, by successive processes such as (11.13) and (11.15). They couple two stages of ionization which differ by two electrons – O^{+2} and O^{+4} in this example – rather than two stages that differ by one electron, as in normal photoionization. It is straightforward to write down the generalizations of equations (2.30) to this situation.

Furthermore, the electrons liberated by high-energy photons themselves naturally have high energy. For instance, a photon with energy $h\nu = 750$ eV that ionizes O^{+2} produces a free electron with energy 695, 686, or 180 eV by one of the three processes (11.11), (11.12), and (11.13), respectively, while the free electron produced by (11.15) and corresponding other similar Auger processes (the excited $1s2s^22p^2$ levels are actually better described by Jl coupling and are not really pure ^4P, ^2P, ^2D, ^2S terms) has an energy between 438 eV and 460 eV.

TABLE 11.6
High-energy photoionization cross section parameters

Ion	Shell	E_0 (eV)	a_T (10^{-18} cm^2)	s
C^0	$1s$	280	1.06	2.47
	$2s$	100	0.593	2.61
	$2p$	100	0.158	3.62
C^+	$1s$	296	0.997	2.48
	$2s$	100	0.651	2.65
	$2p$	100	0.100	3.66
C^{+2}	$1s$	317	0.930	2.49
	$2s$	100	0.476	2.62
C^{+3}	$1s$	347	0.850	2.51
	$2s$	100	0.297	2.57
C^{+4}	$1s$	392	0.526	2.76
C^{+5}	$1s$	490	0.194	2.95
O^0	$1s$	533	0.554	2.58
	$2s$	100	1.00	2.37
	$2p$	100	1.15	3.36
O^+	$1s$	550	0.537	2.57
	$2s$	100	1.06	2.39
	$2p$	100	1.09	3.38
O^{+2}	$1s$	570	0.518	2.59
	$2s$	100	1.22	2.44
	$2p$	100	0.93	3.44
O^{+3}	$1s$	595	0.496	2.60
	$2s$	100	1.61	2.56
	$2p$	100	0.54	3.45
O^{+4}	$1s$	627	0.470	2.61
	$2s$	114	1.10	2.49
O^{+5}	$1s$	672	0.439	2.62
	$2s$	138	0.36	2.41
O^{+6}	$1s$	739	0.275	2.81
O^{+7}	$1s$	870	0.109	2.95

In H II regions and planetary nebulae the free electrons produced by normal photoionization processes are rapidly thermalized by Coulomb collisions, as described in Section 2.2, and give up all their energy to heating the gas. For the high-energy free electrons produced in AGNs the situation is more complex. The Coulomb-scattering cross section decreases with increasing energy, approximately as E^{-2} or v^{-4}, so the high-energy electrons have longer mean free paths before they are thermalized. Furthermore, an appreciable amount of H^0 can be present in the large partially ionized or transition region in the AGN. Hence electrons also lose energy by collisional excitation,

$$H^0(1s\ ^2S) + e \rightarrow H^0(nl\ ^2L) + e, \tag{11.16}$$

of which by far the largest cross section is for collisional excitation of Lα,

$$H^0(1s\ ^2S) + e \rightarrow H^0(2p\ ^2P) + e, \tag{11.17}$$

and also by collisional ionization

$$H^0(1s\ ^2S) + e \rightarrow H^+ + e + e. \tag{11.18}$$

The cross sections for both processes (11.17) and (11.18) increase rapidly with energy to maxima near 50 eV of about 6.3×10^{-17} cm^2 and 7.0×10^{-17} cm^2 respectively, followed by slow declines at higher energy. Thus some of the energy of the fast, newly created electrons goes to increasing the ionization of H I and thus producing still more secondary electrons, and some to Lα line emission, rather than to heating the gas. Similar processes can also occur with He^0 and He^+; they are much less important than the interactions with H^0 because of the lower abundance of He.

Calculations can be made of the relative amounts of energy lost by each process, and the number of secondary electrons produced, for primary electrons of any energy in a gas of a given degree of ionization. For instance, a free electron with initial energy 500 eV in a gas in which H is 10 percent ionized to H^+ will create on the average 6.0 secondary electrons by collisional ionization while slowing down, will cause on the average the emission of 7.3 Lα photons, and will lose 320 eV to heat in Coulomb collisions (of its own and of the secondary electrons it created). In addition, on the average about 20 eV will go to similar ionization and excitation processes in He, assumed to be 10 percent as abundant as H.

All these opacity, ionization, and heating processes must be taken into account in calculating photoionization models of AGNs, in addition to the processes already described in the earlier chapters in the context of model H II regions and planetary nebulae.

11.8 Collisional Excitation of H^0

The large partially ionized region in AGNs, produced by photoionization by high-energy photons, contains H^0, H^+, and e^-. Collisional excitation of neutral atoms hence can occur, and is observed in [O I] $\lambda\lambda6300$, 6364, and [N I] $\lambda5199$. In addition, collisional excitation of H^0 by thermal electrons produces strong Lα emission, and makes a significant contribution to Hα.

Collisional excitation of Lα at low densities arises very largely by direct excitation of the $2\ ^2P$ level

$$\mathrm{H}^0(1s\ ^2S) + e \rightarrow \mathrm{H}^0(2p\ ^2P) + e, \tag{11.19}$$

with a threshold 10.2 eV. The excitation cross-section is zero at the threshold but rises rapidly with energy. For this strongly allowed transition the emission coefficient for collisionally excited Lα emission is given directly by the relevant form of equation (3.23),

$$4\pi j_{\mathrm{L}\alpha} = N_e\, N_{\mathrm{H}^0}\, q_{1\ ^2S,2\ ^2P}\, h\nu_{\mathrm{L}\alpha}, \tag{11.20}$$

where $q_{1\ ^2S,2\ ^2P}$ can be calculated from equations (3.19) and (3.21). Values of the required collisional strength $\Omega(1\ ^2S, 2\ ^2P)$ are listed for a few temperatures in Table 3.12. As can be seen there, the collision strengths decrease rapidly with increasing principal quantum number n. In addition, the threshold energy increases with n, making the contribution to Lα emission from collisional excitation to higher levels followed by cascading down to $2\ ^2P$ quite small. Nevertheless, this contribution can be included, to the extent that excitation cross sections to higher levels n are available. Good values have been calculated for excitation to $n = 3$. Note that every excitation to $3\ ^2S$ or $3\ ^2D$ is followed by emission of Hα and then Lα, but every excitation to $3\ ^2P$ leads, under Case B conditions, to emission of Hα and population of $2\ ^2S$, which does not emit Lα. Thus, taking account of excitation to the levels with $n = 2$ and 3, we can write the collisional excitation contribution to the Lα emission coefficient as

$$4\pi j_{\mathrm{L}\alpha} = N_e N_{\mathrm{H}^0}(q_{1\ ^2S,\ 2\ ^2P} + q_{1\ ^2S,\ 3\ ^2S} + q_{1\ ^2S,3\ ^2D})h\nu_{\mathrm{L}\alpha}. \tag{11.21}$$

Numerical values calculated from this equation are listed in the left-hand side ("low-density") of Table 11.7. The total contribution from the two additional terms that are in equation (11.21) but not (11.20), the cascading following collisional excitation to $n = 3$, is small, ranging from 3 percent at $T = 10,000^\circ$ K to 11 percent at $T = 20,000^\circ$ K. Probably including the still smaller contributions of excitation to the levels $n \geq 4$ would result in further increases in the Lα emission by comparable percentages.

TABLE 11.7
Lα emission coefficients in partly ionized region[a]

T	Low density $N_e \ll 1.5 \times 10^4$ cm^{-3} Collisional	Recombination	High density $N_e \gg 1.5 \times 10^4$ cm^{-3} Collisional	Recombination
(°K)	$\frac{4\pi j_{L\alpha}}{N_e N_{H^0}}$	$\frac{4\pi j_{L\alpha}}{N_e N_p}$	$\frac{4\pi j_{L\alpha}}{N_e N_{H^0}}$	$\frac{4\pi j_{L\alpha}}{N_e N_p}$
10,000	2.10×10^{-24}	2.87×10^{-24}	3.51×10^{-24}	4.25×10^{-24}
12,500	2.15×10^{-23}	2.39×10^{-24}	3.56×10^{-23}	3.53×10^{-24}
15,000	1.02×10^{-22}	2.01×10^{-24}	1.67×10^{-22}	3.06×10^{-24}
20,000	7.31×10^{-22}	1.51×10^{-24}	1.19×10^{-21}	2.34×10^{-24}

[a] All in erg cm^3 sec^{-1}.

In Table 11.7 the recombination emission coefficients for Lα are also listed, for comparison with the collisional-excitation coefficients. The recombination coefficient immediately calculated as

$$\begin{aligned} 4\pi j_{L\alpha} &= N_e N_p \alpha^{eff}_{2\,^2P} h\nu_{L\alpha} \\ &= N_e N_p (\alpha_B - \alpha^{eff}_{2\,^2S}) h\nu_{L\alpha}, \end{aligned} \tag{11.22}$$

and values of $\alpha^{eff}_{2\,^2S}$ are listed in Table 4.11.

Note how rapidly collisionally excited emission of Lα increases with energy because of its high threshold, 10.2 eV. For a half-ionized gas, with $N_{H^0} = N_p$ the collisional and recombination contributions to Lα emission are roughly the same at $T = 10{,}000$ ° K, while at $T = 12{,}500$ ° K the collisional excitations would be almost ten times more important. Actually, it would be almost impossible to heat such a gas to this high a temperature, precisely because of the strong collisional cooling by Lα; a better comparison is that for a mostly ionized gas with $N_{H^0} \approx 0.09 N_p$ the collisional and recombination contributions to Lα emission would be approximately equal at $T = 12,500°$ K.

At higher densities H^0 atoms that arrive in $2\,^2S$ as a result of collisional excitation, of recombination, or of cascading, are collisionally shifted to $2\,^2P$ and also emit Lα. The critical density for this process, as derived in Section 4.3, is

$$N_c = \frac{A_{2\,^2S,\,1\,^2S}}{q^p_{2\,^2S,\,2\,^2P} + q^e_{2\,^2S,\,2\,^2P}} \approx 1.5 \times 10^4 \text{ cm}^{-3} \tag{11.23}$$

for $N_e \approx N_p$. Thus in this high-density limit the first approximation to the collisional-excitation emission coefficient, replacing equation (11.20), is

$$4\pi j_{L\alpha} = N_e N_{H^0}(q_{1\,^2S,\,2\,^2S} + q_{1\,^2S,\,2\,^2P}) h\nu_{L\alpha}, \tag{11.24}$$

while a better approximation, replacing (11.21), is

$$4\pi j_{L\alpha} = N_e N_{H^0} \sum_{n=2}^{3} \sum_{L=0}^{n-1} q_{1\,^2S,n\,^2L} h\nu_{L\alpha}. \tag{11.25}$$

Values calculated from equation (11.25) are listed in the right-hand side of Table 11.7, where they are compared with the corresponding high-density-limit recombination emission coefficients

$$4\pi j_{L\alpha} = N_e N_p \alpha_B h\nu_{L\alpha} \tag{11.26}$$

instead of (11.22). It can be seen that the behavior is very similar to the low-density case, except that both the collisional and the recombination Lα emission coefficients are larger by factors of approximately 1.5.

Collisional excitation to any of the levels $3\,^2S, 3\,^2P$ or $3\,^2D$ leads to Hα emission, under standard Case B conditions. Thus the first approximation to the collisional-excitation emission coefficient for Hα is

$$4\pi j_{H\alpha} = N_e N_p \sum_{L=0}^{2} q_{1\,^2S,3\,^2L} h\nu_{H\alpha}. \tag{11.27}$$

Values calculated from this equation are listed in Table 11.8, where they are compared with the recombination emission coefficient, taken or interpolated from Table 4.4 for $N_e = 10^4$ cm^{-3}. Again the even stronger temperature dependence of the collisionally excited Hα emission can be seen. For a situation in which $N_{H^0} = N_p$, the collisionally excited Hα emission is about 8 percent as large as the recombination emission at $T = 10,000\ ^\circ$ K, while at $T = 12,500\ ^\circ$ K if $N_{H^0} = 0.09\ N_p$, the collisional Hα emission would be about 15 percent as large as the recombination contribution.

Collisional excitation of Hβ and higher Balmer lines is even smaller with respect to the recombination contributions, because of the higher thresholds and smaller cross sections, which, however, are less accurately calculated. Both the degree of ionization of H and the temperature vary strongly as the ionization approaches zero in the transition region of an AGN. Quantitative statements about the total collisionally excited contributions to the various H I line emissions therefore depend upon detailed model calculations, which will be described in Chapter 12. Many such calculations show that collisionally excited Lα emission is quite important, and that collisional excitation adds a small but significant contribution to the Hα recombination emission, while the collisional contributions to Hβ and higher Balmer lines are nearly negligible. These models agree that overall, for the entire AGN NLR, the intrinsic Balmer decrement is approximately Hα/Hβ = 3.1, with the excess over the recombination value of 2.85 resulting from the effects of collisional excitation of Hα.

TABLE 11.8
Hα emission coefficients in partly ionized region[a]

T (°K)	Collisional $\frac{4\pi j_{H\alpha}}{N_e N_{H^0}}$	Recombination $\frac{4\pi j_{H\alpha}}{N_e N_p}$
10,000	3.03×10^{-26}	3.54×10^{-25}
12,500	4.95×10^{-25}	2.89×10^{-25}
15,000	3.21×10^{-24}	2.46×10^{-25}
20,000	3.22×10^{-23}	1.81×10^{-25}

[a] All in erg cm^3 sec^{-1}.

References

The early papers mentioned in the text, in which the objects we now know as Seyfert galaxies were first studied and gradually recognized as a distinct group, are

Fath, E. A. 1909. *Lick Observatory Bulletin* **5,** 71.
Slipher, V. M. 1917. *Lowell Observatory Bulletin* **3,** 59.
Hubble, E. 1926. *Ap. J.* **54,** 369.
Seyfert, C. K. 1943. *Ap. J.* **97,** 28.

The identification of Cygnus A is in

Baade, W., and Minkowski, R. 1954. *Ap. J.* **119,** 206.

The papers mentioned on the identification of quasars are

Schmidt, M. 1963. *Nature* **197,** 1040.
Greenstein, J. L., and Matthews, T. A. 1963. *Nature* **197,** 1041.

These and many other early papers on quasars are reprinted in

Robinson, I., Schild, A., and Schucking, E. L. 1965. *Quasistellar Sources and Gravitational Collapse.* Chicago: University of Chicago Press.

A very good book on the whole subject is

Weedman, D. W. 1986. *Quasar Astronomy.* Cambridge: Cambridge University Press.

The source of Table 11.1 is

Osterbrock, D. E. 1982. *Extragalactic Radio Sources* (IAU Symposium No. 97), D. S. Heeschen and C. M. Wade, eds. Dordrecht: Reidel, p. 369.

There are many published papers that discuss the emission-line spectra of Seyfert galaxies, quasars, QSOs, and AGNs. Four review articles that contain references to many of these papers are

Weedman, D. W. 1977. *Ann. Rev. Astr. Ap.* **15,** 69.

Osterbrock, D. E. 1979. *A. J.* **84,** 901.

Sargent, W. L. W. 1980. *Scientific Research with the Space Telescope,* M. S. Longair and J. W. Warner, eds. U.S. Government Printing Office, pp. 197-214.

Osterbrock, D. E. 1984. *Q.J.R.A.S.* **25,** 1.

The discussion of Cygnus A is based upon the papers

Osterbrock, D. E., and Miller, J. S. 1975. *Ap. J.* **197,** 535.

Osterbrock, D. E. 1983. *P.A.S.P.* **95,** 12.

Tables 11.2, 11.3 and 11.4 are based on both these papers, chiefly the former.

Table 11.5 is based on

Koski, A. T. 1978. *Ap. J.* **223,** 56.

The evidence for photoionization of the BLR is discussed in many papers, of which three are

Searle, L., and Sargent, W. L. W. 1968. *Ap. J.* **153,** 1003.

Yee, H. K. C. 1980. *Ap. J.* **241,** 894.

Shuder, J. M. 1981. *Ap. J.* **244,** 12.

The data of Figure 11.6 are taken from the third of these papers.

The most thorough discussion of the presence of weak [O III] emission from the BLR, and the resulting lower limit to the electron density there, is

Crenshaw, D. M., and Peterson, B. M. 1986. *P.A.S.P.* **98,** 185.

Several general references on ionization by high-energy photons or X-rays are

Tarter, C. B., and Salpeter, E. E. 1969. *Ap. J.* **156,** 953.

Hatchett, S., Buff, J., and McCray, R. 1976. *Ap. J.* **206**, 847.

Halpern, J. P., and Grindlay, J. E. 1980. *Ap. J.* **242,** 1041.

Ferland, G. J., and Truran, J. W. 1981. *Ap. J.* **244,** 1022.

Kallman, T. R., and McCray. R. 1982. *Ap. J. Suppl.* **50,** 263.

The first of these gives a good description of the main physical differences between nebulae or AGNs that are photoionized by X-rays and by normal stellar photons. The third and fourth references give especially clear, systematic descriptions of the present state of this subject. Numerical values of photoionization cross sections are given by

Daltabuit, E., and Cox, D. P. 1972. *Ap. J.* **177,** 855.
Weisheit, J. C. 1974. *Ap. J.* **190,** 735.
Reilman, R. F., and Manson, S. T. 1979. *Ap. J. Suppl.* **40,** 815.

The numerical data listed in Table 11.6 have been taken from the first two of these references.

The slowing down and energy loss by fast electrons in a partly ionized region is worked out by

Bergeron, J., and Collin-Souffrin, S. 1973. *Astr. Ap.* **25,** 1.
Shull, J. M. 1979. *Ap. J.* **234,** 761.
Shull, J. M., and Van Steenberg, M. J. 1985. *Ap. J.* **298,** 268.

The numerical values quoted in the text are taken from complete curves, tables, and fitting formulae in the third of these.

Early calculations of collisional excitation of H I line radiation were made by

Chamberlain, J. W. 1953. *Ap. J.* **117,** 387.
Parker, R. A. R. 1964. *Ap. J.* **139,** 208.

They give the main physical ideas, but have been completely superseded by more accurate cross sections and calculations made with larger computers available today. The best cross sections for H^0 available at the time of writing are the close-coupling calculations for levels up through $3\ ^2D$ of

Callaway, J. 1985. *Phys. Rev. A.* **32,** 775.
Callaway, J., Unnikrishnan, K., and Oza, D. H. 1985. *Phys. Rev. A.,* **36,** 2576.

The numerical values listed in Tables 11.7 and 11.8 are based on these cross sections. More complete lists of collisional-excitation cross sections to larger values of n have been collected by

Drake, S. A., and Ulrich, R. K. 1980. *Ap. J. Suppl.* **42,** 351.

The importance of collisional excitation of Lα and Hα line emission in the NLRs of AGNs, and especially for the intrinsic Hα/Hβ ratios was pointed out by several authors, including

Netzer, H. 1982. *M.N.R.A.S.* **198,** 589.
Ferland, G. J., and Netzer, H. 1983. *Ap. J.* **264,** 105.
Halpern, J. P., and Steiner, J. E. 1983. *Ap. J.* **269,** L37.
Gaskell, C. M. 1983. *Ap. Letters* **24,** 43.
Gaskell, C. M., and Ferland, G. J. 1984. *P.A.S.P.* **96,** 393.

12

Active Galactic Nuclei: Results

12.1 Introduction

In the preceding chapter we have discussed the observed optical spectra of AGNs, the diagnostic information that can be drawn from them, and the basic physical ideas by which they can be understood. The main one is photoionization by a spectrum that extends to high energies, and includes a relatively large proportion of high-energy photons. The source of this ionizing spectrum cannot be a star. The best current working hypothesis is that it is emitted in the immediate surroundings of a black hole, as a consequence of the release of gravitational energy by matter in the process of falling into the hole and ultimately disappearing. This is discussed in the next section.

Following it, models of the narrow-line regions of AGNs are discussed, that is, of Seyfert 2 galaxies, narrow-line radio galaxies, the NLR regions of Seyfert 1s and QSOs, and the lower-ionization LINERs. In all these NLRs the electron densities are comparable with those in planetary nebulae and dense H II regions. In the BLRs however, the densities are much higher. As a result, collisional and radiative processes from excited levels are not negligible. Very large optical depths in resonance lines further complicate the situation. These processes are discussed next. Then come sections on dust and on the velocity fields on AGNs. Finally, the chapter and book conclude with a discussion of current ideas of an overall physical picture of these objects.

12.2 Energy Source

The luminosity of a typical AGN, of order $10^{12}\ L_\odot$, is far too large for its source to be a star. The most massive stars are of order $10^2\ M_\odot$, and have luminosities of order $10^5\ L_\odot$. In such massive stars the radiation pressure dominates over the gas pressure and as a result they are close to the limit of instability. More massive stars, producing (or actually, liberating) energy by thermonuclear reactions, cannot exist. No matter what the energy-production mechanism, any spherically symmetric object whose gravity holds it together against radiation pressure must satisfy the Eddington condition

$$L \leq L_E = \frac{4\pi c G m_H M}{\sigma_T} = 1.26 \times 10^{38} \frac{M}{M_\odot}, \tag{12.1}$$

where σ_T is the electron-scattering or Thomson cross section. This is the minimum opacity and therefore the minimum radiation pressure; any larger opacity would correspond to a smaller upper limit to the luminosity. This equation can also be written

$$\frac{L}{L_\odot} \leq \frac{L_E}{L_\odot} = 3.22 \times 10^4 \frac{M}{M_\odot}.$$

According to it, for instance, the central source in an AGN with $L = 10^{12}\ L_\odot$ must have $M \geq 3 \times 10^7\ M_\odot$. Furthermore, it must be quite small, for the broad-line region which presumably surrounds it has a size of order 0.07 pc $\approx$ 0.2 light year, from the estimates in the previous chapter. The continuum variations observed in many AGNs, discussed there, suggest that the central continuum sources may be even smaller, ranging down in some cases to at most 1 light-week (optical limit) or even 1 light-day (X-ray limit). Note, however, that the Eddington limit applies strictly only to spherically symmetric objects; for more complicated geometries it is no more than a rough estimate. Also, if the object is not in equilibrium or a steady state the limit may be surpassed; an outstanding example is a supernova explosion.

Thus large energies are released in very small volumes in the neighborhood of large masses. Thermonuclear reactions cannot do it. However, gravitational energy release can. The most promising physical picture is an accretion disk around a massive black hole. In such a situation the rest-mass energy of infalling material can be converted into radiation or fast particles with greater efficiency than seems achievable by any other processes we know. The luminosity produced may be written

$$L = \eta \dot{M} c^2, \tag{12.2}$$

with $\dot{M}$ the accretion rate, and η the efficiency of process, or the fraction of the mass that is converted into energy and does not fall into the black hole.

For instance, if $\eta = 10$ percent, for an AGN with $L = 10^{12}\ L_\odot$ the necessary accretion rate is $\dot{M} = 0.7\ M_\odot\ \mathrm{yr}^{-1}$.

Simplified models of such accretion disks exist; the simplest is a thin disk which is optically thick at all radii. It emits a continuum with spectrum

$$L_\nu = C\nu^{1/3} \tag{12.3}$$

over a limited range of frequency, with a high-energy exponential cutoff corresponding to a Planck function with $T = 10^5$ to 10^6 °K for typical accretion rates and masses. This spectrum is quite unlike that observed, and in particular does not account for the X-ray extension. This is not surprising, for the radio observations show that relativistic plasma is continuously being generated, no doubt near the black hole, as a consequence of electromagnetic fields connected with rotation. High-resolution radio measurements often show narrow jet-like plasma structures extending from close to the source out to large distances; they appear to be in the axis of rotation of the accretion disk (which is often not the same as the axis of rotation of the galaxy in which it is located). Undoubtedly the generation of the high-energy photons in the ionizing spectrum is intimately connected with the generation and properties of the relativistic plasma.

Although many attempts have been made, and limited understandings have been reached, to date we do not yet have a complete physical picture of the central energy source, or central "engine" of an AGN. We cannot calculate the emergent spectrum of ionizing photons; indeed, we do not fully understand all the physical parameters on which it depends, quite in contrast to what we know about how the emergent spectrum of a hot star depends ultimately upon its mass, composition, and age (or previous evolutionary history). Thus we are forced to use an assumed power law, of the form of equation (11.2), or a more complicated broken-power law, similar to equation (10.2), perhaps modified by an exponential cut off at high energy, or perhaps a blackbody form

$$L_\nu = AB_\nu(T_*), \tag{12.4}$$

with an assumed very high T_*. Such assumed forms of input spectra are analogous to the blackbody models used to represent the ionizing radiation from hot OB stars and planetary-nebula central stars in the early stages of nebular astrophysics. Ultimately, just as for stars, we must hope to understand the nature of the AGN "engine" well enough to calculate, from first principles, all its observable properties, including its emergent spectrum.

12.3 Narrow-Line Region

To go beyond the diagnostic methods, the next step is to calculate model AGNs. The method is exactly the same as that used in calculating models of planetary nebulae and H II regions, as described in Section 5.10, but including the additional physical processes relevant for high-energy photons that were discussed in the previous chapter. In contrast to planetary nebulae and H II regions, the structures of the AGNs are not resolved, and there is almost no direct observational information on their forms, shapes, symmetries, degree of fine structure, etc. The simplest way to proceed is to assume spherical symmetry for a complete model, or plane-parallel symmetry for a representative dense condensation or *cloud,* small in comparison with its distance from the central source, and therefore illuminated by essentially parallel radiation from it. The actual situation, however, is probably far more complicated.

In Table 11.2 the observed relative line strengths in Cyg A are compared with the predictions of a simple spherical model. In it a power-law input spectrum (11.2) with $n = 1.2$ was assumed. The constant C was chosen to fit the observed featureless continuum of a representative AGN. The assumed abundances (by number) are listed in Table 12.1. The assumed mean density in the clouds was taken as $N_e = 10^4$ cm^{-3}, and the filling factor as $\epsilon = 10^{-2}$. In this model the collisional excitation of the H I Balmer lines was not included in the calculations, though cooling by collisional excitation of Lα was, and the extra ionization by Auger electrons was also omitted. These effects are routinely included in present model calculations, but their importance was not realized at the time when this model was calculated.

TABLE 12.1
Assumed relative abundances in Cyg A model

Element	Abundance	Element	Abundance
H	1×10^4	Ne	0.79
He	6×10^2	Mg	0.25
C	3.5	Si	0.35
N	1.3	S	0.79
O	7.9	Fe	0.14

Nevertheless, as can be seen from Table 11.2, the model gives a good representation of the observed spectrum of Cyg A. The characteristic emission lines covering a wide range of ionization, from [O I] and [S II] through strong [O III] to [Ne V] and [Fe X] are reproduced by the model. A closer comparison of the observed and model columns shows that both He I and He II are predicted too weak, by a factor of about 1.4; to the first order this could be rectified by increasing the helium abundance by this same factor, from 0.06 to 0.09. The latter value is closer to our present ideas of the correct abundance value than the smaller value that was assumed to apply to quasars (and by extension, to all AGNs) when this early model was calculated. Also, the observed [N II] lines are stronger than predicted; [N I] is also measured to be rather strong though no calculated value is available from this model for comparison with it. Again to a first order, the discrepancy can be corrected by increasing the assumed nitrogen abundance about fivefold, which also agrees better with recent ideas about the overabundance of N in AGNs.

In the best models available at the time of writing, all the processes described in Chapter 11 are taken into account. These include Auger transitions in heavy ions, and collisional excitation and ionization by the fast electrons produced by them and by photoionization by high-energy photons. Also, line excitations by photoionization, leaving the residual ion in an excited level of the ground configuration, are taken into account. Charge-exchange reactions are, of course, especially important in AGN models because of the wide range of ionization.

Once the relative abundances and the electron density N_e are specified, the cloud model is completely specified by the form of the photoionizing spectrum and the value of the ionization parameter

$$\Gamma = \frac{1}{4\pi r^2 c N_e} \int_{\nu_0}^{\infty} \frac{L_\nu}{h\nu} d\nu = \frac{Q(H^0)}{4\pi r^2 c N_e}, \tag{12.5}$$

(as previously defined) at the inner face of the cloud. The integration is carried forward from there into the cloud until the ionization has dropped so low, and with it the heating rate and the temperature, that further contributions to the emission lines are negligible. This is the assumption of optically thick clouds; we can instead terminate the integration at any specified physical dimension or optical depth, if these are taken as known or specified. Because of collisional-deexcitation effects, the computed structure and emission depend not only upon Γ but upon N_e as well.

The simplest type of model is then one specified by the input spectrum (for instance, the exponent n of an assumed power law), Γ, N_e, and a set of assumed abundances. A more sophisticated model can be built up as a weighted sum of such simple models, with different values of Γ and/or N_e, representing a distribution of clouds at different distances r from the central source and/or densities. There is thus a great range of possible models, which would be considerably reduced if we could adopt a definite physical picture, for instance, a cylindrically symmetric distribution of clouds with an exponential height distribution, a power-law radial distribution, and a power-law spectrum of densities. To date nothing this complicated has been attempted, because we do not yet have a physical basis for such assumptions.

These types of models may be used to analyze and interpret the observational results. As a sample, Figures 12.1, 12.2 and 12.3 show measured diagnostic line-intensity ratios, corrected for reddening, for a large sample of emission-line galaxies. The open circles are H II regions in external galaxies, starburst, or H II region galaxies, objects known to be photoionized by OB

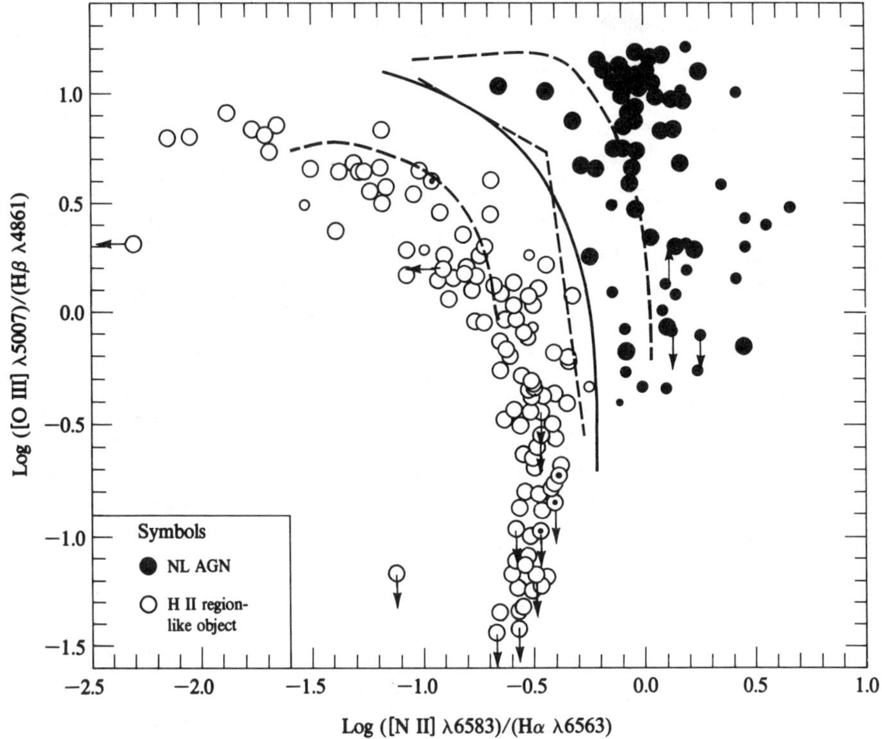

FIGURE 12.1
[O III]/Hβ *vs.* [N II]/Hα diagnostic diagram for emission-line galaxies. Observed galaxies shown by symbols indicated at lower left. The solid line is the dividing line between active galactic nuclei (*upper right*) and H II-region galaxies (*lower left*). The dashed lines are calculated results from models, as described in the text.

stars. The black circles are AGNs. The ratios were chosen to give the best separation of the two classes of objects; essentially the [O III]/Hβ ratio is mainly an indicator of the mean level of ionization and temperature, while the [O I]/Hα and [S II]/Hα ratios are indicators of the relative importance of a large partially ionized zone produced by high-energy photoionization. The significance of the [N II]/Hα ratio is not so immediately obvious, but it also gives a good separation between H II region nuclei and AGNs. The solid curve on each diagram is the best empirical dividing line between the two types of objects, as deduced from these data. No doubt there is some observational scatter, but note that the ratios have been chosen to minimize the effects of dust extinction. Some of the galactic nuclei close to the dividing line probably contain both OB stars and an active-nucleus hard-photon source.

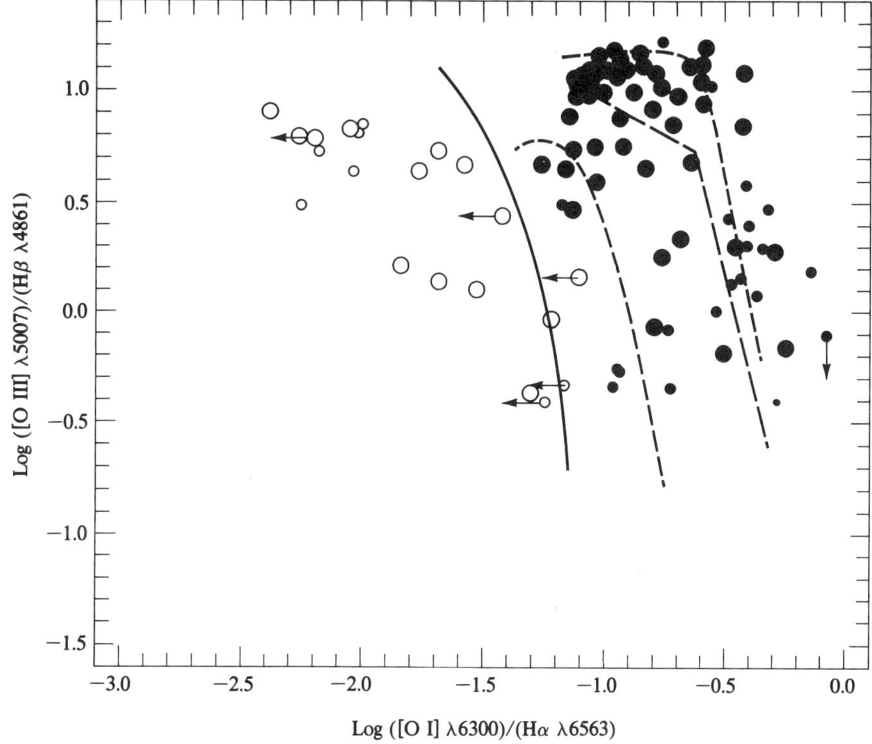

FIGURE 12.2
[O III]/Hβ *vs.* [O I]/Hα diagnostic diagram for emission-line galaxies. Symbols and lines as in Figure 12.1.

Several sets of computed models are also shown on these three diagrams. Two sets, indicated by short dashed lines, are simple models with assumed power-law spectra with exponent $n = 1.5$, electron density essentially $N_e = 10^3$ cm^{-3}, and either essentially solar abundances (upper right curve on all three diagrams), or abundances of all the heavy elements reduced tenfold relative to H and He (lower left curve on all three diagrams). The ionization parameter varies from $\Gamma = 10^{-1.5}$, at the upper left end of each curve, to $\Gamma = 10^{-4}$ at the lower right. The third set, indicated by the long dashed line segments, consists of three composite models, also with an assumed power-law spectrum with exponent $n = 1.5$ and with the same solar abundances, but containing two types of clouds, with densities $N_e = 10^6$ and 10^2 cm^{-3}, respectively. Both types of clouds are taken to have the same ionization parameter which varies along the line segments from $\Gamma = 10^{-2}$ at the upper left to $\Gamma = 10^{-4}$ at the lower right. In this model the denser clouds are therefore assumed to be, on average 10^2 times closer to the ionizing source than the lower density clouds. The assumed solar abundances used in calculating these models are listed in Table 12.2.

TABLE 12.2
Assumed relative abundances in AGN models

Element	Abundance	Element	Abundance
H	10^4	Mg	0.26
He	10^3	Si	0.40
C	4.7	S	0.33
N	1.0	Al	0.063
O	8.3	Fe	0.16
Ne	1.1		

These diagrams show that the AGN models predict line ratios in the general area in which the measured ratios lie. The general picture of photoionization by a spectrum to high energies is consistent with the observational data. On the other hand, models calculated for photoionization by OB stars (not plotted on these diagrams) do not agree with the observed AGN ratios, but do agree with the H II region and starburst-galaxy measurements. Looking more closely at the AGNs, we see that their measured ratios on the [S II] and [O I] diagrams mostly fall between the solar-abundance and 0.1 solar abundance simple-model sequences. They would roughly agree with abundances averaging about 0.3 of solar abundances, with considerable scatter about this mean. On the other hand, in the [N II] diagram many of the observed ratios indicate abundances higher than solar; perhaps (by extrapolation) a factor 1.5 times solar would represent a good average. To a first approximation, since O and H, not N, dominate the heating and cooling, the observed and predicted line ratios could be brought into agreement by increasing the N abundance by a factor of 1.5. Thus it appears that N may be somewhat overabundant relative to the other heavy elements in these narrow-line regions of typical AGNs.

However, note that the two-component models, which are plotted only for solar abundances, are displaced from the corresponding simple models in the direction of lower heavy-element abundances. This is a consequence of collisional deexcitation, which tends to weaken many of the forbidden lines at densities $N_e \approx 10^6$ cm^{-3}, and thus requires a higher abundance to reproduce the same ratio of a heavy-element line relative to an H I line. Differences in line profiles, to be discussed in Section 12.7, show that these collisional deexcitation effects do occur, and that the composite models containing gas at densities 10^6 cm^{-3} as well as at lower densities are relevant. Hence it is clear that the simple one-component low-density models underestimate the abundances. More sophisticated models with a wider range of densities and of elemental abundances are certainly required, but very crudely these two-

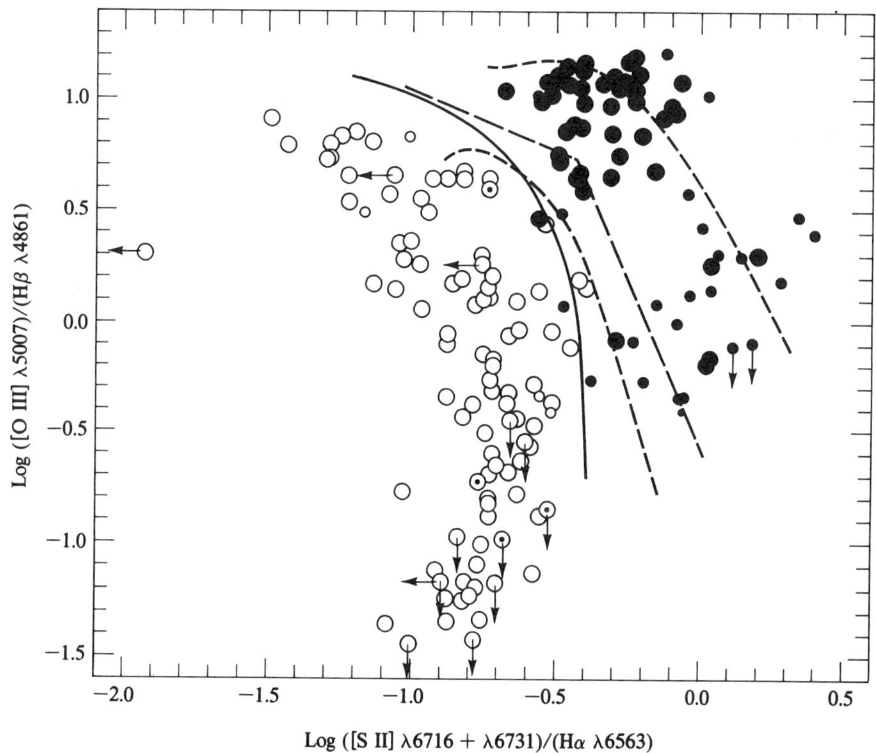

FIGURE 12.3
[O III]/Hβ *vs.* [S II]/Hα diagnostic diagram for emission-line galaxies. Symbols and lines as in Figure 12.1.

component models suggest approximately normal abundances of O and S, but an over-abundance perhaps three times normal of N. However, these abundances are still highly model-dependent; other analyses of a few specific AGNs, based on still more sophisticated models, have given essentially solar abundances for all three of these elements.

The two-component models do not predict [Fe VII] λ6087 and [Fe X] λ6375 to be as strong as observed in many Seyfert 2 nuclei, no doubt because they are too simplified. A relatively small amount of gas, close to the ionizing source and hence with a relatively large Γ, is probably present in these real galaxies, but not in the models, which have only two "average" types of clouds. Models with a continuous distribution of gas, extending in close to the ionizing source, such as the model used for comparison with Cyg A in Table 11.2, do in fact predict [Fe VII] and [Fe X] with roughly the observed intensities. Photoionization by an assumed hard spectrum seems to explain approximately the observed emission-line intensities. It is the best hypothesis to follow in seeking a complete physical picture of the nature and structure of AGNs.

12.4 LINERs

Seyfert 2 galaxies have relatively high ionization. All the classical objects of this type have ([O III] $\lambda5007$)/Hβ ≥ 3, and for most of them it is ≥ 5. This criterion alone is not enough to make an observed galaxy a Seyfert 2, however, for as Figures 12.1, 12.2 and 12.3 show, many H II-region galaxies also satisfy it. Most starburst and H II region galaxies do have lower ionization, with typically ([O III] $\lambda5007$)/Hβ < 3. Again, however, not all galactic nuclei that satisfy this criterion are photoionized by OB stars. Many low-ionization galaxies have stronger [O I] $\lambda6300$ and [S II] $\lambda\lambda6716$, 6731 than do H II-region or starburst galaxies. These objects have been named "Low-Ionization Nuclear Emission-line Regions" or "LINERs," and they are the subject of this section.

The original observational definition of a LINER was a galaxy nucleus with emission-line ratios ([O II] $\lambda3727$)/([O III] $\lambda5007$) ≥ 1, and ([O I] $\lambda6300$)/([O III] $\lambda5007$) $\geq 1/3$. The first of these criteria is satisfied by many H II-region galaxies, but the second is not. Because of the difficulties in comparing lines over a wide range of wavelengths, and particularly in making comparisons with $\lambda3727$, which can be strongly affected by interstellar extinction, perhaps a better definition, which is usually but not always equivalent, is ([O III] $\lambda5007$)/Hβ < 3 and ([O I] $\lambda6300$)/Hα > 0.05, ([S II] $\lambda6716 + \lambda6731$)/H$\alpha$ $>$ 0.4, and ([N II] $\lambda6583$)/Hα > 0.5.

In many LINERs and H II-region galaxies the emission lines are quite faint, and therefore difficult to see, badly affected by the underlying integrated absorption-line spectrum of the stars in and near the nucleus. This, of course, is especially true at Hβ and Hα, which are seen as absorption lines in almost all galaxies, their strengths depending on the spectral type. Hence to be certain whether or not the emission lines are present, and to measure their strengths at all accurately, we must correct the observed spectrum for the underlying galaxy spectrum. This can be done by subtracting a template spectrum of a galaxy without emission lines. All galaxies do not have identical absorption-line spectra; so the real problem is to subtract the spectrum the galaxy would have if it had the same population of stars, but no ionized gas. We cannot do this exactly, and probably cannot even define the operation logically, since the stellar population, the presence and amount of gas, the elemental abundances in it, and the ionization conditions are all linked through the past evolutionary history of the galaxy. The best approximations are to take the spectrum of another galaxy with very weak emission lines, or to add the spectra of representative stars of various spectral types to form a weighted average absorption-line spectrum template. With either method, we know that the template is approximately correct if it cancels out the absorption lines that are not at the positions of the emission lines, leaving only the latter and an essentially featureless continuum or nearly zero flux.

Surveys of this type have been made of a good sample of bright spiral galaxies. They show that essentially every spiral galaxy nucleus has at least Hα and [N II] emission lines in its spectrum. Among the earlier-type galaxies of the Hubble sequence, Sa and Sb, a large fraction, perhaps 80 percent, are LINERs, and the remainder are H II-region galaxies; there is an abrupt change at Sc, to about 20 percent LINERs and 80 percent H II.

When the class of LINERs was first isolated, it was suggested that they might be objects in which shock-wave heating rather than photoionization is the main energy-input mechanism. Since only a few emission lines are observed in them, this possibility cannot be proved or ruled out. However, it now seems far more likely that LINERs are simply the extension of Seyfert 2 nuclei to lower luminosities, smaller ionization parameters, and somewhat larger exponents n in the representative power-law spectrum. Figures 12.1, 12.2, and 12.3 extend smoothly to LINERs. Several objects just a little above the LINER class, with [O III] $\lambda 5007/\mathrm{H}\beta \approx 3$, have been observed to have reasonably strong He II $\lambda 4686$, which can only be explained in terms of photoionization. Shocks with velocity low enough to produce such a weak [O III] line would have too low a temperature to ionize He^{+} to He^{++}, and would not produce observable He II. No examples are known of higher-ionization, higher-temperature nuclei of spiral galaxies that are unequivocally shock-wave heated. Thus continuity argues strongly that the LINERs in spiral galaxies are also objects photoionized by a spectrum extending to high energies.

The computed models of Figures 12.1, 12.2 and 12.3 extend down to the regions of the observed LINER spectra for ionization parameters around $\Gamma \approx 10^{-3}$ to 10^{-4}. Another set of models, in which a power-law spectrum with exponent $n = 2$ was assumed, and a distribution of number of clouds with distance from the ionizing source was also assumed, gives a good match to the observed LINER spectra. Thus evidently many spiral-galaxy nuclei contain weak AGNs, in which a relatively low-luminosity central source produces a LINER.

Although emission lines are on the average much weaker in elliptical galaxies than in spirals, some nuclei of ellipticals do have LINER spectra. The two best known, bright examples with the strongest emission line spectra are NGC 1052 and NGC 4278. Even in them, however, the emission lines are not strong, and are badly blended with the absorption lines of the integrated stellar continuum. In NGC 1052 the [O II] $\lambda\lambda 3726, 3729$ lines are partially resolved; they and [S II] $\lambda\lambda 6716, 6731$ agree with an electron density $N_e \approx 2-3 \times 10^2$ cm^{-3}, near the low density limit for both ions. This corresponds, from the observed Hα luminosity, to a mass of ionized gas of only $5 \times 10^5\ M_{\odot}$.

The source of ionization in these elliptical-galaxy LINERs is not as clear.

The continuity arguments do not apply. The relative strengths of the few lines seen can be matched either by photoionization, as in the spiral-galaxy nuclei, or by shock-wave heating. Attempts to detect a weak featureless continuum in the satellite ultraviolet spectral region of NGC 1052 have been unsuccessful, and point toward shock heating, but are somewhat inconclusive. The crucial test should be the temperature indicated by the [O III] ($\lambda 4959 + \lambda 5007)/\lambda 4363$ ratio, but the measured strength of the very faint $\lambda 4363$ is critically dependent upon the correction for underlying absorption lines in the underlying galaxy spectrum. Four groups attempted to measure this ratio in the years 1976 through 1984, presumably each time with improved techniques. Their results for the derived [O III] temperature in the nucleus of NGC 1052 were 33,000°, 39,000°, 26,000°, and $\leq 17,600°$ K. The first two papers concluded that shock heating was the mechanism; the last two, that it was photoionization. However, as will be discussed in Section 12.8, a very weak broad Hα emission component has been detected in the elliptical galaxy NGC 1052. It thus has a low-luminosity Seyfert-1-type nucleus within it. Undoubtedly photoionization is the input source in it, and probably in other elliptical LINERs as well.

12.5 Broad-Line Region

Broad permitted emission lines are the characteristic feature of Seyfert 1 galaxy and BLRG nuclei, quasars and QSOs. They arise in the small, dense regions close to the central ionizing source. Thus they contain very important information on the structure right at the heart of the AGN. But because the density within them is so high, typically $N_e \approx 10^9$ cm^{-3} or, as we shall see, even higher, the physics is far more complicated than in the narrow-line regions of AGNs, or in planetary nebulae and H II regions. In many aspects, the BLRs are physically as closely related to stellar atmospheres as to traditional nebulae. Thus the conclusions that can be drawn from the observations are more highly model-dependent than for nebulae, and hence less certain.

Consider a dense cloud close to the photoionizing source. If the electron density within it is $N_e = 10^9$ cm^{-3} rather than, say, 10^5 cm^{-3}, as in a typical narrow-line cloud, but the dense cloud is 0.3 pc rather than, say, 30 pc from the source, the ionization parameter Γ will be the same. However, the cloud is so dense that the forbidden lines are all greatly weakened by collisional deexcitation. The temperature is therefore raised to the point at which the energy is radiated away by collisionally excited permitted and semiforbidden lines, mostly in the ultraviolet, such as C IV $\lambda 1549$ and C III] $\lambda 1909$. One important coolant is, of course, H I Lα itself. The calculated temperatures in the BLR models are somewhat higher than in NLR models, because of this increased collisional deexcitation, but $T \approx 15,000°$ K is perhaps a typical value, since the radiative cooling rises steeply with temperature.

Optical-depth and radiative-transfer effects are very important in the BLRs. Since much of the ionization occurs by high-energy photons, the optical depth τ_0 at the Lyman limit may be as large as 10^2, giving an optical depth τ_{0l} in the center of Lα of order 10^6. (This is correct for thermal Doppler broadening only, but overestimates the optical depth over larger distances in which the velocity field varies.) Hence, a Lα line photon emitted in the cloud is scattered many times before it escapes or is absorbed in some other physical process. The full physical problem, taking into account the variation of emission and absorption with depth into the ionized cloud, and with frequency in the line profile, is complicated . The simplest way to handle it is by the escape-probability formalism sketched in Section 4.5. For optical depths $\tau_{0l} \le 10^4$, we can calculate analytically the mean number of scatterings a typical resonance-line photon suffers before it escapes, and the mean distance it travels; for larger optical depths Monte Carlo calculations are necessary. For instance, for photons emitted uniformly in a plane parallel slab of optical thickness τ_{0l}, good fits to $N_{\rm esc}$, the mean number of scatterings before escape, are given by

$$N_{esc} = \begin{cases} 1.11 \;\; \tau_{0l}^{0.826} & (\tau_{0l} < 1) \\ \\ \dfrac{1.11 \;\; \tau_{0l}^{1.071}}{1 + (log_{10}\tau_{0l}/5)^5} & (\tau_{0l} > 1). \end{cases} \tag{12.6}$$

Thus, for instance, in a cloud with an optical depth $\tau_{0l}(\mathrm{L}\alpha) = 10^6$, $N_{\rm esc} = 1.2 \times 10^6$ is the average number of times a Lα photon is scattered before escaping; to put this another way, the escape probability of each Lα photon emitted is $1/N_{\rm esc} = 8.6 \times 10^{-5}$.

In an optically thin nebula, every time an H^0 atom reaches the 2 ^2P level as a result of recombination, cascading down from higher levels, or collisional excitation from the ground level, it spends a mean lifetime $\tau_{2\,^2P} = 1/\mathrm{A}_{2\,^2P,\,1\,^2S} = 1.6 \times 10^{-9}$ sec in the excited level before decaying. But in an optically thick nebula the photon emitted does not escape directly, but instead is absorbed, leads to another radiative excitation to 2 ^2P, is emitted again, and so on. Thus the average time that *some* atom spends in this excited level as a result of each recombination, cascading or collisional excitation is $N_{\rm esc}\tau_{2\,^2P} = 1.8 \times 10^{-3}$ sec, a very large increase. Hence the population in the 2 ^2P level is quite significant. Since the electron density in the BLR region is high, collisional excitation to other levels $n\ ^2L$ can occur, leading to collisionally excited Balmer, Paschen, etc. lines. Since transitions to the 3 ^2L levels have the smallest threshold and largest cross sections, Hα is especially favored. Angular momentum-changing collisional transitions to 2 ^2S have zero threshold energy and are even more favored, coupling the populations of it and 2 ^2P. The finite populations in these two levels make for non-negligible optical depths in the Balmer lines, and hence lead to radiative-transfer effects on them of the type discussed in Section 4.5.

Note further that a Lα photon, with energy $(3/4)h\nu_0 = 10.2$ eV, is an ionizing photon for H^0 atoms in the excited $2\,^2S$ and $2\,^2P$ levels. Since they have significant populations, this process occurs, "destroying" at least the Lα photon absorbed, and often also the one that originally led to the population of the $2\,^2P$ level that absorbed it. One further process that can destroy Lα photons is collisional deexcitation of $2\,^2P$. If the Lα photon escaped freely, the critical density for this process would be, by equation (3.31)

$$N_c(2^2P) = \frac{A_{2\,^2P,\,1\,^2S}}{q_{2\,^2P,\,1\,^2S}} = 8.7 \times 10^{16}\ \mathrm{cm}^{-3}$$

at $T = 10,000°$ K, using the collision strength listed in Table 3.12. But in the optically thick case, the mean lifetime in the excited state is increased by a factor N_{esc}, corresponding to decreasing the effective transition probability, and hence the critical density, by this same factor. Thus for the example with $\tau_{0l} = 10^6$,

$$N_c(2\,^2P) = \frac{A_{2\,^2P,\,1\,^2S}}{N_{\mathrm{esc}}q_{2\,^2P,\,1\,^2S}} = 7.5 \times 10^{10}\ \mathrm{cm}^{-3}.$$

This density is relatively high in comparison with the mean density $N_e = 10^9$ cm^{-3} derived for a typical BLR in Section 11.6, but may well be reached in some parts of some BLRs.

All these processes must be taken into account in calculating the broad-line H I spectrum of a model AGN. As an illustration, Table 12.3 gives results calculated from assumed homogeneous pure H "models," with constant electron density $N_e = 10^{10}$ cm^{-3}, assumed constant temperature as tabulated, and optical depth $\tau_{0l}(L\alpha) = 5 \times 10^6$. The relative intensities of the lowest three Balmer lines, as well as of Lα, all with respect to Hβ, are listed. The optical depths in Hα range from $\tau_{0l}(\mathrm{H}\alpha) = 64$ for $\Gamma = 10^{-6}$ to $\tau_{0l}(\mathrm{H}\alpha) = 4.5 \times 10^4$ for $\Gamma = 10^{-2}$. Taking the entire slab as homogeneous, rather than integrating the ionization thermal-equilibrium and radiative-transfer equations through it point by point is quite unrealistic, but perhaps illustrates some of the effects involved. At low T and Γ, most of the H is neutral, collisional excitation dominates, and therefore Hα/Hβ and Lα/Hβ are both very large. At larger ionization parameters, recombination becomes more important, but the higher temperatures (which were chosen to roughly mimic equilibrium values) enhance collisional excitation also. The competition between these

TABLE 12.3
Relative emission-line intensities in homogeneous BLR models

Γ	10^{-6}	10^{-5}	10^{-4}	10^{-3}	10^{-2}
T^a	8,000	10,000	12,000	14,000	16,000
Hα	28.3	8.78	3.61	2.71	2.38
Hβ	1.00	1.00	1.00	1.00	1.00
Hγ	0.54	0.18	0.30	0.36	0.38
Lα	58.3	13.0	6.7	9.8	30.3

^aT in ° K.

two effects, as well as the radiative-transfer effects in the Balmer lines, make the Lα/Hβ ratio first decrease, then increase, along the sequence as listed.

This ratio is important, because it is a straightforward indication of deviations from a pure recombination H I spectrum. Under Case B conditions, in the low density limit approximately 2/3 of all recombinations go through $2\ ^2P$ and lead to Lα emission (the actual number at $T = 10^4$ °K is 0.677), while the remainder go through $2\ ^2S$ and emit the two-photon continuum. The ratio of recombination-line intensities in this limit is thus

$$\frac{j_{\mathrm{L}\alpha}}{j_{\mathrm{H}\beta}} = 0.677 \frac{\alpha_B}{\alpha^{eff}_{\mathrm{H}\beta}} \frac{h\nu_{\mathrm{L}\alpha}}{h\nu_{\mathrm{H}\beta}} = 23.1.$$

In the high-density limit, collisions transfer atoms in the $2\ ^2S$ level to $2\ ^2P$ before they emit the two-photon continuum, and the intensity ratio is

$$\frac{j_{\mathrm{L}\alpha}}{j_{\mathrm{H}\beta}} = \frac{\alpha_B}{\alpha^{eff}_{\mathrm{H}\beta}} \frac{h\nu_{\mathrm{L}\alpha}}{h\nu_{\mathrm{H}\beta}} = 34.2.$$

Collisional excitation can only increase these ratios.

One of the first indications that the pure recombination conditions do not apply in BLRs was the discovery in the 1970s that Lα/H$\beta \approx 10$ is a more characteristic observed value for quasars and QSOs than 23 or 34. This conclusion has been abundantly verified since. How much of the discrepancy is due to the high-density effects discussed above, and how much is due to extinction by dust (see Section 12.6), is still not clear.

In addition to H I, He I, and He II (in which similar effects occur), permitted broad Fe II lines are also observed in the optical spectra of many Seyfert

1s and QSOs. An example is the spectrum of Mrk 376, shown in Figure 12.4. The strongest Fe II features are marked $\lambda\lambda$4570, 5190, 5320; they are unresolved blends of lines of several multiplets. They are shown with better resolution in Figure 12.5 in the spectra of Mrk 486 and I Zw 1, which have significantly narrower line widths. The wavelengths of the individual Fe II lines, grouped by multiplets, are drawn below the spectrum of I Zw 1. Note that the few individually resolved Fe II lines have essentially the same line widths as Hβ.

These Fe II permitted emission lines are not observed in planetary nebulae or H II regions. They are observed in T Tauri stars. Fe II is a low stage of ionization; the ionization potential of Fe^0 is only 7.9 eV, while the ionization potential of Fe^+ is 16.2 eV, between those of N^0 and Ne^0. These lines thus arise in the large partly ionized transition region of the BLR. The energy-level diagram in Figure 12.6 shows that the strong optical lines of Fe II observed in Seyfert 1 nuclei come from the energy levels of the terms $z\,^6D^o$, $z\,^6F^o$, $z\,^6P^o$, $z\,^4D^o$, $z\,^4F^o$, $z\,^4P^o$ between 4.8 and 5.6 eV above the ground $a\,^6D$ term. All the upper levels of the observed optical Fe II lines are connected with the ground term or the metastable $a\,^4F$ and $a\,^4D$ terms by strong permitted lines in the ultraviolet spectral region, in the range $\lambda\lambda2300 - 2800$. Thus the observed optical Fe II lines are similar to the Balmer lines of H I, while the ultraviolet resonance lines of Fe II are similar to the Lyman lines. The optical depths of the Fe II resonance lines are large in any reasonable model of the BLR, so any photons originally emitted in them are converted by multiple scattering to the longer-wavelength optical lines, with more highly excited lower levels. This analysis is confirmed by the relative strengths of the individual Fe II lines within the optical multiplets, which have clearly been evened out by multiple scattering and fluorescence.

The primary excitation mechanism of the Fe II emission is not completely understood. There is a range of strength of Fe II features in Seyfert 1 galaxies; Mrk 376, Mrk 486, and I Zw 1 are three of the objects in which they are strongest. Collisional excitation alone does not seem sufficient to produce such strong Fe II emission, at least with what are thought to be reasonable Fe abundances. Resonance fluorescence, if it were the primary mechanism, would deplete the continuum in the $\lambda\lambda2300 - 2800$ region, but this is not observed in QSOs with redshifts sufficient to bring it into the optical region. In fact, the Fe II resonance multiplets are observed in emission there, with approximately the strengths expected from multiple scattering and fluorescence for optical depths that fit the longer wavelength-multiplets. A combination of collisional excitation and resonance scattering seems to give the overall best fit to the observed spectrum, but this problem is unresolved at the time of writing. It is an extremely complex one, because Fe II has so many energy levels, and there are so many collisional and radiative transitions that connect them. Also, the collision strengths and (to a lesser extent) the transition probabilities are not at all accurately known.

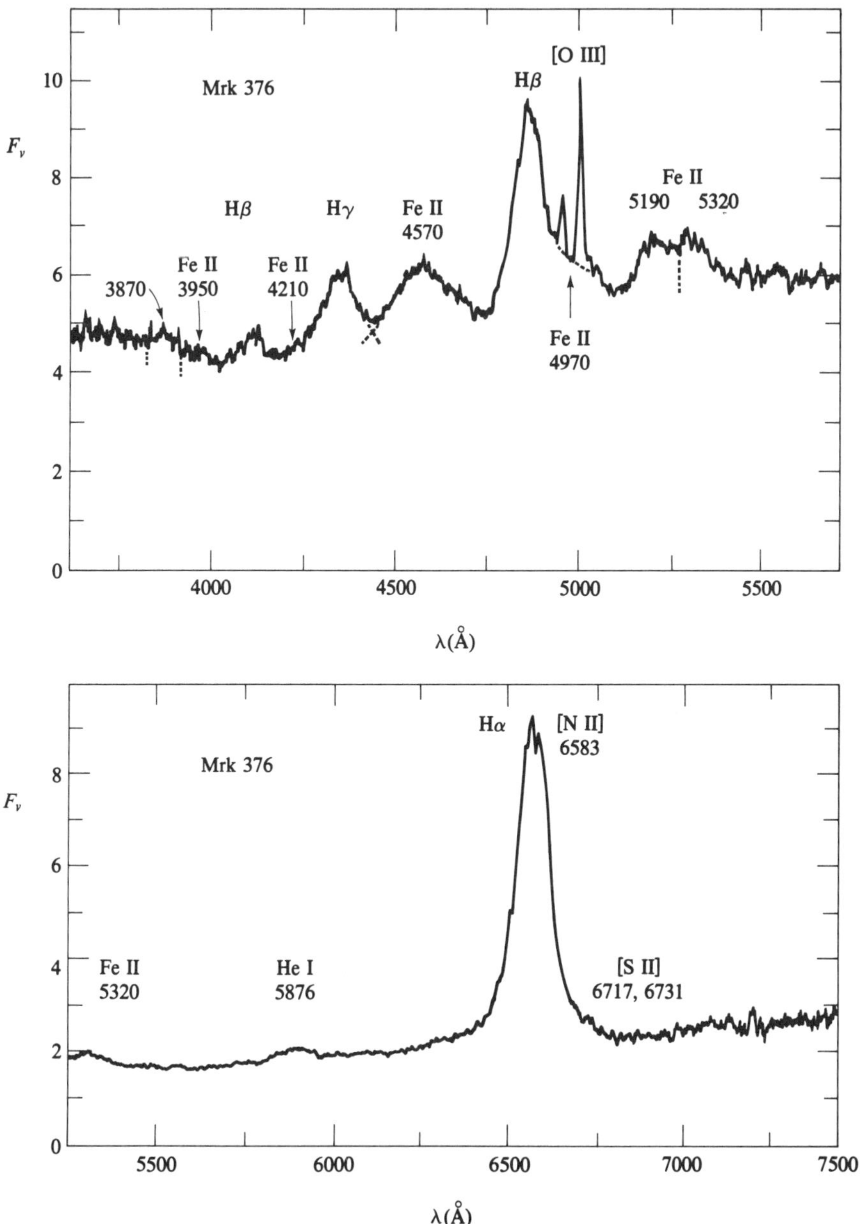

FIGURE 12.4
Spectral scans of Mrk 376, a Seyfert 1 galaxy with strong, broad Fe II and H I emission lines. Relative flux per unit frequency interval plotted *vs.* wavelength in the rest system of the object.

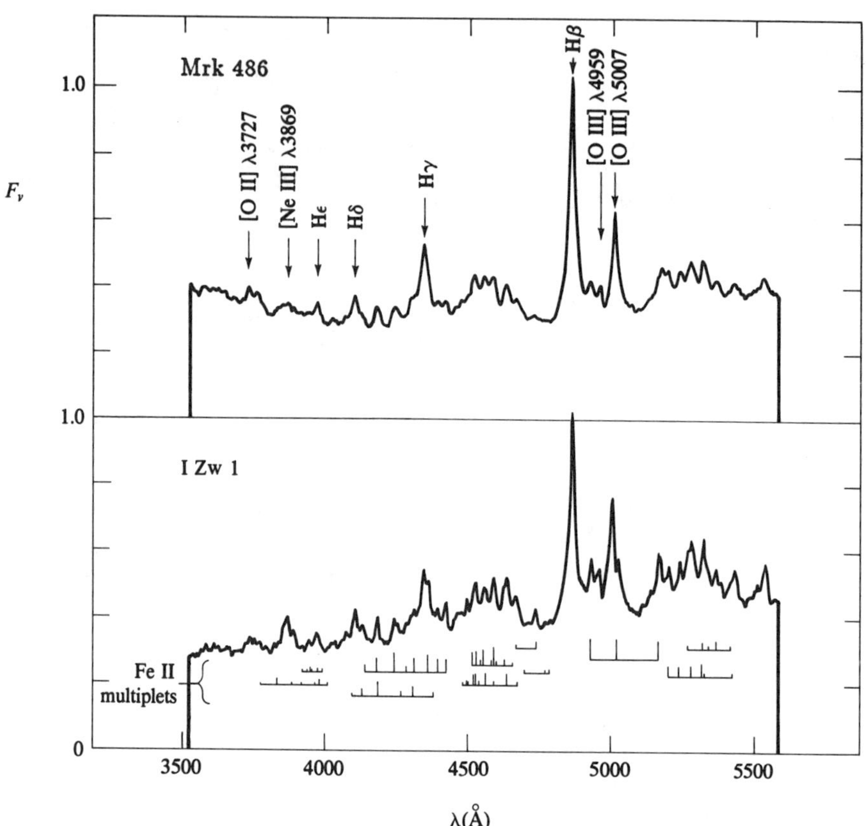

FIGURE 12.5
Spectral scans of Mrk 486 and I Zw 1, two Seyfert 1 galaxies with successively narrower Fe II and H I emission lines. Scales as in Figure 12.4.

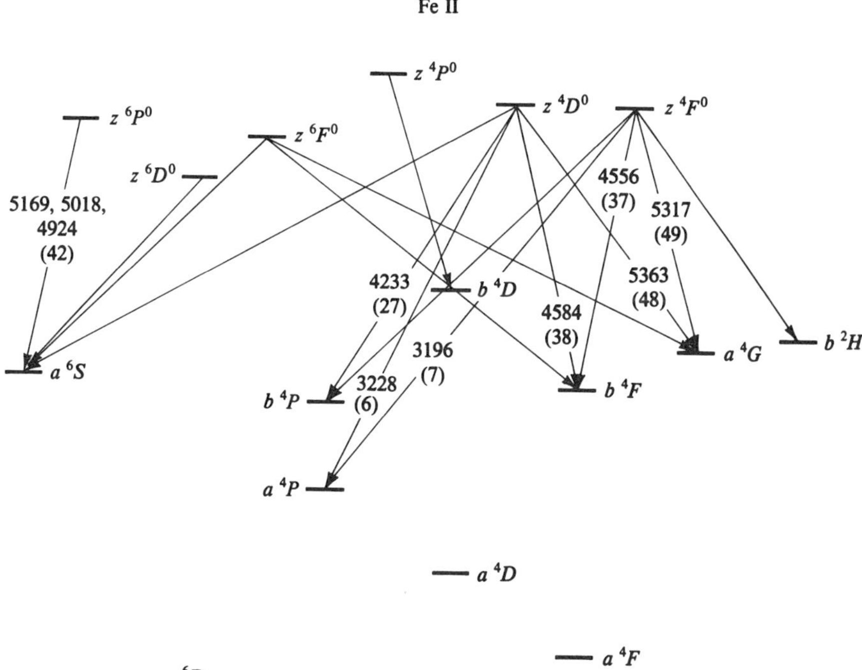

FIGURE 12.6
Schematic energy-level diagram of Fe II, with strongest observed optical emission-line multiplets indicated by wavelength and multiplet number. The still stronger ultraviolet resonance multiplets connect the lower $a\ ^6D$, $a\ ^4F$, and $a\ ^4D$ terms with the six higher z terms.

Since much of the cooling of the BLR occurs by collisionally excited lines in the ultraviolet, especially Fe II, C IV λ1549 and H I Lα, that spectral region is far more relevant for understanding and analyzing the BLR than the optical region. For Seyfert 1 galaxies this region is unobservable from the ground, and the IUE satellite, although it has provided very valuable data, is only a small telescope (28-inch aperture), whose observations are restricted to strong emission lines in bright objects. The Hubble Space Telescope, when it is launched, will undoubtedly provide very important new physical information on the nature of the BLRs and central sources in these objects. The large redshifts of distant quasars and QSOs make the ultraviolet spectra of these objects accessible to ground-based telescopes, but move their optical spectra into the near infrared region. Most of these objects are quite faint, but great progress has been made with large telescopes, particularly with fast linear detectors such as Reticons and CCDs. However, our knowledge of the BLRs in AGNs is still far behind our knowledge of H II regions and planetary nebulae because of these instrumental limitations.

The ultraviolet spectral region is most strongly affected by extinction by dust. Hence we will postpone the discussion of model predictions for this region until the end of the next section, following a discussion of dust in AGNs.

12.6 Dust in AGNs

Dust is undoubtedly present in AGNs. Its extinction effects can easily be seen in the Balmer decrements of observed NLRs, for instance, in the heavily reddened spectrum of Cyg A listed in Table 11.2. For a few objects there are measurements of the [S II] ratios $I(^4S - ^2P)/I(^2D - ^2P)$, mentioned in Section 7.2, comparing multiplets of the same ion with the same upper term, one in the violet spectral region and one in the near infrared. They give very similar extinctions to the values derived from the H I Balmer decrement. For well-observed NLR spectra, if we assume the average interstellar extinction curve of Section 7.2, calculate values of the extinction constant c from several different pairs of H I lines, and average them, it nearly always turns out that for dereddened NLR spectra the intrinsic Hα/Hβ ratio is larger than the recombination value 2.85. A more typical value for AGNs derived in this way is 3.1. As was explained in Section 11.8, this can be understood to result from the additional contribution of collisional excitation of Hα in the partly ionized transition region. It is the best overall average value to adopt for the intrinsic Hα/Hβ ratio in NLRs, along with values of the ratios of the higher Balmer lines to Hβ the same as in the case B recombination spectrum.

Dust is also present in the BLRs of Seyfert 1 galaxies and many quasars and QSOs, though its effects are not as directly obvious as in the NLRs. As explained above, in BLRs the H I Balmer-line ratios are modified by optical-depth and collisional-excitation effects, and are not known almost directly from physics alone as in NLRs, planetary nebulae and H II regions. Thus in a plot of measured $H\alpha/H\beta$ *vs* $H\beta/H\gamma$, the Seyfert 1 and BLRG nuclei do not cluster tightly around a reddening line, with slope determined by the standard interstellar extinction curve. Instead the observed ratios exhibit a general tendency to lie in the direction of this line, but with a relatively large scatter about it. In the Seyfert 1.8 and 1.9 nuclei, the observed ratios of broad $H\alpha/H\beta$ are very large, indicating strong dust extinction. For QSOs and quasars with emission lines measured over a wide range of wavelength, extending far into the ultraviolet spectrum, approximate agreement with models requires that the observed line strengths be corrected for extinction. This is particularly shown by the small observed $L\alpha/H\beta$ ratio mentioned in the previous section, and by similarly small observed He II $\lambda1640/\lambda4686$ ratios.

As a final example, in Table 12.4 a typical composite spectrum for QSOs is listed, obtained by averaging many observed spectra. They have then been corrected for interstellar extinction, using a highly simplified extinction law

$$\tau_\lambda = Cf(\lambda) = D\lambda^{-1} \tag{12.7}$$

in terms of equation (7.3), that is, the dust optical depth for extinction has been taken to be proportional to $1/\lambda$. The constant D, that is, the amount of dust, has been adjusted to give the best overall agreement with the model listed in the last column of the table. This model was calculated with an assumed power-law input photoionization spectrum that increases toward high energies, with an exponential cutoff

$$L_\nu = B\nu^{1/2}e^{-\nu/\nu_c}, \tag{12.8}$$

with $h\nu_c = 200$ eV. It has a weak maximum at $h\nu = 67$ eV, and was chosen so that when reddened by the same extinction law, it approximately matches the observed continuum in a typical QSO.

Obviously this whole procedure is highly schematic, because we know almost nothing about the properties and spatial distribution of the dust in AGNs. The assumed λ^{-1} dependence of the effective extinction is convenient and over a wide range of wavelength is probably approximately the correct form, but no doubt is quite wrong in detail. The only well-determined extinction curve we know was derived from stars in our Galaxy near the Sun, and as can be seen from Figure 7.3 or Table 7.2, it only loosely approximates equation (12.7). Furthermore, as the discussion of Section 7.3 emphasizes, observations of distant nebulae (including AGNs) are affected differently by dust extinction than observations of distant stars. Scattering by dust along the line of sight removes light from the beam; scattering by dust within the object does not have the same effect, although it does lengthen the paths of the photons and thus gives true absorption more chances to be effective.

TABLE 12.4
Composite QSO BLR Spectrum and Model

Ion	λ (Å)	Observed Relative Intensity	Corrected for Reddening	Model
O VI	1034	20	31	38
Lα	1216	100	100	100
N V	1240	25	24	27
O IV]	1407	10	7	2
N IV]	1488	3	2	3
C IV	1549	40	23	30
He II	1640	5	2.6	3.5
O III]	1663	3	1.5	1
C III]	1909	18	7	5
Mg II	2798	23	5	4
He II	4686	2.5	0.37	0.44
Hβ	4861	17	2.6	1.7
He I	5876	4	0.5	0.1
Hα	6563	77	9.6	7.7
Pα	18751	6	0.5	0.4

Furthermore, it is not clear at all that the same extinction correction should be applied to the observed continuum as to the emission lines. If the dust is concentrated in or near the same clouds as the gas, as seems likely, and the latter fill only a small fraction of the volume with the AGN and have a covering factor less than one, as is quite possible, then the continuum observed at the Earth may be much less subject to extinction than the radiation from the emission-line clouds. On the other hand, the ionizing continuum incident on the gas in the emitting clouds may be more affected by extinction than the continuum observed at the Earth. This extinction has not been taken into account in the model of Table 12.4.

Finally, a very important effect of dust is the destruction of Lα line photons, and to a lesser extent other resonance-line photons, such as C IV λ1549. As discussed in Section 7.4, the great lengthening of the paths of these photons by resonance scattering makes their absorption by dust much more probable than for neighboring continuum photons. This effect has certainly been observed for C IV λ1549 in planetary nebulae. (The bright geocoronal Lα emission has to date frustrated attempts to measure quantitatively the fluxes of Lα in planetary nebulae.) The amount of destruction depends critically upon the path length of the scattered photon, the pure absorption of the dust at the wavelength of the line, and the distribution of dust and scattering atoms or ions within the AGN. This process has not been taken into account in the model of Table 12.4. It is certain to be important for Lα, and probably for C IV λ1549 as well, but its amount is difficult to calculate accurately.

Very probably the dust in AGNs is different from the dust in our Galaxy near the Sun. The radiation field, particularly of high-energy photons which are effective in destroying dust, is much more intense in AGNs. The evolutionary history of the dust in AGNs is also no doubt very different from the evolutionary history of dust in our part of the Galaxy.

At the time of writing the amount, nature and space distribution of the dust in AGNs are thus among the most important unknowns in this subject. Progress in understanding the effects of this dust will no doubt come from theory and observations combined. One important result is that in AGNs refractory elements, such as C and Fe, generally appear to be relatively more abundant than in the gas in planetary nebulae and H II regions. In the latter, and in interstellar matter in our Galaxy as observed in absorption lines, the abundances of these refractory elements are low relative to their abundances in stars. In nebulae and interstellar matter, appreciable fractions of these elements are apparently locked up in dust grains and are therefore unobservable. Probably in AGNs there is less dust, the dust particles are smaller, and they are composed of the more refractory elements. All these effects would be expected to result from the more extreme radiation fields in AGNs.

Probably there are differences in the properties of dust in the NLRs and the BLRs. There may be differences in the amount or arrangement of the dust in Seyfert 1 and BLRG nuclei, for the latter have on the average greater amounts of reddening in their broad emission lines.

New information on the dust in AGNs will come from infrared observational data. AGNs are strong infrared sources. Part of this radiation comes from the extension of the featureless continuum into the infrared, part from heated dust. Separating the two components is difficult. However, the IRAS measurements of very many galaxies to wavelengths as long as 100 μ clearly show excess heated dust in AGNs. Similar excesses are seen in starburst or H II-region galaxies. The whole subject of the dust in AGNs will no doubt be further clarified as these observational data are interpreted; certainly much remains to be learned.

12.7 Internal Velocity Field

The characteristic feature of the spectra of AGNs, in addition to the wide range of ionization covered by their emission lines, is that these lines are broad. The broadening clearly results from the velocity field in the ionized gas within the AGNs. Understanding this velocity field and how it arises is crucial for understanding the nature of AGNs. At the time of writing, such understanding does not exist, and research in this field is of very great importance.

Observationally, the line profiles in Seyfert 2 galaxies typically have full widths at half maximum of about 500 km sec^{-1}, noticeably wider than the emission lines in the nuclei of starburst or H II galaxies, or the absorption lines of the integrated stellar spectra of normal galaxies. To a first approximation, all the emission lines in a given Seyfert 2 nucleus have the same FWHM, but the values cited usually refer to [O III] $\lambda\lambda 4959$, 5007. The line widths in different Seyfert 2 galaxies range from about 250 km sec^{-1} to about 1,200 km sec^{-1}. NGC 1068, which has often been called a "typical" Seyfert 2, in fact has an extreme FWHM = 1,200 km sec^{-1}.

The line profiles can be fitted to a first rough approximation by a Gaussian, but to a better approximation the observed profiles do not fall to zero that rapidly and have more extensive wings. Furthermore, they are often asymmetric, with the wing almost always extending further to the blue (shorter wavelength) than to the red. All these features can be seen, for example, in the profiles of four emission lines in the spectrum of IRAS 1319 − 164, plotted in Figure 12.7. These profiles, as is generally the case in Seyfert nuclei narrow-line profiles, are quite similar in form to one another, differing only in velocity scale. Different galaxies differ from one another in degree of asymmetry, and in the FWHMs of the various lines.

Furthermore, in most well-studied Seyfert 2s, there are regularities in the line widths. For most, but not all, there is a correlation of FWHM with critical density for collisional deexcitation, in the sense that the lines with higher critical densities tend to have larger FWHMs. For instance, in one well-observed sample of Seyfert 2 nuclei, [O III] $\lambda 4363$, with critical density $N_c(^1S) = 3 \times 10^7$ cm^{-3} is broader in 70 percent of the objects than [O I] $\lambda 5007$, with $N_c(^1D) = 7 \times 10^5$ cm^{-3}. Likewise [O I] $\lambda 6300$ with $N_c(^1D) = 2 \times 10^6$ cm^{-3} is broader than [S II] $\lambda\lambda 6716$, 6731, with $N_c(^2D) = 2 \times 10^3$ cm^{-3}, in 66 percent of the objects.

The narrow-line profiles in Seyfert 1 and 1.5 profiles are quite similar to those in Seyfert 2s, covering essentially the same range of widths and exhibiting very similar asymmetries. However, in Seyfert 1 and 1.5 nuclei there is more tendency for the FWHMs to be correlated with ionization potential, in the sense that ions with higher ionization potential have larger FWHMs. However, none of these tendencies are universal. A minority of Seyfert 1 nuclei show a correlation of FWHM with critical density, similar to the one described for Seyfert 2s above, and a minority of Seyfert 2s have the correlation of FWHM with ionization potential. In some members of both groups there is little correlation of width with either critical density or ionization potential, or the range in FWHM is small.

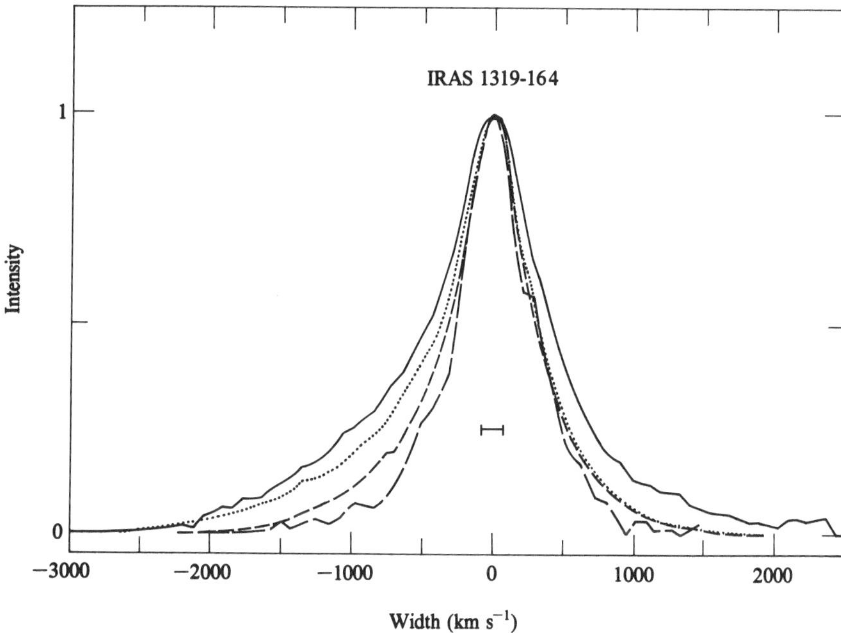

FIGURE 12.7
Line profiles of several lines in IRAS 1319-164, all plotted in velocity units and normalized to same peak intensity. Solid line [O I] λ6300; dotted, Hα; short dashed, [N II] λ6583; long dashed, [S II]λλ6716, 6731; average. Instrumental profile FWHM indicated at zero velocity.

given AGN. Presumably the highest ionization occurs closest to the central source; the correlation of line width with ionization potential shows that the highest internal velocities also occur there. In a structure with a wide range of electron densities, the region or regions with density near the critical density of any energy level is most effective in the emission of the line (or lines) arising in that level. The correlation with line width seems to show that these regions also are nearest the central source. At least in the AGNs with broad [O I] λ6300 profiles, the emission-line clouds must be optically thick to ionizing radiation, so that both high-ionization lines, like [O III] λ4959, 5007, and low-ionization lines, like [O I] λλ6300, 6364, are emitted by the same high-velocity clouds.

The asymmetric profiles of the emission lines, with wings almost always extending to the blue, can be understood as resulting from extinction by dust if the ionized gas is flowing outward more or less radially, or more or less perpendicularly to the central plane of the AGN. Then if the dust is mixed with the ionized gas, or concentrated to the center or central plane of its distribution, in almost any fashion except completely outside the ionized volume, an asymmetry of this type will be introduced. Line photons emitted on the more distant side of the structure will pass through more dust on their way to us, and will suffer more extinction. If the further side is moving away from us, there will be fewer photons observed from the red side of the profile than the blue side. Thus an outward flow, with extinction, seems the best working hypothesis suggested by the observed form of the profiles. On the other hand, if the dust is assumed to be concentrated on the least-ionized side of the cloud, furthest from the central source, then the ionized part of the clouds on the *near* side of the structure suffer the most extinction. On this rival picture, the same line profiles thus indicate infall.

Note that if the electron density in the clouds is assumed, as an example, to decrease radially outward as a power law in distance,

$$N_e \propto r^{-m}, \tag{12.9}$$

then since the flux of ionizing photons decreases as r^{-2}, the ionization parameter at the face of the cloud is

$$\Gamma = \frac{Q(\mathrm{H}^0)}{4\pi r^2 c N_e} \propto r^{m-2}, \tag{12.10}$$

where $Q(\mathrm{H}^0)$ is the number of ionizing photons emitted by the central source per unit time. If $m = 2$, the ionization parameter and hence the degree of ionization at the front surface of all the clouds is independent of distance. If each cloud is optically thick, the degree of ionization decreases to zero with increasing optical depth within it, and all clouds have, to a first approximation, the same ionization and thermal structure. Hence, even though the velocities of the clouds depend on distance, the line profiles would not depend on ionization potential at all. On the other hand, the density variation would lead to collisional deexcitation in clouds close to the photoionization source and thus to a dependence of line profile on critical density. This is observed in many objects, as was stated above, in the sense that higher critical densities are correlated with larger FWHMs. We conclude that these objects have density variations fitted approximately by the exponent $m \approx 2$.

If, on the other hand, $m > 2$, the ionization parameter Γ would increase outward, and high stages of ionization like Ne^{+4} and Fe^{+6} would not exist in clouds close to the central source. If the velocity field were the same as in the previous case, the FWHM would *decrease* with increasing ionization potential. This situation has not been observed in any AGN; we conclude that $m > 2$ does not occur in nature. If $m < 2$, Γ decreases outward, and high stages of ionization can occur only in clouds close to the central source. For the same type of velocity field, the FWHM would therefore increase with increasing ionization potential. This is observed in many Seyfert galaxy nuclei. In the extreme case $m = 0$, there would be no density dependence on distance, and hence no correlation of FWHM with critical density.

Observationally, AGNs with FWHMs correlated with both N_c and ionization potential exist; they can be understood as objects with $0 < m < 2$, say $m \approx 1$. More Seyfert 1s have FWHMs correlated with ionization potential, that is, in simple terms, they tend to have $m \approx 0$, while more Seyfert 2s tend to have FWHM correlated with critical density; that is, to have $m \approx 2$.

What drives the flow out? One possibility is radiation pressure. The main effect is from continuum absorption. The main problem is that if radiation pressure is strong enough to accelerate the clouds, it would also disrupt them. Another possibility is an invisible, high-temperature, low-density wind flowing outward from the central source. Its pressure could explain how the emission-line clouds are confined to a fractional volume of order $\epsilon \sim 10^{-3}$ of the entire volume of the AGN, as estimated from the luminosity, electron density, and ionization parameter Γ. Whatever the mechanism, it seems from the observational data that the velocity must be directed outward and must decrease outward. Very probably gravitational deceleration is involved.

The broad lines in Seyfert 1 nuclei typically have FWHMs from 500 to 5,000 km sec^{-1}, with full widths at zero intensity (or actually, as close to zero as the profiles can be defined) ranging from 5,000 to nearly 30,000 km sec^{-1}. These are the mean FWHMs and FW0Is of the H I Balmer lines; generally Hβ is slightly broader than Hα. He I λ5876 tends to be broader than the H I lines (on the average by a factor about 1.3), He II still broader, and Fe II the same as H I or slightly narrower (to perhaps a factor about 0.75).

In some Seyfert 1 nuclei the broad lines are symmetric, in others they are asymmetric with a stronger blue wing, and in others they are asymmetric with a stronger red wing. There is no common asymmetry in the broad lines, as there is in the narrow ones. Hence the physical explanation of the velocity field in the BLR cannot be symmetric radial flow alone, with extinction. If the expansion picture is adopted, in some of the objects there must be significantly more material flowing away from us than toward us. The objects themselves must be asymmetric.

Another possible velocity field for the BLR is rotation under the gravitational field of the central black hole. This will give a symmetric line profile whether there is extinction or not; asymmetric profiles, some to the red and some to the blue, must be modeled on this picture again by deviations from symmetry in the object, either in the distribution of extinction or in the distribution of the gas in the BLR. A very attractive feature of the rotational interpretation is that the observed broad-line velocities are about what would be expected for rotation in the gravitational field of a black hole of the anticipated mass. For instance, an *average* rotational velocity 2,500 km sec^{-1}, at an *average* radius 0.035 pc, corresponds to a mass

$$M = \frac{rv^2}{G} = 10^8 \, M_\odot,$$

which is about the value expected for a black hole with the Eddington luminosity $L_E = 3 \times 10^{12}\ L_\odot$. Assumed models with dense clouds arranged in rotating disks, plus outflow centered about the axis of rotation, will give profiles similar to the observed ones.

Note further that on the rotational picture the highest velocities occur closest to the central object. Thus even though the BLR clouds are optically thick, the highest level of ionization, He II, is expected to occur preferentially closest to the source, and to have the broadest lines, He I, on the average somewhat further out, H I still further out, and Fe II can be emitted even in the most distant, only partly ionized clouds. The differences between Hα and Hβ must result from optical depth and collisional effects; detailed models show that in general Hβ/Hα strengthens either with increasing ionization parameter, Γ, or electron density, N_e, as does He I λ5876/Hβ. Thus the H I line profiles also agree with the higher velocity regions (which preferentially emit the wing of Hβ, since it is wider than Hα) having larger Γ (that is, smaller r) or higher N_e (which presumably also corresponds to smaller r). Thus the rotational picture, possibly with outflow also, is very attractive. However, a complete physical basis of the nature of such a rotating structure, the cloud structure within it, the temperature and density structure, their stability, etc., does not exist at the time of writing. No doubt research in this direction in coming years will be very productive.

12.8 Physical Picture

As this and the previous chapter have made clear, though much observational material and diagnostic data on AGNs is at hand, we do not have a complete physical picture of them as we do, for instance, of stars. What we need is a well-defined physical model, from which we can derive definite predictions that can be tested by observational data. Though such a model does not now exist, we have made steps toward it through partial understanding gained from the observations themselves. No doubt in years to come, much more progress will be made toward a complete model.

The observational data clearly show the importance of photoionization. We visualize it as arising in a central source, most likely powered by the rotating accretion disk around a black hole. The broad-line region is small (diameter $d \sim 0.1$ pc) and dense (mean $N_e \sim 10^9\ \text{cm}^{-3}$). Its velocity field seems to have a significant rotational component. It may be more nearly a disk than a spherically symmetric object, though there is partial evidence favoring either picture. The narrow-line region is much larger ($d \sim 10^2$ pc) and less dense (mean $N_e \sim 10^4\ \text{cm}^{-3}$). Its velocity field certainly has a large radial component. It may also be disk-like in form, or it may be more nearly spherically symmetric.

A highly schematic drawing of a possible arrangement is shown in Figure 12.8. It is not to scale and cannot be, since it shows both the BLR and the NLR. The central source, shown in black, emits ionizing photons. The BLR is a disk, much larger than the central source, which is optically thick to ionizing radiation near its equatorial plane, and optically thin near its poles. The NLR, which is much larger than the BLR, is drawn here as spherically symmetric. The regions near the pole are ionized by photons that have penetrated through the BLR, but the regions near the equator are neutral because they are shielded by the outer parts of the BLR.

Note that there is no reason to assume that the angular momentum vector, or axis of rotation, of the BLR disk is in the direction of the axis of rotation of the galaxy in which it lies. The angular momentum per unit mass in the BLR is very small in comparison with that of typical stars or interstellar matter at distances of the order of several kpc from the nucleus. The angular momentum of the very small, low-mass BLR may depend strongly on the specific events by which the mass in it arrived at the center. Observationally, there is no correlation between the widths of the broad emission lines (resulting chiefly from the component of rotational velocity in the direction of the observer) and the axial ratio of the observed galaxy (depending on its orientation with respect to the line of sight). High-resolution radio observations of jets, which are axial structures in the ionized plasma closely connected with the central source, show that they typically are not aligned with the overall structure of the galaxy, but more often are in arbitrary directions. These jets are much more likely to indicate the axis of the central source, and probably of the BLR also, than the axis of the galaxy is. Hence all of Figure 12.8 should be imagined as inclined by an unknown, perhaps almost random, angle to the overall plane of the galaxy.

Note also that although the drawing shows homogeneous regions, this is an extreme oversimplification. All the available information on luminosities, densities, and dimensions, whether inferred for the BLR from observed variations in the luminosities and profiles of the broad emission lines, inferred for the BLRs and NLRs from Γ, N_e, and L, or seen directly for a few resolved NLRs in the nearest AGNs, indicate that the filling factor is very small. Perhaps $\epsilon \sim 10^{-3}$ is a good overall estimate to keep in mind. Thus within the BLR and within the NLR there are strong variations in density, which can be thought of as density condensations, or "clouds." It is hard to imagine that these clouds are immersed in a vacuum, for they would very quickly expand to fill it and thus dissipate. Very probably there is a much more dilute, high-temperature medium, most likely an outflowing wind, between the clouds and in approximate pressure equilibrium with them. Thus in both the BLR and the NLR the dense, "cool" ($T \sim 10^4$ °K), ionized gas that is observed should be thought of as concentrated in clouds, as is emphasized for the NLR in the highly schematic Figure 12.9.

Here the NLR is shown as a disk, rather than as spherically symmetric, as in Figure 12.8. Either representation is consistent with the observational data. If the material in the NLR comes from the BLR, or goes into it, they probably are thick, and the degree of ionization therefore decreases nearly to

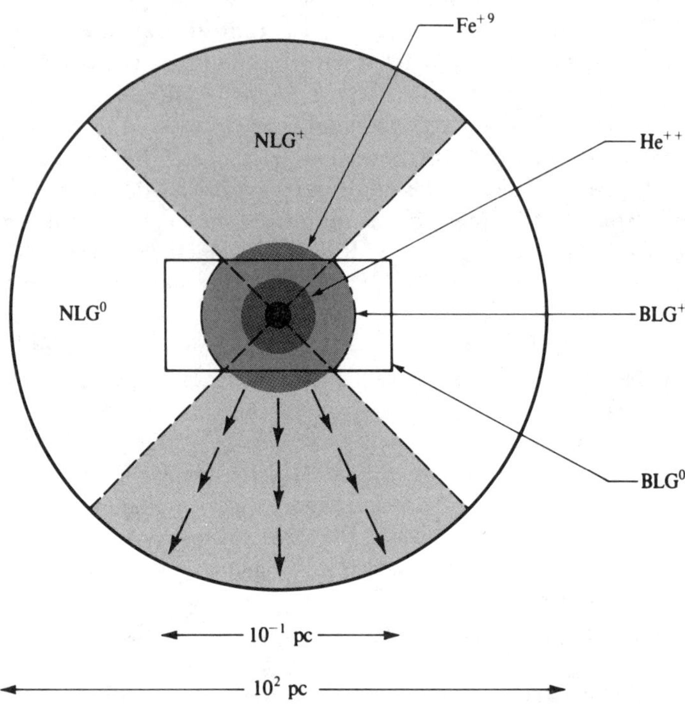

FIGURE 12.8
Schematic AGN model, showing central photoionization source as black filled circle, broad-line gas (BLG) in a disk ionized near source and neutral further away, and narrow-line gas in much larger sphere, ionized in the cone in which ionizing photons can penetrate the broad-line gas and escape. Note that the scale is distorted in the interest of legibility.

zero deep within each one. Since the BLR is not homogeneous, but also has a cloud structure, along some rays, even in the central plane of the BLR, there are no clouds, so ionizing photons escape in those directions also, but only over limited solid angles.

In this picture Seyfert 1, 1.5 and 2 galaxies differ in the amount of dense, broad-line gas near their nuclei. Objects with large, thick BLRs allow practically no ionizing photons to penetrate out into the NLR. They have almost pure broad line spectra, and are extreme Seyfert 1 galaxies. Examples are Mrk 376 and Mrk 42. Objects with no broad-line gas have completely ionized NLRs, and are pure Seyfert 2s. Examples are Mrk 1157 and Akn 347. Objects with intermediate amounts of broad-line gas, perhaps in a thinner disk, or perhaps in one scaled down in all directions, have mixed broad plus narrow-line spectra. An example (always according to this picture, it must be understood) is Mrk 926.

As indicated in Figure 12.8, the high-ionization lines, such as [Fe X] λ6375 and [Fe XI] λ7892, are emitted in the narrow-line region closest to the ionizing source, where Γ is largest. Actually it is not certain that these ions are produced by photoionization. The alternative is collisional ionization, which would require coronal temperatures, $T \sim 10^6$ °K. Since only one line of each is observed, there is no observational information on the temperature of the region in which they are emitted. However, there is a very good observational correlation between the strengths of [Ne V] and [Fe VII], between [Fe VII] and [Fe X], and between [Fe X] and [Fe XI], exactly of the type predicted by photoionization models. If [Fe X] and [Fe XI] are produced by collisional ionization, the amount of high-temperature coronal gas in AGNs must be closely linked to the high-energy end of the ionizing spectrum. The observed line profiles of [Fe X] λ6375 and [Fe XI] λ7892 also fit well with the correlations observed in the other narrow lines in Seyfert 1 galaxies. They have the highest ionization potentials, and generally have the largest FWHMs. Although their ionization mechanism is not certain, the best present working hypothesis is that it is photoionization. Note that only one Seyfert 1 galaxy (III Zw 77) is known to have [Fe XIV] λ5303 in its emission-line spectra; the ionization potential to produce this ion is 361 eV. III Zw 77 has quite strong [Ne V], [Fe VII], [Fe X] and [Fe XI] emission lines also.

The high-energy ionizing photons in AGNs are X-rays. Correspondingly, essentially all bright Seyfert 1 and 1.5 galaxies are observed X-ray sources, and most luminous X-ray sources that are galaxies are Seyfert 1 galaxies. The X-rays from some objects vary on time scales of a few days, indicating that the region in the central source that emits them is very small. Only a few Seyfert 2 galaxies have been observed as X-ray sources, but their nuclei are generally much less luminous in the featureless continuum than Seyfert 1 nuclei. The Seyfert 2s may have a smaller ratio of X-ray to optical featureless-continuum luminosities, that is, a larger value of the power-law index n, if equation (11.2) is used to fit their continuous spectra between optical and X-ray regions. However, very few Seyfert 2s have as yet been measured as X-ray sources; so this result is not certain.

The high-ionization line profiles, with FWHMs larger than those of the other narrow lines, but smaller than those of the broad lines, indicate that the BLR-NLR dichotomy is too extreme a simplification. Undoubtedly there is a smooth transition from one to the other, and a range of densities bridging the two. The velocity field must also have some continuity between the BLR and NLR. Likewise, many apparent problems remain in understanding the physics of the rotational and radial flows.

The picture sketched above, though attractive, is not the whole story, as has emerged in recent years from measurements of the polarization of the emission lines and the continuum in Seyfert galaxies. The best-studied object by far is NGC 1068, the bright, somewhat atypical Seyfert 2 mentioned previously and shown on the dust cover of this book. These measurements show that its forbidden lines are only slightly polarized, of the order of 1 percent but

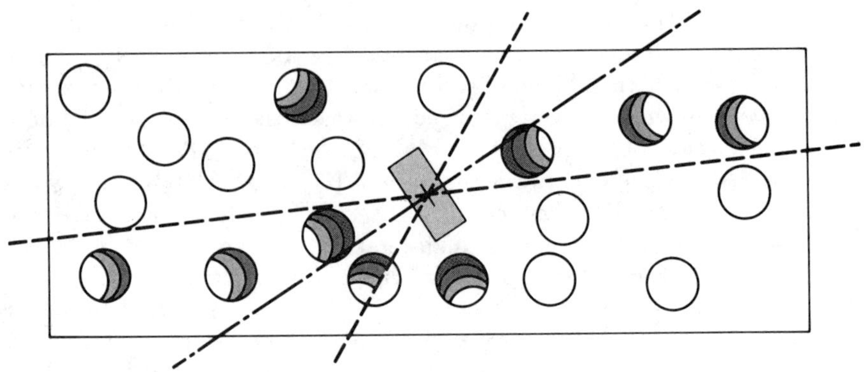

FIGURE 12.9
Schematic AGN model, showing tipped broad-line disk in larger narrow-line disk, with cloud structure indicated schematically. The ionizing photons mostly escape through the broad-line disk in the cone around its axis, but a few routes for escape also may exist even near its equators. Highest degree of ionization, indicated by darkest shading, occurs only in clouds nearest central source (for uniform density clouds).

the featureless continuum, after the galaxy integrated stellar absorption-line spectrum has been carefully removed, is strongly plane-polarized. The degree of polarization is approximately 16 percent, independent of wavelength and in a constant direction. Furthermore, in the polarized-light spectrum, broad Hα, Hβ, and Hγ emission lines can clearly be seen, with FWHM $\approx$ 3,500 km sec^{-1} and FW0I $\approx$ 7,500 km sec^{-1}. These are much wider than the strong narrow lines seen in the Seyfert 2 spectrum of NGC 1068, but are comparable with the widths of the H I lines in many Seyfert 1s. Furthermore, the polarized-light spectrum shows weakly, but definitely present, the broad unresolved Fe II $\lambda\lambda$4570, 5190, 5320 features that are characteristic of Seyfert 1 spectra. In other words, there is a "hidden" Seyfert 1 nucleus in the Seyfert 2 galaxy NGC 1068, that is clearly seen only in plane polarized light. Knowing they are there, we can just barely see the strongest of these broad polarized features in the total-light spectrum (with the integrated stellar spectrum removed), but they are so weak that they are difficult to recognize against the strong narrow-emission-line spectrum. They have essentially the same plane polarization, in amount and degree, as the featureless continuum.

If only the continuum were observed to be polarized, it could be attributed to emission as synchrotron radiation. However, this mechanism will not explain the equally strong polarization of the broad emission lines. They can only be polarized by scattering. Since the featureless continuum has the same polarization in direction and amount, it must also be polarized by the same scattering process. The polarization is independent of wavelength over the observed range $\lambda\lambda3500 - 7000$. Electron scattering has this property; interstellar dust particles as we know them do not. Very probably the scattering is by electrons. It would be natural to connect them with the hot low-density wind postulated above to explain the radial flow and the confinement of the clouds. However, because of their small mass, electron scattering broadens lines; the temperature of the electrons must be $< 10^6$ °K for otherwise the lines would be thermally broadened by more than their observed widths. Finally, in NGC 1068 the observed plane of polarization is not obviously connected with the optical appearance of the galaxy. Instead, it has the E vector perpendicular to the direction of the observed radio jet in this object.

A straightfoward interpretation of these polarization measurements is that the nucleus of NGC 1068 has a structure much like those shown in Figures 12.8 or 12.9, but the disk BLR has so much dust around its periphery, in a ring or torus, that it is optically thick. This ring is thick enough, and so oriented, that no light gets out directly in our direction either from it or from the central source. Thus we do not see directly either the broad emission lines or the strong featureless continuum, but only the narrow lines from the surrounding NLR. However, photons emitted by the central source and the BLR that escape more or less along the axis can be scattered by electrons above and below the thick BLR disk, and thus observed by us.

It is not far from this picture to imagine that not only NGC 1068, but all Seyfert 2s, may contain hidden BLRs, and that the true physical difference between Seyfert 1s and Seyfert 2s is not in the amount of broad-line gas present in the object, but rather in its orientation with respect to the observer. Only a few other Seyfert 2 galaxies have been observed spectropolarimetrically to the excellent level of signal-to-noise ratio necessary to detect similar weak Seyfert 1 features in their polarized light spectra, and to date none of these results have been published. This physical picture has many attractive features, and no doubt will be explored thoroughly, both observationally and theoretically, in coming years. Note that one of its implications is that the obscuring ring or torus should be optically thick to X-rays as well as optical radiation from the central source, since the Seyfert 2 nuclei are so much less luminous than Seyfert 1s in this region also. For hard X-rays the Compton cross section of equation (11.10) is the main opacity source. For example, the optical depth $\tau_X > 2$ requires

$$N_H \; a_{C\nu} \; \Delta > 2$$

or for $N_H = 10^9$ cm^{-3}, the required thickness of the opaque torus is $\Delta > 10^{-3}$ pc, which is not unreasonable at all.

If we understand AGNs correctly, the central energy source is the rotating accretion disk around a black hole. A massive black hole is required, plus a supply of mass to fuel it at a rate 1 to 10 $M_\odot$ yr^{-1}. If such a black hole is present in a galaxy, as a result of previous dynamical evolution of the stars in the nucleus, it will still not be an AGN without the reservoir of mass to continue to replenish the disk. The question of how this fuel gets nearly to the nucleus, with essentially zero angular momentum on the scale of the galaxy, is an important one. Seyfert galaxies as a group are spiral galaxies, chiefly of types Sa and Sb. Many of them are somewhat distorted, and a considerable fraction of the nearby ones whose images are large and well-resolved show bars or outer rings (or both). Recent statistical studies reveal that a significantly larger fraction of the Seyferts than of non-Seyferts have nearby "companions," or galaxies passing by near enough that they are presumably interacting gravitationally. All these clues seem to show that gravitational perturbations, either from outside or from a 2θ (bar-like) internal potential, can induce flows in which some interstellar matter is delivered close to the nucleus with nearly zero angular momentum. On the other hand, violently distorted galaxies, though they often clearly have had recent star formation, usually are not Seyfert galaxies; evidently in them the perturbations are too extreme. Likewise in clusters of galaxies the fraction of galaxies that contain AGNs is much smaller than in the general field. No doubt the perturbations are too frequent or strong, and perhaps most of the interstellar matter has been stripped from the galaxies. This whole topic will no doubt also be studied much more deeply in coming years.

A very interesting study, at the frontier of present-day research, is the number of low-luminosity AGNs. Among a well-defined sample of 101 bright, nearby spiral galaxies, four either had no nucleus or were edge-on so that their nuclei were obscured. Of the remaining 97 whose nuclei could be observed, one is a Seyfert 1, four are Seyfert 2s, and 52 are LINERs. All the rest have emission lines characteristic of H II regions. In many of these objects the emission lines are quite faint, and were detected only after careful subtraction of template spectra derived from observational data obtained with equally high signal-to-noise ratio of absorption-line galaxies. Thus there is a very high percentage of low-level AGNs, but many of these AGNs would be missed, and in fact were missed, on lower-quality spectral data.

Likewise many galaxies previously classified as Seyfert 2s, LINERs, or simply emission-line galaxies show, on equally good data, weak broad Hα emission-line components, indicating the presence of a faint, underlying Seyfert 1 galaxy. A single example, Mrk 883, is shown in Figure 12.10. Another outstanding example is the elliptical galaxy NGC 1052, discussed as possibly having a shock-wave heated or collisionally ionized nucleus in Section 12.4. The very weak, broad Hα emission observed in NGC 1052 shows that it has, at a low level of luminosity, an AGN quite similar to those in Seyfert 1 galaxies. This is the strongest evidence that it is in fact photoionized, not shock-heated.

Perhaps at a low-enough level of activity every galaxy, or at least every spiral galaxy, is an AGN. Our own Galaxy has been fairly securely classified as an Sbc, even though we cannot observe it from the outside, on the basis of the surface brightness in its central bulge and the relative importance of its spiral arms. This morphological type makes it a prime candidate to be a Seyfert galaxy. We cannot observe the nucleus of our Galaxy in the optical

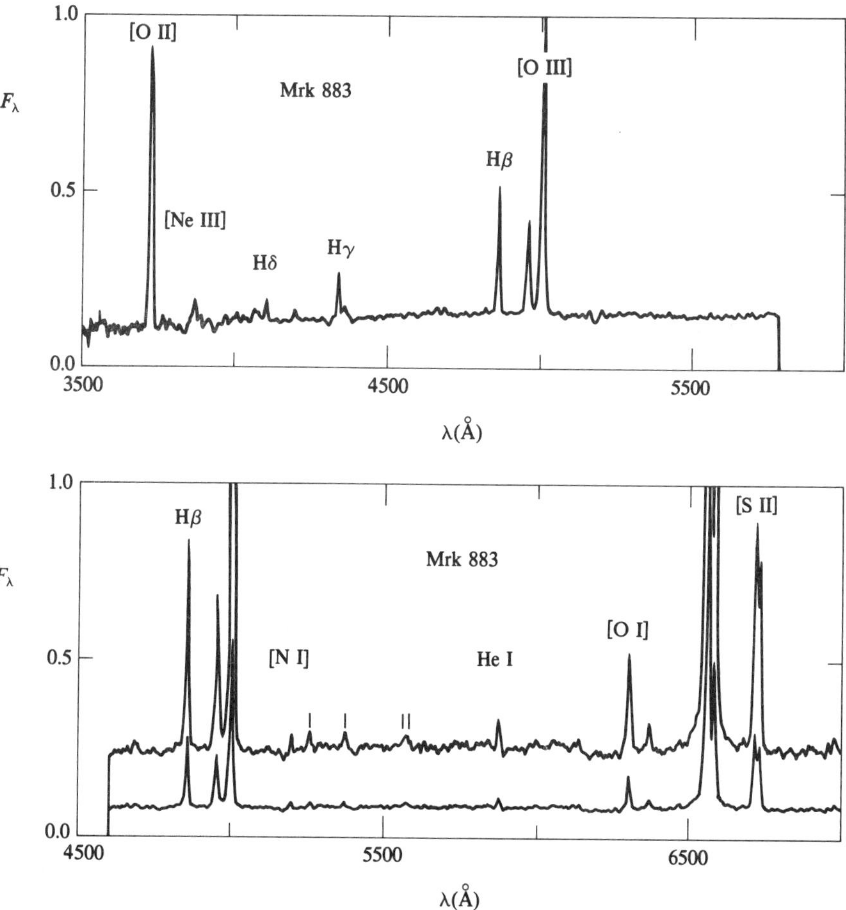

FIGURE 12.10
Spectral scans of Mrk 883, a LINER with a weak, broad emission-line component of Hα. It can only barely be seen in the lower scan plotted at normal scale, but is easily apparent at the three times enlarged scale. The four weak emission lines marked with vertical ticks in the scan at three times enlarged scale are from the sky, not from Mrk 883.

spectral region, because there are approximately 25 magnitudes of extinction between us and it. But there is a well observed compact nonthermal radio source at the center, Sgr A*. Furthermore, in the far infrared, ionized gas condensations have been observed in emission in [Ne II] $\lambda 12.8\ \mu$, and also in neutral, slightly ionized, and molecular lines at longer wavelength. Their measured velocities seem to indicate a tilted thick ring with a radius of about 2 pc, and inside it a central point mass with $M \approx 5 \times 10^6 M_\odot$. This is presumably the black hole at the center of our Galaxy. It is much smaller than the black holes of $M \approx 10^8 - 10^9 M_\odot$ believed to be present in Seyfert galaxy nuclei. Correspondingly, the radio and infrared emission-line luminosity of the center are far too small for it to be considered a Seyfert galaxy, but nevertheless it is a low-level AGN. In the nearby M 31 there is no sign of nuclear activity, either radio or optical, but the observed rotational velocity curve very near the nucleus seems to indicate a black hole with $M \approx 10^7\ M_\odot$.

In coming years AGNs, from the most luminous quasars and QSOs to the barely detectable, will no doubt be very intensively studied. Great strides can be expected especially from combining data which bridges a wide range in wavelength. Guided by sound theoretical analysis, we may be sure that a complete physical picture of these most interesting objects will gradually but surely emerge.

References

A good overall reference for the entire subject of AGNs is

Miller, J. S., ed. 1985. *Astrophysics of Active Galaxies and Quasi-Stellar Objects.* Mill Valley: University Science Books. (This volume is called "Miller" in the specific references, below.)

Two excellent early discussions of the physical nature of the central sources or "engines" in AGNs, and two more recent summaries are

Rees, M. J. 1977. *Q. J. R. A. S.* **18,** 429.

Rees, M. J. 1978. *Observatory* **98,** 210.

Rees, M. J. 1984. *Ann. Rev. Astr. Ap.* **22,** 471.

Begelman, M. C. 1985. Miller, p. 411.

A very good overall treatment of accretion disks, both around stars, which are relatively well-observed and understood, and in AGNs, which are not, is the book

Frank, J., King, A. R., and Raine, D. J. 1985. *Accretion Power in Astrophysics.* Cambridge: Cambridge University Press.

Two papers in which calculated high-energy spectra from simple accretion-disk models are explicitly calculated, or used as input photoionizing spectra, are

Aldrovandi, S. M. V. 1981. *Astr. Ap.* **97,** 122.

Reynolds, S. P. 1982. *Ap. J.* **256,** 13.

The early photoionization model for a power-law spectrum with exponent $n = 1.2$ is contained in a thesis,

MacAlpine, G. M. 1971. Ph.D. Thesis, University of Wisconsin.

It was never published in full, but a description of the program and some other models from it are in

MacAlpine, G. M. 1972, *Ap. J.* **175,** 11.

Two other very important early papers on photoionization models are

Davidson, K. 1972. *Ap. J.* **171,** 213.

Davidson, K. 1973. *Ap. J.* **181,** 1.

The more recent models described in the text and used in the diagrams of Figures 12.1, 12.2, and 12.3 are from

Ferland, G. J., and Netzer, H. 1983. *Ap. J.* **264,** 105.

Stasinska, G. 1984. *Astr. Ap. Suppl.* **55,** 15.

Stasinska, G. 1984. *Astr. Ap.* **135,** 341.

The observational data of Figures 12.1, 12.2, and 12.3 and the discussion in terms of these models are based upon

Veilleux, S., and Osterbrock, D. E. 1987. *Ap. J. Suppl.* **63,** 295.

LINERs are discussed from an observational point of view in

Heckman, T. M. 1980. *Astr. Ap.* **87,** 152.
Stauffer, J. R. 1982. *Ap. J.* **262,** 66.
Keel, W. C. 1983. *Ap. J.* **269,** 466.
Keel, W. C. 1985. Miller, p. 1.

The first of these defined the class, while the next two give data showing how many spiral galaxies have this type of emission-line spectrum. The third paper presents particularly convincingly the evidence favoring photoionization as the energy-input mechanism to LINERs, and the fourth is a general review. Analyses in terms of photoionization models are given by Ferland and Netzer (1983), and by

Halpern, J. P., and Steiner, J. E. 1983. *Ap. J.* **269,** L37.
Pequignot, D. 1984. *Astr. Ap.* **131,** 159.
Binette, L. 1985. *Astr. Ap.* **143,** 334.

Good overall references for models and analyis of BLRs are

Davidson, K., and Netzer, H. 1979. *Rev. Mod. Phys.* **51,** 715.
Osterbrock, D. E. 1985. Miller, p. 111.
Ferland, G. J., and Shields, G. A. 1985. Miller, p. 157.

Table 12.4 is taken from the first of these references.

Discussions of resonance-line scattering, particularly Lα, in the context of nebulae and AGNs, include

Osterbrock, D. E. 1962. *Ap. J.* **135,** 195.
Adams, T. F. 1972. *Ap. J.* **174,** 439.
Ferland, G., and Netzer, H. 1979. *Ap. J.* **229,** 274.
Bonilha, J. R. M., Ferch, R., Salpeter, E. E., Slater, G., and Noerdlinger, P. D. 1979. *Ap. J.* **233,** 649.
Eastman, R. G., and MacAlpine, G. M. 1985. *Ap. J.*, **299**, 785.

The interpolation formulae of equation (12.6) are taken from the third of these references. The last is a good discussion of the Bowen resonance-fluorescence process in O III in the context of AGNs.

Much theoretical work has been done on the H I line spectrum of a dense, optically thick BLR. All the models are necessarily highly simplified. Some of the most important papers, all of which give references to earlier work, are

Krolik, J. H., and McKee, C. F. 1978. *Ap. J. Suppl.* **37,** 459.
Kwan, J., and Krolik, J. H. 1979. *Ap. J.* **233,** L91.
Drake, S. A., and Ulrich, R. K. 1980. *Ap. J. Suppl.* **42,** 351.
Canfield, R. C., and Puetter, R. C. 1981. *Ap. J.* **243,** 390.
Collin-Souffrin, S., Dumont, S., and Tully, J. 1982. *Astr. Ap.* **106,** 362.
Kwan, J. 1984. *Ap. J.* **283,** 70.

The data of Table 12.3 came from the first of these papers. The models of the last paper are the most sophisticated, and include the full ionization, thermal equilibrium and radiative-transfer (for the ionizing continuum) equations into the gas cloud, rather than assuming homogeneous temperature or ionization.

A few of the many papers that summarize large amounts of observational data on BLRs are

Weedman, D. W. 1977. *Ann. Rev. Astr. Ap.* **15,** 69.
Osterbrock, D. E. 1977. *Ap. J.* **215,** 733.
Osterbrock, D. E. 1984. *Q. J. R. A. S.* **25,** 1.

A very important paper with a lot of early observational data on quasars and QSOs of different redshifts, which were linked up to provide a composite spectrum over a wide wavelength range, including the first realization of the small value of the $L\alpha/H\beta$ ratio, is

Baldwin, J. A. 1975. *Ap. J.* **201,** 26.

A more recent paper with a lot of good observational data, particularly concentrated in the ultraviolet spectra of quasars and QSOs, is

Wills, B. J., Netzer, H., and Wills, D. 1985. *Ap. J.* **288,** 94.

A good reference for the ultraviolet spectra of Seyfert 1 galaxies, taken with the IUE satellite telescope, is

Wu, C.-C., Boggess, A., and Gull, T. R. 1983. *Ap. J.* **266,** 28.

A few of the many papers on the Fe II emission features in Seyfert 1 galaxies and low redshift QSOs are

Wampler, E. J., and Oke, J. B. 1967. *Ap. J.* **148,** 695.
Sargent, W. L. W. 1968. *Ap. J.* **203,** 329.
Phillips, M. M. 1977. *Ap. J.* **215,** 746.
Phillips, M. M. 1978. *Ap. J. Suppl.* **38,** 187.
Phillips, M. M. 1978. *Ap. J.* **226,** 736.
Netzer, H., and Wills, B. J. 1983. *Ap. J.* **275,** 445.

Two reviews of the subject of dust and extinction in AGNs are

Osterbrock, D. E. 1979. *A. J.* **84,** 901.
MacAlpine, G. M. 1985. Miller, p. 259.

The second of these gives many references to published papers on the subject.

Good overall reviews of the velocity fields in AGNs, and their structure, include

Mathews, W. G., and Capriotti, E. R. 1985. Miller, p. 185.
Wilson, A. S., and Heckman, T. M. 1985. Miller, p. 39.
Mathews, W. G., and Osterbrock, D. E. 1986. *Ann. Rev. Astr. Ap.* **24,** 171.

All three contain many references to previous work. Some specific references to the observational data on line profiles are

Pelat, D., Alloin, D., and Fosbury, R. A. E. 1981. *M.N.R.A.S.* **195,** 787.
Osterbrock, D. E., and Shuder, J. M. 1982. *Ap. J. Suppl.* **49,** 149.
Wilkes, B. J. 1984. *M.N.R.A.S.* **207,** 73.
Wilkes, B. J. 1986. *M.N.R.A.S.* **218,** 331.
Crenshaw, D. M. 1986. *Ap. J. Suppl.* **62,** 821.
Whittle, M. 1985. *M.N.R.A.S.* **213,** 33 and **216,** 817.
Vrtilek, J. M., and Carleton, N. P. 1985. *Ap. J.* **293,** 106.

Vrtilek, J. M. 1985. *Ap. J.* **294,** 121.
DeRobertis, M. M., and Osterbrock, D. E. 1986. *Ap. J.* **286,** 171, and **301**, 727.

Some references on the disk model for the BLR, and the concept that it may be (and probably is) tipped with respect to the overall structure of the galaxy in which it is located are

Shields, G. A. 1977. *Ap. Lett.* **18,** 119.
Osterbrock, D. E. 1978. *Proc. Natl. Acad. Sci.* **75,** 540.
Tohline, J. E., and Osterbrock, D. E. 1982. *Ap. J.* **252,** L49.

Two good papers on possible connections between obscuration and observed Seyfert galaxy type according to the disk model of AGNs are

Lawrence, A., and Elvis, M. 1982. *Ap. J.* **256,** 410.
Lawrence, A. 1987. *P.A.S.P.* **99,** 309.

A very good study of the theoretical and observational constraints on a hot wind in the BLR, as well as on rotational flow models, is

Mathews, W. G., and Ferland, G. J. 1987. *Ap. J.*, **323**, 456.

The very high-ionization Seyfert 1 galaxy III Zw 77, and the quite-similar Seyfert 2 galaxy Tololo 0109–383, respectively, are described in

Osterbrock, D. E. 1981. *Ap. J.* **246,** 696.
Fosbury, R. A. E., and Sansom, A. E. 1983. *M.N.R.A.S.* **204,** 1231.

Some of the important papers on spectropolarimetry of the emission lines and continuum in AGNs are

Angel, J. R. P., Stockman, H. S., Woolf, N. J., Beaver, E. A., and Martin, P. G. 1976. *Ap. J.* **206,** L5.
McLean, I. S., Aspin, C., Heathcote, S. R., and McCaughrean, M. J. 1983. *Nature* **296,** 331.
Miller, J. S., and Antonucci, R. R. J. 1983. *Ap. J.* **271,** L7.
Antonucci, R. R. J., and Miller, J. S. 1985. *Ap. J.* **297,** 621.

The last two of these present the measurements and interpretation of the spectra of NGC 1068, including the "hidden Seyfert 1 nucleus," and the suggested ring model for all or most Seyfert galaxies. Some analysis of this model is presented in

Krolik, J. H., and Begelman, M. C. 1986. *Ap. J.* **308,** L55.

Several papers on the interaction picture of Seyfert galaxies, from early papers to the most recently published statistics, plus two attempts at a theoretical interpretation, are

Adams, T. F. 1977. *Ap. J. Suppl.* **33,** 19.
Simkin, S. M., Su, H. J., and Schwarz, M. P. 1980. *Ap. J.* **237,** 404.
Kennicutt, R. C., and Keel, W. C. 1984. *Ap. J.* **279,** L5.
Dahari, O. 1984. *A. J.* **89,** 966.
Keel, W. C., Kennicutt, R. C., Hummel, E., and van der Hulst, J. M. 1985. *A. J.* **90,** 708.
Dahari, O. 1985. *A. J.* **90,** 1772.

Dahari, O. 1985. *Ap. J. Suppl.* **57,** 643.
Roos, N. 1981. *Astr. Ap.* **104,** 218.
Lin, D. N. C., Pringle, J. E., and Rees, M. J. 1988. *Ap. J.* **328**, 103.

The low-luminosity, previously undetected BLRs in many Seyfert 2 nuclei and LINERs, are described, with excellent spectral data, by

Filippenko, A. V., and Sargent, W. L. W. 1985. *Ap. J. Suppl.* **57,** 503.

Three very good references on the activity at the center of our Galaxy, and the detection of a black hole there from measured rotational velocities, are

Lacy, J. H., Townes, C. H., Geballe, T. R., and Hollenbach, D. J. 1980. *Ap. J.* **241,** 132.
Lacy, J. H., Townes, C. H., and Hollenbach, D. J. 1982. *Ap. J.* **262,** 120.
Genzel, R., Watson, D. M., Crawford, M. H., and Townes, C. H. 1985. *Ap. J.* **297,** 766.

The observational results on the detection of a black hole in the nucleus of M 31 are given in

Dressler, A., and Richstone, D. O. 1988. *Ap. J.*, **324**, 701.
Kormendy, J. 1988. *Ap. J.*, **325**, 128.

APPENDIX 1

Milne Relation Between Capture and Photoionization Cross Sections

The Milne relation expresses the capture cross section to a particular level (with threshold ν_T) in terms of the absorption cross section from that level, and is based on the principle of detailed balancing or microscopic reversibility. According to this principle, in thermodynamic equilibrium each microscopic process is balanced by its inverse. Thus, in particular, recombination (spontaneous plus induced) of electrons with velocity in the range between $v + dv$ is balanced by photoionization by photons with frequencies in the range between $\nu + d\nu$, where

$$\frac{1}{2}\ mv^2\ +\ h\nu_T\ =\ h\nu,$$

so

$$mv\ dv\ =\ h\ d\nu.$$

The rate of induced downward radiative transitions (induced recombinations in this case) in thermodynamic equilibrium is always just $e^{-h\nu/kT}$ times the rate of induced upward transitions (photoionizations in this case), so the equilibrium equation may be written

spontaneous recombination rate = $(1 - e^{-h\nu/kT})$ photoionization rate

or, in the notation of Chapter 2,

$$N_e N(X^{+i+1})\ v\sigma(v)f(v)dv = \\ (1 - e^{-h\nu/kT})N(X^{+i})\frac{4\pi B_\nu(T)}{h\nu}a_\nu d\nu.$$

Substituting the thermodynamic equilibrium relations for the Maxwell-Boltzmann distribution function,

$$f(v)\ =\ \frac{4}{\sqrt{\pi}}\left(\frac{m}{2kT}\right)^{3/2}v^2e^{-mv^2/2kT},$$

the Planck function,

$$B_\nu(T) = \frac{2h\nu^3}{c^2} \frac{1}{e^{h\nu/kT} - 1},$$

and the Saha equation

$$\frac{N(X^{+i+1})N_e}{N(X^{+i})} = \frac{2\omega_{i+1}}{\omega_i} \left(\frac{2\pi mkT}{h^2}\right)^{3/2} e^{-h\nu_T/kT},$$

we obtain the Milne relation

$$\sigma(v) = \frac{\omega_i}{\omega_{i+1}} \frac{h^2\nu^2}{m^2c^2v^2} a_\nu.$$

Though the preceding equation was derived using arguments from thermodynamic equilibrium, it is a relation between the recombination cross section at a specific v and the absorption cross section at the corresponding ν. It can also be derived quantum mechanically and depends only on the fact that the matrix elements between two states are independent of their order.

In particular, if the photoionization cross section a_ν can be represented by the interpolation formula

$$a_\nu = a_T \left[\beta\left(\frac{\nu}{\nu_T}\right)^{-s} + (1 - \beta)\left(\frac{\nu}{\nu_T}\right)^{-(s+1)}\right],$$

it follows that the recombination cross section can be written

$$\sigma(v) = \frac{\omega_i}{\omega_{i+1}} \frac{h^2\nu^2}{m^2c^2v^2} a_T \left[\beta\left(\frac{\nu}{\nu_T}\right)^{-s} + (1 - \beta)\left(\frac{\nu}{\nu_T}\right)^{-(s+1)}\right].$$

Substituting, the recombination coefficient to the level X^{+i},

$$\alpha(X^{+i}, T) = \int_0^\infty v\, f(v)\, \sigma(v) dv,$$

becomes

$$\alpha(X^{+i}, T) = \frac{4}{\sqrt{\pi}} \frac{\omega_i}{\omega_{i+1}} \left(\frac{m}{2kT}\right)^{3/2} e^{h\nu_T/kT} \frac{h^3\nu_T^3}{m^3c^2} a_T$$

$$\times \left[\beta E_{s-2}\left(\frac{h\nu_T}{kT}\right) + (1-\beta)E_{s-1}\left(\frac{h\nu_T}{kT}\right)\right],$$

if s is an interger, where E_n is the exponential integral function. If s is nonintegral, $E_n(x)$ must be replaced in this formula by $x^{n-1}\Gamma(1-n, x)$.

APPENDIX 2

Escape Probability of a Photon Emitted in a Spherical Homogeneous Nebula

Let $\tau = \kappa R$ be the optical radius of the nebula (see Figure A2.1), where R is the linear radius. Consider a ray making an angle θ to the outward normal; the total optical length along this ray is $\tau_\theta = 2\tau \cos\theta$, so if ϵ is the (constant)-emission coefficient per unit volume per unit solid angle per unit time, the emergent intensity in this direction is

$$I(\theta) = \int_0^{\tau_\theta} \epsilon\, e^{-t} ds = \frac{\epsilon}{\kappa}(1 - e^{-2\tau \cos\theta}),$$

where s is the linear coordinate along the ray, and $t = \kappa s$ is the optical-length coordinate. Therefore, the outward flux per unit area per unit time is

$$\pi F = 2\pi \int_0^{\pi/2} I(\theta) \cos\theta \sin\theta \, d\theta$$

$$= \frac{\pi\epsilon}{2\kappa\tau^2}[2\tau^2 - 1 + (2\tau + 1)e^{-2\tau}].$$

To find the escape probability, this expression may be compared with the flux the nebula would emit in the absence of any absorption, which is simply given by the total emission in the volume divided by the area:

$$\pi F(0 \text{ absorption}) = \frac{4\pi\epsilon(4\pi/3)R^3}{4\pi R^2} = \frac{4\pi\epsilon R}{3}.$$

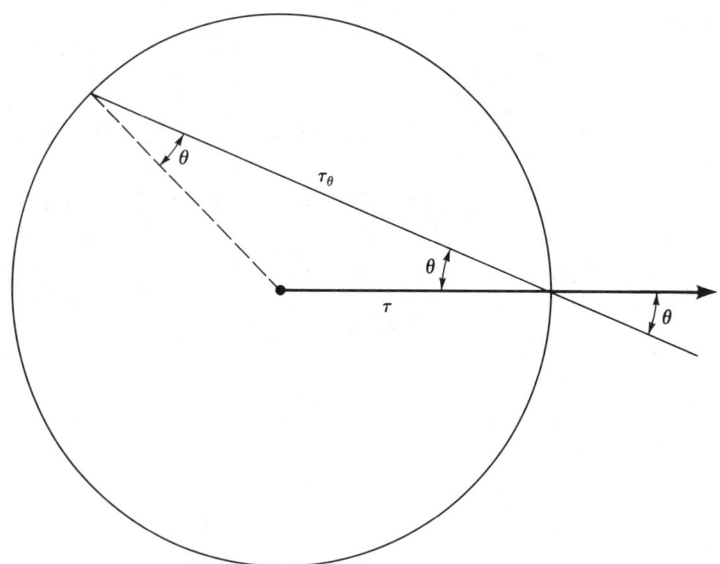

FIGURE A2.1
Cross section through the center of a spherical homogeneous nebula of radius r, optical radius $\tau = \kappa R$, showing a ray of optical length τ_θ making an angle θ with respect to the outward normal.

Thus the escape probability is

$$p(\tau) = \frac{\pi F}{\pi F(0 \text{ absorption})}$$

$$= \frac{3}{4\tau}\left[1 - \frac{1}{2\tau^2} + \left(\frac{1}{\tau} + \frac{1}{2\tau^2}\right)e^{-2\tau}\right].$$

It may be verified that as $\tau \to 0, p(\tau) \to 1$, but as $\tau \to \infty, p(\tau) \to 3/4\tau$, which is the contribution due to the escape of photons from the surface layer only.

APPENDIX 3

Names and Numbers of Nebulae

Though most nebulae are ordinarily referred to by their numbers in the New General Catalogue (NGC) or the Index Catalogues, a few of the brighter ones are sometimes referred to in published papers by their Messier numbers. NGC numbers are used most throughout the present book. The H II regions and planetary nebulae with Messier numbers (M) are listed in Table A3.1, together with their NGC numbers and their common names, which are also sometimes used in published papers.

TABLE A3.1
Identification of Messier numbers

	M	NGC	Common names(s)
H II Regions	8	6523	Lagoon
	16	6611	–
	17	6618	Omega, Horseshoe
	20	6514	Trifid
	42	1976	Orion
	43	1982	Part of Orion complex
Planetary nebulae	27	6853	Dumbbell
	57	6720	Ring
	76	650-651	–
	97	3587	Owl
Supernova remnant	1	1952	Crab
Galaxies	31	224	Andromeda
	32	221	Round companion to Andromeda
	33	598	Triangulum
	51	5194	Whirlpool
	77	1068	Seyfert galaxy
	81	3031	–
	101	5457	–

APPENDIX 4

Emission Lines of Neutral Atoms

Collisionally excited emission lines of several neutral atoms are observed in gaseous nebulae, particularly [O I] $\lambda\lambda6300$, 6364, [N I] $\lambda\lambda5198$, 5200, [Mg I] $\lambda\lambda4562$, and Mg I] $\lambda4571$. These emission lines arise largely in the transition regions or ionization fronts at the boundaries between H^+ and H^0 regions. As was described in Chapter 5, the relative strengths of the neutral-atom lines therefore provide information on the transition regions and on the dense neutral condensations in nebulae. These lines are relatively strong in AGN spectra, as explained in Section 11.5.

The emission coefficients for these lines may be calculated by the methods described in Section 3.5, using the transition probabilities and collision strengths in the following tables. The transition probabilities for [O I] are listed in Table 3.10. Note, however, that the collision strengths of neutral atoms are zero at the threshold and vary rapidly with energy; their mean values, defined by equation (3.20), must therefore be used. These mean values for the most important collision strengths are listed in Tables A4.2 and A4.3.

TABLE A4.1
Transition probabilities for [N I], [Mg I], *and* Mg I]

Atom	Transition	Transition probability (sec^{-1})	Wavelength (Å)
[N I]	$^4S_{3/2} - {}^2D_{3/2}$	2.0×10^{-5}	5197.9
[N I]	$^4S_{3/2} - {}^2D_{5/2}$	7.3×10^{-6}	5220.4
[Mg I]	$^1S_0 - {}^3P_2^0$	2.8×10^{-4}	4562.5
Mg I]	$^1S_0 - {}^3P_1^0$	$4.3 \times 10^{+2}$	4571.1

TABLE A4.2
Collision strengths for O^0

T (° K)	$\Omega(^3P, {}^1D)$	$\Omega(^3P, {}^1S)$	$\Omega(^3P_2, {}^3P_1)$	$\Omega(^3P_2, {}^3P_0)$	$\Omega(^3P_1, {}^3P_0)$
6,000	0.15	0.019	0.057	0.018	0.014
8,000	0.21	0.026	0.078	0.024	0.020
10,000	0.27	0.032	0.099	0.029	0.027
12,000	0.31	0.038	0.120	0.034	0.034
15,000	0.38	0.046	0.152	0.042	0.046
20,000	0.50	0.061	0.206	0.054	0.069

TABLE A4.3
Collision strengths for N^0 *and* Mg^0

T(° K)	N^0		Mg^0
	$\Omega(^4S, {}^2D)$	$\Omega(^2D_{3/2}, {}^2D_{5/2})$	$\Omega(^1S, {}^3P)$
6,000	0.31	0.16	2.4
8,000	0.41	0.22	2.4
10,000	0.48	0.27	2.4
12,000	0.55	0.32	2.5
15,000	0.62	0.38	2.5
20,000	0.79	0.46	2.6

References

The transition probabilities for [N I] in Table A4.1 are taken from

Zeippen, C. J. 1982. *M.N.R.A.S.* **198,** 111.

The transition probabilities for Mg I] and [Mg I] (this book follows the convention of using a single bracket for electric-dipole intercombination lines, and both brackets for electric-quadrupole and magnetic-dipole and magnetic-quadrupole lines) in Table A4.1 are from

Weise, W. L., Smith, M. W., and Miles, B. 1969. *Atomic Transition Probabilities* **2,** *Sodium through Calcium,* Washington, D. C.: Government Printing Office, p. 25.

The mean values of the collision strengths in Tables A4.2 and A4.3. are based upon

Fabrikant, I. I. 1974. *J. Phys. B.* (Atom. Molec. Phys.) **7,** 91 (Mg^0).

Dopita, M. A., Mason, D. J., and Robb, W. D. 1976. *Ap. J.* **207,** 102. (N^0).

LeDournef, M., and Nesbet, R. K. 1976. *J. Phys. B.* **9,** 1241. (O^0).

Berrington, K. A., and Burke, P. G. 1981. *Planetary & Space Science* **29,** 377. (N^0 and O^0).

Glossary of Physical Symbols

The symbols used in this book are listed here with their physical significance and the section (indicated without parentheses) or equation (indicated by parentheses) in which they first appear. Dummy mathematical variables and symbols used only once are not listed. Note that in some cases the same symbol has been used for two widely differing quantities, usually one from quantum mechanics and the other from stellar or nebular astronomy.

ROMAN

a	Radius of a dust particle	7.3
$A_{i,j}$	Radiative transition probability between upper level i and lower level j	2.2
a_0	Bohr radius $h^2/4\pi^2me^2$	(2.4)
a_C	Compton scattering cross section	(11.10)
a_T	Threshold absorption cross section for an arbitrary atom or ion	(2.31)
A_λ	Albedo of a dust particle at wavelength λ	(7.9)
a_ν	Absorption cross section per atom	(2.1)
b	Fraction of excitations to a level leading to emission of a particular line photon	5.9
b	Minimum distance (or impact parameter) of a light ray from the central star of a nebula	(7.10)
b_j	Deviation from thermodynamic equilibrium factor	(4.6)
$B_\nu(T)$	Planck function at frequency ν	(4.34),(4.36)
C	Constant giving amount of interstellar extinction along a ray for use with natural logarithms	(7.3)
c	Constant giving amount of interstellar extinct-tion along a ray for use with logarithms to base 10	(7.6)
c	Velocity of light	(2.14)

$C(i,j)$	Collisional transition rate per ion level i from level i to level j, with temperature exponential factored out	(5.3)
$C_{i,j}$	Probability population of upper level i is followed by population of lower level j	4.2,(4.9)
C_λ	Extinction cross section of a dust particle at wavelength λ	(7.9)
c_0	Isothermal velocity of sound in H^0 region	6.3, (6.18)
c_1	Adiabatic velocity of sound in H^+ region	6.3
D_i	Distance of a star i	(7.2)
D/Dt	Time derivative following an element of volume	(6.1)
E	Emission measure	(4.32)
E	Internal kinetic or thermal energy per unit mass	(6.5)
E_0	Initial energy of supernova outburst	10.4
e	Electron charge (absolute value)	2.2
E_p	Proton-emission measure	(5.14),(5.15)
f	Fraction of excitations to a level leading to emission of a particular line photon	5.9
f_{ij}	f-value of a line between a lower level i and an upper level j	(4.45)
$f(v)$	Maxwell-Boltzmann distribution function	(2.6)
$f(\lambda)$	Wavelength dependence of interstellar extinction by dust	(7.3)
F_ν	Flux of radiation divided by π; πF_ν = flux of radiation	(2.2)
$F_{\nu s}$	Flux of stellar radiation divided by π; $\pi F_{\nu s}$ = stellar flux	(2.11)
G	Energy input rate due to photoionization	(3.1),(3.7)
g_{ff}	Gaunt factor for free-free emission	(4.22)
g_ν	Frequency dependence of H^0 2-photon emission coefficient	(4.29)
h	Planck's constant	(2.1)
I_ν	Specific intensity of radiation	(2.9)
$I_{\nu d}$	Specific intensity of diffuse radiation field	(2.10)
$I_{\nu s}$	Specific intensity of stellar radiation field	(2.10)
J	Rotational quantum number of a molecule	9.7
j_{ij}	Emission coefficient in a line resulting from a radiative transition from an upper level i to a lower level j	(4.12)
J_ν	Mean specific intensity of radiation; $J_\nu = 1/4\pi \int I_\nu d\omega$	(2.1)

$J_{\nu d}$	Mean specific intensity of diffuse radiation field	2.3
$J_{\nu s}$	Mean specific intensity of stellar radiation field	(2.11)
j_ν	Emission coefficient in continuum	(2.9)
k	Boltzmann constant	(2.6)
k_{0l}	Line absorption coefficient at center of a line	(4.44)
$k_{\nu L}$	Line absorption coefficient at frequency ν corrected for stimulated emission	(4.42)
$k_{\nu l}$	Line-absorption coefficient at frequency ν	(4.42)
L	Luminosity of star; $L = \int L_\nu d\nu$	(7.23)
L	Orbital angular momentum quantum number	2.2
L_C	Energy loss rate due to collisionally excited radiation	(3.23),(3.30)
L_E	Eddington luminosity	(12.1)
l	Mean free path of an ionizing photon	(11.1)
L_{FF}	Energy loss rate due to free-free emission	(3.14)
L_n	Lyman line from upper level n to lower level *1*	(4.13)
L_R	Energy loss rate due to recombination	(3.3),(3.8)
L_ν	Luminosity of star per unit frequency interval	(2.2)
M	Mach number	(6.19)
M	Mass of line-emitting region in AGN	(11.8)
M_B	Absolute magnitude in the B photometric system	11.3
M_{bol}	Absolute bolometric magnitude	10.3
$\dot{M}$	Mass accretion rate per unit time	(12.2)
m	Electron mass	2.2
m_D	Mass of a dust particle	7.3
m_H	Mass of proton or hydrogen atom	(4.13)
M_V	Absolute visual magnitude	1.6
$M_\odot$	Solar mass	1.5
n	Principal quantum number	2.2
n	Exponent in power law continuous spectrum	(10.1)
$N_c(i)$	Critical electron density for collisional de-excitation of level i	(2.23),(3.31)
n_{cL}	Principal quantum number above which collisions dominate distribution of atoms among levels with different angular momentum L	4.2
N_D	Number density of dust particles per unit volume	(7.9)
N_e	Electron density	1.4
N_{esc}	Mean number of scatterings of a photon before escape	(12.6)
N_H	Hydrogen density; $N_H = N_{H^0} + N_p$	2.1
N_{H^0}	Neutral hydrogen atom density	(2.1)
N_{He}	Helium density; $N_{He} = N_{He^0} + N_{He^+} + N_{He^{++}}$	(2.28)

N_{He^0}	Neutral helium atom density	(2.21)
N_{He^+}	Singly ionized helium density	2.4
N_j	Density of atoms in level j	(4.1)
N_p	Proton density	(2.1)
N_{ph}	Density of ionizing photons	(8.1)
N_s	Particle density just behind shock front	(10.6)
N_0	Ambient particle density	10.3
p	Gas pressure	(6.1)
$P_{i,j}$	Probability that population of upper level i is followed by a direct radiative transition to lower level j	(4.8)
$q_{i,j}$	Collisional transition rate from level i to level j per particle in level i per colliding particle per unit volume per unit of time	(2.22)
$Q(X)$	Number of ionizing photons for element X emitted by star; for example: $$Q(H^0) = \int_{\nu_0}^{\infty} \frac{L_\nu}{h\nu} d\nu$$	(2.19)
R	Stellar radius	(2.2)
R	Radius of line-emitting region in AGN	(11.9)
r	Distance from star	(2.2)
r	Ratio of line to continuum brightness temperature	(5.17)
r_N	Radius of nebula	(9.1)
R_s	Radius of shock front	(10.3)
r_1	Strömgren radius or critical radius of H^+ zone	(2.19)
r_2	Radius of He^+ zone corresponding to Strömgren radius of H^+ zone	(2.27)
r_3	Radius of He^{++} zone corresponding to Strömgren radius of H^+ zone	(2.29)
S_ν	Source function in equation of transfer	(5.19)
T	Absolute thermodynamic temperature	1.4
$T_{b\nu}$	Brightness temperature at frequency ν	(4.37)
T_C	Brightness temperature in the radio-frequency continuum	5.6
T_c	Color temperature of continuum radiation	7.4
T_D	Temperature of a dust particle	7.5
T_i	Initial temperature of newly created photoelectrons	(3.2)
T_L	Brightness temperature at the center of a radio-frequency line	5.6
T_s	Temperature just behind shock front	(10.4)
T_*	Stellar effective temperature; $L = 4\pi R^2 \sigma T_*^4$	1.4

U	Internal kinetic or thermal energy per unit volume	(6.4)
V	Volume of line-emitting region	(11.7)
v	Electron velocity	(2.5)
v	Vibrational quantum number of a molecule	9.7
$\mathbf{v}$	Velocity vector	(6.1)
v_D	D-critical velocity	(6.24)
v_R	R-critical velocity	(6.23)
v_s	Shock velocity	(10.3)
W_0	Equivalent width of an emission line	(11.4)
x	$N_{He^{++}}/N_{He^+}$	9.2
X_n	Ionization potential of level n	(4.4),(4.5)
y	N_{He}/N_H	9.2
Z	Charge on a dust particle in units of electron charge	7.5
Z	Nuclear charge in units of charge of proton	(2.4)
z	Redshift $(\lambda/\lambda_0 - 1)$	11.2

GREEK

α	Recombination coefficient	(2.1)
α_A	Recombination coefficient summed over all levels	(2.7)
α_B	Recombination coefficient summed over all levels above ground level, $\alpha_B = \alpha_A - \alpha_1$	(2.18)
α_i	Recombination coefficient to level i	(2.5)
α_{ij}^{eff}	Effective recombination coefficient for emission of a line resulting from a radiative transition from upper level i to lower level j	(4.14)
$\alpha(X, T)$	Recombination coefficient to the species X at temperature T	2.1
β	Correction to line-absorption coefficient due to maser effect	(5.18)
$\beta_A(X, T)$	Effective recombination coefficient for recombination energy loss of species X at temperature T	(3.4)
β_i	Effective recombination coefficient to level i for recombination energy loss	(3.4)
Γ	Ionization parameter	(8.1)
γ_ν	Frequency dependence of continuum-emission coefficients	(4.23),(4.24)

$\delta(T)$	Charge exchange rate per unit volume per unit time per unit number density of the two reacting species	(2.36),(2.37)
Δ	Thickness of opaque torus in AGN model	12.8
$\epsilon(x)$	Escape probability of a photon emitted at a point x in a nebula	(4.47)
ϵ	Filling factor	9.2
η	Efficiency of energy release in accretion process	(12.2)
κ_C	Absorption coefficient per unit volume in continuum	5.6
κ_L	Line absorption coefficient per unit volume at center of a line	(5.11)
λ	Wavelength	1.2
λ_{ij}	Wavelength corresponding to transition from upper level i to lower level j	(4.13)
λ_0	Threshold wavelength for ionization of H^0; $\lambda_0 = c/\nu_0$	2.2
μ	Mean atomic or molecular weight for particle	(6.2)
ν	Frequency; $\nu = c/\lambda$	(2.1)
ν_{ij}	Frequency corresponding to a transition from upper level i to lower level j	3.5
ν_T	Threshold frequency for ionization of an arbitrary atom or ion	(2.31)
ν_0	Threshold frequency for ionization of H^0; $h\nu_0 = 13.6$ eV	2.1
ν_1	Threshold frequency for ionization of H-like ion of change Z; $\nu_1 = Z^2\nu_0$	(2.4)
ν_2	Threshold frequency for ionization of He^0; $h\nu_2 = 24.6$ eV	2.4
ξ	Fraction of neutral H; $\xi = N_{H^0}/(N_{H^0} + N_p)$	2.1
ρ	Gas density	(6.1)
$\sigma_{i,j}(v)$	Collisional cross section for transition from level i to level j for relative collision velocity v	(2.22)
$\sigma_i(X,v)$	Recombination cross section to level i of species X for electrons with velocity v	(2.5)
σ_T	Thomson electron scattering cross section	11.6
τ_C	Optical depth in the continuum	(5.10)
τ_{cL}	Optical depth at the center of a line corrected for stimulated emission	(5.10)
τ_{0l}	Optical radius at the center of a line	4.5
τ_i	Mean lifetime of excited level i	(2.3)
τ_L	Contribution to the optical depth at the center of a line due to the line alone	(5.10)
τ_0	Optical depth at frequency ν_0	2.3

$\tau_\lambda(i)$	Optical depth at wavelength λ along a ray to star or nebula i	(7.1)
τ_ν	Optical depth at frequency ν	(2.12)
ϕ	Angular radius of nebula	9.2
ϕ_i	Flux of ionizing photons	(6.13),(6.14)
χ	Threshold energy for an excitation process	(2.22)
ω_i	Statistical weight of level i	(3.15)
$\Omega(i,j)$	Collision strength between levels i and j	(3.15)

MISCELLANEOUS

*	Used as a superscript to denote a quantity under conditions of thermodynamic equilibrium	5.7
$\partial/\partial t$	Partial time derivative at a fixed point in space	(6.1)

Index

Note: Page numbers in italics refer to entries in tables on those pages. When the subject appears in an illustration, the abbreviation (fig.) follows the page number.